Complete PCB Design Using OrCAD® Capture and PCB Editor

Complete PCB Design Using OrCAD® Capture and PCB Editor

Kraig Mitzner

AMSTERDAM • BOSTON • HEIDELBERG • LONDON
NEW YORK • OXFORD • PARIS • SAN DIEGO
SAN FRANCISCO • SINGAPORE • SYDNEY • TOKYO
Newnes is an imprint of Elsevier

Newnes is an imprint of Elsevier
30 Corporate Drive, Suite 400, Burlington, MA 01803, USA
Linacre House, Jordan Hill, Oxford OX2 8DP, UK

Library of Congress Cataloging-in-Publication Data
Mitzner, Kraig.
    Complete PCB design using OrCAD Capture and PCB editor / Kraig Mitzner.
        p. cm.
    Includes bibliographical references and index.
    ISBN 978-0-7506-8971-7 (pbk.)
1. Printed circuits—Design and construction. 2. OrCAD SDT. I. Title.
TK7868.P7.M562 2009
621.3815'31—dc22

2009003604

British Library Cataloguing-in-Publication Data
A catalogue record for this book is available from the British Library.

For information on all Newnes publications,
visit our Web site at www.elsevierdirect.com

Printed and bound by CPI Group (UK) Ltd, Croydon, CR0 4YY
Transferred to digital print 2012

Working together to grow
libraries in developing countries

www.elsevier.com | www.bookaid.org | www.sabre.org

ELSEVIER   BOOK AID
             International   Sabre Foundation

Contents

OrCAD PCB Editor is a relatively new printed circuit board (PCB) design application that has replaced OrCAD Layout. The purpose of this book is to provide new users of the software with a basic introduction to OrCAD PCB Editor and the design of PCBs.

OrCAD PCB Editor is based on Allegro PCB Editor, so this book will be useful to new Allegro PCB Editor users as well. Allegro PCB Editor is a powerful, full-featured design tool. While OrCAD PCB Editor has inherited many of those features, it does not possess all of the tools and features. Consequently most of the basic tools and features are described here, but only a few of the more-advanced tools are covered, as outlined later.

Chapter 1 introduces the reader to the basics of PCB design. The chapter begins by introducing the concepts of computer-aided engineering, computer-aided design, and computer-aided manufacturing. The chapter then explains how these tools are used to design and manufacture multilayer PCBs. Many 3-D pictures are used to show the construction of PCBs. Topics such as PCB cores and layer stack-up, apertures, D codes, photolithography, layer registration, plated through holes, and Gerber files are explained.

Chapter 2 leads new users of the software through a very simple design example. The purpose of the example is to paint a "big picture" of the design flow process. The example begins with a blank schematic page and ends with the Gerber files. The circuit is ridiculously simple, so that it is not a distraction to understanding the process itself. Along the way some of PCB Editor's routing tools are briefly introduced along with some of the other tools, which sets the stage for Chapter 3.

Chapter 3 provides an overview of the OrCAD project files and structure and explains PCB Editor's tool set in detail. The chapter revisits and explains some of the actions performed and tools used during the example in Chapter 2. Gerber files are also explained in detail.

Chapter 4 introduces some of the industry standards organizations related to the design and fabrication of PCBs (e.g., IPC and JEDEC). PCB performance classes and producibility levels are also described, along with the basic ideas behind standard fabrication allowances. These concepts are described here to help the reader realize some of the fabrication issues up front to help minimize board failures and identify some of the guides and standards resources that are available for PCB design.

Chapter 5 addresses the mechanical aspect of PCB design—design for manufacturability. The chapter explains where parts should be placed on the board, how far apart, and in what orientation from a manufacturing perspective. OrCAD PCB Editor's design rule checker is then considered relative to the manufacturing concepts and IPC's courtyard concepts. To aid in understanding the design issues, manufacturing processes such as reflow and wave soldering, pick-and-place assembly, and thermal management are discussed. The information is then used as a guide in designing plated through holes, surface-mount lands, and PCB Editor footprints in general. Tables summarize the information and serve as a design guide during footprint design and PCB layout.

Chapter 6 addresses the electrical aspect of PCB design. Several good references are available on signal integrity, electromagnetic interference, and electromagnetic compatibility. Chapter 6 provides an overview of those topics and applies them directly to PCB design. Topics such as loop inductance, ground bounce, ground planes, characteristic impedance, reflections, and ringing are discussed. The idea of "the unseen schematic" (the PCB layout) and its role in circuit operation on the PCB is introduced. Look-up tables and equations are provided to determine required trace widths for current handling and impedance as well as required trace spacing for high-voltage designs and high-frequency designs. Various layer stack-up topographies for analog, digital, and mixed-signal applications are also described. The design examples in Chapter 9 demonstrate how to apply the layer stack-ups described in this chapter. A demonstration on how to use PSpice to simulate transmission lines to aid in circuit design and PCB layout is also provided.

Chapter 7 explains how to construct Capture parts using the Capture Library Manager and Part Editor and the PSpice Model Editor. Heterogeneous and homogeneous parts are developed in examples using four methods. Different methods are used depending on whether a part will be used for simple schematic entry, design projects intended for PCB layout, PSpice simulations, or all of these. The chapter also demonstrates how to attach PSpice models to Capture's schematic parts using PSpice models downloaded from the Internet and basic PSpice models developed from functional Capture projects. The Capture parts can then be used for both PSpice simulations and PCB layout as demonstrated in Chapter 9.

Detailed coverage of padstacks and footprints is covered in Chapter 8. The chapter begins with an overview of PCB Editor's symbols library, describes the various types of symbols, and explains the anatomy of a footprint. Then a detailed description of the padstack (as it relates to PCB manufacturing described in Chapters 1 and 5) is given, as it is the foundation of both footprint design and PCB routing. Design examples are provided to demonstrate how to design discrete through-hole and surface-mount devices and how to use the footprint design wizard. The IPC Land Pattern Viewer is also introduced in this chapter.

Chapter 9 provides four PCB design examples that use the material covered in the previous chapters. The first example is a simple analog design using a single op amp. The design shows how to set up multiple plane layers for positive and negative power supplies and ground. The design also demonstrates several key concepts in Capture, such as how to connect global nets, how to assign footprints, how to perform design rule checks, how to use the Capture part libraries, how to generate a bill of materials (BOM), and how to use the BOM as an aid in the design process in Capture and PCB Editor. The design also shows how to perform important tasks in PCB Editor, such as how to set up a board outline, place parts, and modify padstacks. Intertool communication (such as annotation and back annotation) between Capture and PCB Editor is also demonstrated. The second design is a mixed digital/analog circuit. In addition to the tasks demonstrated in the first example, the design also demonstrates how to set up and use split planes to isolate analog and digital power supplies and grounds. Other tasks include using copper areas on routing layers to make partial ground planes, setting up split power and ground planes, and defining anticopper areas on plane and routing layers. The third example uses the same mixed digital/analog circuit from the second example but demonstrates how to use multiple-page schematics and off-page connectors to add PSpice simulations to a Capture project used for PCB layout, all within a single project design. It also demonstrates how to construct multiple, separated power and ground planes and a shield plane to completely isolate analog from digital circuitry. The use of guard rings and guard traces is also demonstrated.

The fourth example is a high-speed digital design, which demonstrates how to design transmission lines, stitch multilayer ground planes, perform pin/gate swapping, place moated ground areas for clock circuitry, and design a heat spreader.

The last part of Chapter 9 includes a short discussion about the differences between using negative and positive planes in PCB design.

Chapter 10 describes taking the PCB design from the CAD stage through fabrication. A simple design example shows how to produce the artwork (Gerber) files for a PCB design. PCB Editor is then used to review the artwork files before they are sent to a manufacturer. PCB Editor is also used to generate a drawing (dfx) file that can be opened and edited with many drawing applications, so they can be 3-D modeled to review form, fit, and function. The chapter also describes how to create a custom report that can be used for pick and place machines during the assembly process.

At the Web site for this book, www.elsevierdirect.com/companions/ 9780750689717, you should find supplementary content including the demo version of the software and the design files used in writing this book. The design files include the design flow example from Chapter 2, the Capture parts from Chapter 7, the footprint and padstack symbols from Chapter 8, the board layout files from Chapter 9, and the manufacturing files from Chapter 10. The newest version of the demo can also be obtained from the cadence Web site.

During the development of this text, the author received technical assistance from Cadence, so many of the procedures and concepts were written from lessons learned and feedback from Cadence. While every effort has been made to ensure technical accuracy of the material, it is possible that some of the steps outlined in the design examples could be done more efficiently or in other ways. Please work closely with your board manufacturer during the procurement and fabrication process to ensure that the end product will be what you intended.

Acknowledgments

The author would like to thank Cadence for the use of the software and the technical assistance during the development of this book.

The author would also like to thank his family and friends for their support and encouragement without which this book would not have been completed.

And last, but not least, a hardy thank you goes to the readers of the first book. Your enthusiasm and feedback provided the motivation for this book.

Introduction to PCB Design and CAD

COMPUTER-AIDED DESIGN AND THE OrCAD DESIGN SUITE

Before digging into the details of PCB Editor, we take a moment to discuss computer-aided engineering (CAE) tools in general. Computer-aided engineering tools cover all aspects of engineering design from drawings to analysis to manufacturing. Computer-aided design (CAD) is a category of CAE related to the physical layout and drawing development of a system design. CAD programs specific to the electronics industry are known as electronic CAD (ECAD) or electronic design automation (EDA). EDA tools reduce development time and cost because they allow designs to be simulated and analyzed prior to purchasing and manufacturing hardware. Once a design has been proven through drawings, simulations, and analysis, the system can be manufactured. Applications used in manufacturing are known as computer-aided manufacturing (CAM) tools. CAM tools use software programs and design data (generated by the CAE tools) to control automated manufacturing machinery to turn a design concept into reality.

So how does OrCAD/Cadence fit into all of this? Cadence owns and manages many types of CAD/CAM products related to the electronics industry, including the OrCAD design suite. The OrCAD design suite can be purchased through resellers such as EMA Design Automation, Inc., which packages different combinations of CAD/CAM applications, including Capture, PSpice, and PCB Editor, to suit customers' needs. Although these applications can operate individually, bundling the individual tools into one suite allows for intertool communication.

Capture is the centerpiece of the package and acts as the prime EDA tool. Capture contains extensive parts libraries that may be used to generate schematics that stand alone or interact with PSpice, PCB Editor, or both simultaneously. A representation of a Capture part is shown in Figure 1-1.

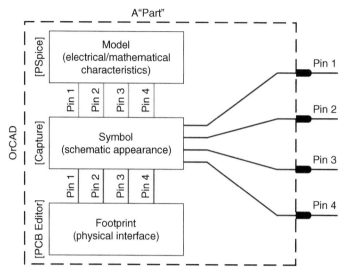

FIGURE 1-1
The pieces of a "part."

The pins on a Capture part can be mapped into the pins of a PSpice model or the pins of a physical package in PCB Editor. PSpice is a CAE tool that contains the mathematical models for performing simulations, and PCB Editor is a CAD tool that converts a symbolic schematic diagram into a physical representation of the design. Netlists are used to interconnect parts within a design and connect each of the parts with its model and footprint. In addition to being a CAD tool, PCB Editor also functions as a front-end CAM tool by generating the data on which other CAM tools operate when manufacturing the printed circuit board (PCB). Combining all three applications into one package produces a powerful set of tools to efficiently design, test, and build electronic circuits. The key to successful project design and production is in understanding the PCB itself and knowing how to use the tools that build the PCB.

PRINTED CIRCUIT BOARD FABRICATION

We now look at how PCBs are manufactured for a better understanding of what we are trying to accomplish with PCB Editor and why. A PCB consists of two basic parts: a substrate (the board) and printed wires (the copper traces). The substrate provides a structure that physically holds the circuit components and printed wires in place and provides electrical insulation between conductive parts. A common type of substrate is FR4, which is a fiberglass/epoxy laminate. It is similar to older types of fiberglass boards but is flame resistant. Substrates are also made from Teflon, ceramics, and special polymers.

PCB Cores and Layer Stack-Up

During manufacturing the PCB starts out as a copper clad substrate as shown in Figure 1-2.

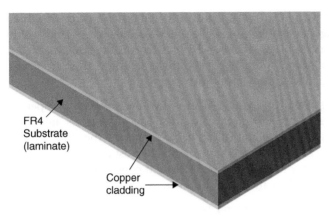

FIGURE 1-2
A double-sided copper clad FR4 substrate.

A rigid substrate is a C-stage laminate (fully cured epoxy). The copper cladding may be copper plated onto the substrate or copper foil glued to the substrate. The thickness of the copper is measured in ounces (oz) of copper per square foot, where $1.0\,oz/ft^2$ of copper is approximately 1.2–1.4 mils (0.0012–0.0014 in.) thick. It is common to drop "$/ft^2$" and refer to the thickness only in oz. For example, you can order 1 oz copper on a $\frac{1}{8}$-in.-thick FR4 substrate.

A substrate can have copper on one or both sides. Multilayer boards are made up of one or more single- or double-sided substrates called *cores*. A core is a copper-plated epoxy laminate. The cores are glued together with one or more sheets of a partially cured epoxy, as shown in Figure 1-3.

FIGURE 1-3
Cores and prepreg.

The sheets are also referred to as *prepreg* or *B-stage laminate*. Once all of the cores are patterned (described next) and aligned, the entire assembly is fully cured in a heated press.

There are three methods of assembling the cores when making a multilayer board. Figure 1-4 shows the first two methods in an example with four routing

FIGURE 1-4
Two stack-up methods for a six-layer board: left, multicore, outer clad; right, multicore, outer foil.

layers and two plane layers. Figure 1-4 (left) shows three (double-sided) cores bonded together by two prepreg layers, while Figure 1-4 (right) shows the same six layers made of two cores, which make up the four inner layers, bonded together by one prepreg layer. The outer layers in this panel are copper foil sheets bonded to the assembly with prepreg.

The routing layers in Figure 1-4 are shown as patterned copper segments, and the plane layers are shown as solid lines. The inner layers are patterned prior to bonding the cores together. The outer layers are patterned later in the process, after the cores have been bonded and cured and most of the holes have been drilled. Because the outer layers are etched later and copper foil is typically less expensive than copper cladding, the stack-up shown in Figure 1-4 (right) is more widely used.

The third method uses several fabrication techniques by which highly complex boards can be fabricated, as illustrated in Figure 1-5. This circuit board may have a typical four-layer core stack-up at its center, but additional layers are built up layer by layer on the top and the bottom, using sequential lamination techniques. The techniques can be used to produce blind and buried vias as well as typical plated through-hole vias, nonplated holes, and back-drilled plated through holes. Resistors and capacitors can also be embedded into the substrate. More about blind vias is discussed in later chapters (Chapters 8 and 9).

FIGURE 1-5
A built-up, multitechnology, PCB stack-up.

PCB Fabrication Process

The copper traces and pads seen on a PCB are produced by selectively removing the copper cladding and foil. Two methods are commonly used for removing the unwanted copper: wet acid etching and mechanical milling. Acid etching is more common when manufacturing large quantities of boards because many boards can be made simultaneously. One drawback to wet etching is that the chemicals are hazardous and must be replenished occasionally, and the depleted chemicals must be recycled or discarded. Milling is usually used for smaller production runs and prototype boards. During milling, the traces and pads are formed by a rotating bit that grinds the unwanted copper from the substrate. With either method, a digital map is made of the copper patterns. The purpose of CAD software like OrCAD PCB Editor is to generate the digital maps.

Note: Only one layer is considered in the following explanation of the fabrication process.

Photolithography and Chemical Etching

Selectively removing the copper with etching processes requires etching the unwanted copper while protecting the wanted copper from the etchant. This protection is provided by a polymer coating (called *photoresist*) deposited onto the surface of the copper cladding, as shown in Figure 1-6.

FIGURE 1-6
A copper clad board coated with photoresist.

The photoresist is patterned into the shape of the desired printed circuit through a process called *photolithography*. The patterned resist protects selected areas of the copper from the etchant and exposes the copper to be etched.

Resist

The two steps to photolithography are patterning the photoresist and developing it. Patterning is accomplished by exposing the resist to light (typically ultraviolet), and developing it is accomplished by washing it in a chemical bath. The two types of photoresist are positive resist and negative resist. When positive resist is exposed to ultraviolet (UV) light, the polymer breaks down and can be removed from the copper. Conversely, negative resist shielded from UV light is removed.

Cu

Substrate

A mask is used to expose the desired part of the photoresist. A mask is a specialized black and white photographic film or glass photoplate on which a picture of the traces and pads is printed with a laser photoplotter. Two types of masks are shown in Figure 1-7.

FIGURE 1-7
Photolithography masks: (left) positive mask, (right) negative mask.

The masks are examples of a trace connected to a pad. Figure 1-7 (left) shows a positive mask used to expose positive photoresist, and Figure 1-7 (right) shows a negative mask used to expose negative photoresist. Masks that will be used repeatedly are sometimes produced on glass photoplates instead of film.

The mask is placed on top of the photoresist, as shown in Figure 1-8, and the assembly is exposed to the UV light. The dark areas block UV light and the white (transparent) areas allow the UV light to hit the photoresist, which imprints the circuit image into the photoresist. A separate mask is used for each layer of a circuit board. OrCAD PCB Editor generates the data that the photoplotter uses to make these masks.

Another way of exposing the photoresist is by using a programmable laser to "draw" the pattern directly onto the photoresist. This is a newer technique, called *laser direct imaging* (LDI). A benefit of the LDI process is that it uses the same data as the photoplotters but no masks are required.

After the photoresist has been exposed (either with the mask and UV or with the laser), it is washed in a chemical called the *developer*. In the case of positive resist, the resist breaks down during exposure and is removed by the developer. In the case of negative resist, the UV light cures the resist, and only the unexposed resist is removed by the developer. Common developers are sodium hydroxide (NaOH) for positive resist and sodium carbonate (Na_2CO_3) for negative resist. Once the resist has been exposed and developed, a circuit image made of the photoresist is left on the copper, as shown in Figure 1-9.

FIGURE 1-8
Positive photomask on photoresist-coated board.

Next, the board is etched in a corrosive solution, such as alkaline ammonia or cupric chloride. The etching solution does not significantly affect the photoresist but attacks the bare copper and removes it from the substrate, leaving behind the resist-coated copper, as shown in Figure 1-10.

Some processes use a plated tin alloy as the etch resist. The tin alloy plating is more resistant to etchants and preprepares the copper surface for solder processes. In this case, photolithography processes are used to selectively plate the circuit pattern onto the copper surfaces prior to etching.

FIGURE 1-9
Developed photoresist on copper.

When polymer etch resists are used, the photoresist is cleaned from the copper with a resist stripper, leaving behind the copper traces. Figure 1-11 shows the final patterned copper. When metal etch resists are used, the plating is typically left in place. Holes for the leads and so forth are not etched into the pads because they are drilled after all of the cores have been glued together (later in the process) to ensure proper alignment of the holes between board layers.

Mechanical Milling

As mentioned previously, milling is an alternative to etching. To mill the board, a computer numerical control (CNC) machine is programmed with the digital map of the board and grinds away the unwanted copper. The unwanted copper can be completely removed (like that in Figure 1-11), or just enough copper may be removed to isolate the pads and traces from the bulk copper, as shown in Figure 1-12. Removing only enough copper to isolate the traces from the bulk copper reduces milling time but can affect the impedance of the traces.

FIGURE 1-10
Unwanted copper removed by etching.

FIGURE 1-11
Copper pad and trace after etching and resist stripping.

FIGURE 1-12
A mechanically milled trace.

Layer Registration

After the inner layers have been patterned, the cores are aligned (called *registration*) and glued together. Registration is critical because the pads on each layer need to be properly aligned when the holes are drilled. Registration is accomplished using alignment patterns (called *fiducials*) and tooling holes in the board, which slide onto guide pins. With the cores in place and properly aligned, a heated press cures the assembly.

After the assembly is cured, holes are drilled for through-hole component leads and vias. The drilling process inevitably heats the laminate due to friction between the laminate and the high-speed drill bit. This tends to soften the laminate and smear it across the walls of the drilled copper. After the drilling processes are finished, the assembly is placed into a bath to etch back the laminate slightly and clean the faces of the copper pad walls. This is called *laminate etchback* or *desmear*.

Once the holes have been drilled and desmeared, a physical path exists between pads on different layers, but as Figure 1-13 (left) shows, there is no electrical connection between them. To make electrical connections between pads on different layers, the board is placed into a plating bath that coats the insides of the holes with copper, which electrically connects the pads, hence, the term *plated* through holes. The plating thickness varies but is typically about 1 mil (0.001 in.) thick. The cutaway view of Figure 1-13 (right) shows a plated through hole on an internal layer of a PCB. The top and bottom copper is actually patterned *after* the plating process is finished because the plating process would replate the areas where copper had been removed.

FIGURE 1-13
Holes are drilled into the board and copper plated: (left) a non-plated through hole, (right) a plated through hole.

FIGURE 1-14
A connection to a plane layer through a thermal relief.

Not all layers have traces. Some layers are planes (see Figure 1-14). Plane layers are typically used to provide low-impedance (resistance and inductance) connections to power and ground and to provide easy access to power and ground at any location on the board. Plane layers and impedance issues are discussed in Chapter 6. The leads of components are connected to ground or power by soldering them into plated through holes. Since copper conducts heat well, soldering to a plane layer could require an excessive amount of heat, which could damage the components or the plating in the hole (called the *barrel*). Thermal reliefs are used, as shown in Figure 1-14, to reduce the path for heat conduction but maintain electrical continuity with the plane.

Since ground and power planes are often inner layers and signal layers will likely be above and below them, in some instances, a via will run through a plane layer but must not touch it. In this case, a "clearance" area (shown in Figure 1-15) is etched into the plane layer around the via to prevent a connection to the plane.

The clearance is larger than the normal pad size to ensure that the plane stays isolated from the plated hole.

After the through holes are plated, the top and bottom layers are patterned using the photolithograpy process as described for the inner layers. After the outer copper has been patterned, the exposed traces and plated through holes can be tinned (although tinning is sometimes deferred until later). Nonplated holes (such as for mounting holes) may be drilled at this time.

FIGURE 1-15
A clearance area provides isolation between a plated hole and a plane.

Next, a thin polymer layer is usually applied to the top and bottom of the board. This layer (shown in dark green in Figure 1-16) is called the *soldermask* or *solder resist*. Holes are opened into the polymer using photolithography to expose the pads and holes where components will be soldered to the board. The soldermask protects the top and bottom copper from oxidation and helps prevent solder bridges from forming between closely spaced pads. Sometimes openings in the soldermask are not made over small or densely placed vias (called *tenting a via*). Tented vias are protected from having chemicals such as flux from becoming trapped inside the hole. Tenting also prevents solder migration into the hole, which could lead to poor solder joints on small components that are close to and connected to the via.

FIGURE 1-16
Final layers are the soldermask (dark green) and silk screen (white).

Finally, markings (called the *silk screen*) are placed on the board to identify where components are to be placed. The silk screen is shown in white in Figure 1-16.

FUNCTION OF OrCAD PCB EDITOR IN THE PCB DESIGN PROCESS

PCB Editor is used to design the PCB by generating a digital description of the board layers for photoplotters and CNC machines, which are used to manufacture the boards. Separate layers are used for routing copper traces on the top, bottom, and all inner layers; drill hole sizes and locations; soldermasks; silk screens; solder paste; part placement; and board dimensions. These layers are not all portrayed identically in PCB Editor. Some of the layers are shown from a positive perspective, meaning what you see with the software is what is *placed onto* the board, while other layers are shown from a negative perspective, meaning what you see with the software is what is *removed from* the board. The layers represented in the positive view are the board outline, routed copper, silk screens, solder paste, and assembly information. The layers represented in the negative view are drill holes and soldermasks. Copper plane layers are handled in a special way, as described next.

Figure 1-17 shows routed layers (top and bottom and an inner, for example) that PCB Editor shows in the positive perspective. The background is black and the traces and pads on each layer are a different color to make it easier to keep track of visually. The drill holes are not shown because, as mentioned already, the drilling process is a distinct step performed at a specific time during the manufacturing process.

FIGURE 1-17
Copper in routed layers.

Figure 1-18 shows examples of drill symbols and silk screen representations used by PCB Editor (traces not shown). The red shapes are examples of drill symbols that indicate locations of the drill holes and are used in conjunction with a drill chart (see Figure 1-19), which produces ID numbers for the different drill tools. The white print is the silk screen discussed previously.

FIGURE 1-18
Circuit layout with drill symbols and silk screen on a PCB.

FIGURE 1-19
Drill chart with drill symbols and specifications.

Figure 1-20(a) shows the soldermask with the patterned holes that allow access to pads, and Figure 1-20(b) shows the negative representation used by PCB Editor. Here, the black background is actually the polymer film and the green circles are the holes in the soldermask.

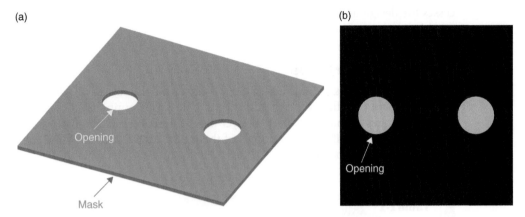

FIGURE 1-20
Soldermask layer and negative view: (a) soldermask, (b) negative view in PCB Editor.

Plane layers are processed as negative layers in PCB Editor and are used to generate negative image Gerber files. However, PCB Editor displays plane layers in the positive view. Figure 1-21(a) shows a physical copper plane layer with a thermal relief for a pin or via, Figure 1-21(b) shows the negative view, and 1–21(c) shows the representation displayed by PCB Editor in "what you see is what you get" (WYSIWYG) mode.

FIGURE 1-21
Copper in a plane layer: (a) copper plane with thermal relief, (b) negative view as processed, (c) displayed in WYSIWYG mode.

DESIGN FILES CREATED BY PCB EDITOR

PCB Editor Format Files

When you are designing a board, PCB Editor works with and saves the design in a format that is efficient for the computer. The PCB Editor design file has a .brd extension. When you are ready to fabricate the board, PCB Editor processes the design and converts it into a format that the photoplotters and CNC machines can use. These files are called *Gerber files*.

Artwork (Gerber) Files

Artwork (Gerber) files are created for each of the etch and mask layers discussed previously, and drill files are created for both plated and nonplated holes. PCB Editor generates as many as 30 or so layer files to describe various manufacturing aspects of the PCB. Some examples of these files, their extensions, and their functions are listed in Table 1-1.

Table 1-1 Drill Files, Their Extensions, and Their Functions	
File name and extension	**Function**
ProjectName.brd	Main board project file
Assembly_Top.art	Top side assembly
Solderpaste_Top.art	Top side solder paste
Silkscreen_Top.art	Top side silk screen
Soldermask_Top.art	Top side soldermask
TOP.art	Top side copper (usually routing)
INNER1.art	Inner layer 1 (usually routing)
INNER2.art	Inner layer 2 (usually routing)
INNERx.art	Inner layer x (usually routing)
PWR.art	Power layer (a plane layer)
GND.art	Ground layer (a plane layer)
BOTTOM.art	Bottom side copper (usually routing)
Soldermask_Bottom.art	Bottom side soldermask
Silkscreen_Bottom.art	Bottom side silk screen
Solderpaste_Bottom.art	Bottom side solder paste
Assembly_Bottom.art	Bottom side assembly
ProjectName.rou	Board outline cutting path
ProjectName.drl	Drill hole data

These and other files that PCB Editor generates are discussed in greater detail in the next two chapters and in the PCB Design Examples.

PCB Assembly Layers and Files

Several layer files generated by PCB Editor are not part of the actual fabrication process. These files are used for automated assembly of a finished board and are mentioned only briefly here. The first layer is the solder-paste layer. It is used to make a contact mask for selectively applying solder paste onto the PCB's pads so that components can be reflow soldered to the board. There may be one solder-paste layer for the top side of the board and one for the bottom side, as indicated in Table 1-1. The second layer file is the assembly layer, which contains information for board assemblers as to the part type, its position, and its orientation on the board. As with the soldermask, there may be one assembly layer for the top side of the board and one for the bottom side. PCB design for the various soldering and assembly processes is discussed in Chapter 5.

The purpose of this chapter has been to introduce you to the process by which PCBs are manufactured. The purpose of the next chapter is to show you how to use OrCAD PCB Editor to design a board and generate the files needed to manufacture the PCB.

Introduction to the PCB Design Flow by Example

Now that we have covered the construction of a PCB and know PCB Editor's role in it, we go through a simple design example so that you get a feel for the overall design process. This simple example sets the stage for Chapter 3, in which we dig deeper into the details of the process and learn more about PCB Editor itself and the PCB Design Examples.

OVERVIEW OF THE DESIGN FLOW

This section illustrates the basic procedure for generating a schematic in Capture and converting the schematic to a board design in PCB Editor. The basic procedure is as follows:

1. Start Capture and set up a PCB project using the PC Board wizard.
2. Make a circuit schematic using OrCAD Capture.
3. Use Capture to generate a PCB Editor netlist and automatically start PCB Editor and open the project as a .brd file.
4. Make a board outline.
5. Position the parts within the board outline.
6. Route the board.
7. Generate manufacturing data files.

Creating a Circuit Design with Capture

If you do not have a full version of OrCAD, you can go to the OrCAD Web site and download the latest demo. If you are using an older version of PCB Editor, most of the following information in this book still applies, but some of the dialog boxes and menu items may be different.

STARTING A NEW PROJECT

Before you make a PCB layout, you need to have a circuit to lay out. You use Capture to make the schematic, so the first step is to start the Capture application by clicking the Windows Start button on your task bar and navigate to All Programs → OrCAD 16.0 Demo → OrCAD Capture. Once Capture is running,

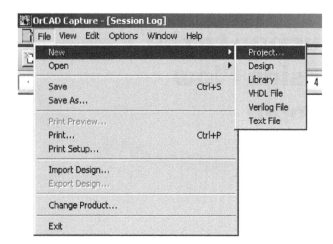

FIGURE 2-1
Starting a new project
in Capture.

FIGURE 2-2
New Project dialog box.

Toggle Grid
button

you should have a blank Capture session frame and a session log. Go to the File dropdown menu and navigate to File → New and click Project as shown in Figure 2-1.

The New Project dialog box in Figure 2-2 will pop up. Type a name for your project, then select the PC Board Wizard radio button. If you feel comfortable selecting your own location to save the project, you can do that (use the Browse... button), or you can use the default location for now (just remember where it is). Click OK.

After you click OK, the PCB Project Wizard dialog box shown in Figure 2-3 (left) will pop up. For now, circuit simulation will not be performed, so leave the Enable project simulation box unchecked (we look at circuit simulation in the PCB Design Examples). Click Next.

After you click Next, the PCB Project Wizard dialog box shown in Figure 2-3 (right) will pop up. This box allows you to add specific libraries to your project. Scroll down until you find the Discrete.olb library, highlight it by clicking on it, then click the Add>> button; then click Finish. This completes the project setup.

You should have a Project Manager window in the left side of the Capture session frame, as shown in Figure 2-4.

You may also have a Schematic window in the work space. If the schematic is not open, expand the *projectname*.dsn directory by clicking the "+" box to the left of the *projectname*.dsn icon (where *projectname* is the name you gave your project while using the project setup wizard). Click the + box next to the Schematics folder, then double click the file called Page1. The Schematic page should open. If you do not see the dots, that means your grid is turned off. *The grid must be turned on to properly place and connect parts. To turn the grid on,* click the Toggle Grid button, ■. If the grid is on, the grid dots are visible and the grid button is gray instead of red.

Note: The menu bar at the top of the Capture session frame changes depending on whether you are working with the Project Manager window or the Schematic

FIGURE 2-3
PCB Project Wizard dialog boxes: (left) simulation selection, (right) parts library selection.

FIGURE 2-4
Example view of a new project.

Page window. If you need to access options or tools for the Project Manager (or Schematic Page), you must have that window active. To make the desired window active, click on its title bar or select the desired window from the Window menu. When inactive, it is gray; and when active, it is blue (or whatever colors you set up in Windows). Also, for projects that have PSpice simulation capabilities, an additional toolbar is displayed, which is not shown in Figure 2-4.

PLACING PARTS

Place Part
tool button

To add parts to your schematic, make the Schematic page active and select Place from the Part dropdown menu, or press the Place Part tool button ▱, or press P on your keyboard. The Place Part dialog box shown in Figure 2-5 will pop up. In the Libraries selection box in the bottom left of the dialog box, click DISCRETE. Then, in the Part List box, click R (for resistor). You should see its symbol in the preview window on the lower right. Click OK. In the Libraries window, you may have libraries different from that shown in Figure 2-5. At the very least, you should have the DISCRETE library, since you had the wizard include it. If, for some reason, you do not see any parts or the DISCRETE library is not there, you can follow along for now to get an overview of the process or you can find and add the library to your project.

FIGURE 2-5
Place Part dialog box.

To add a part library to the project, select Place from the Part dropdown menu as described previously. In the Place Part dialog box shown in Figure 2-5, press the Add Library... button to bring up the Browse File dialog box shown in Figure 2-6.

Find and select the desired library and click Open. You can also find a resistor in the pspice library folder. To add it, double click the pspice folder, select the Analog.olb library, and click Open. You should now be back to the Place Part dialog box, and the library you just added should be shown in the Libraries list box. Find and select R from the Part List selection box and click OK.

FIGURE 2-6
Add a library using the
Browse File dialog box.

After you click OK, you should immediately return to the Schematic page and have a resistor tagging along with your mouse pointer. Left click on the schematic page to place the part as shown in Figure 2-7.

Place one resistor on the page; then hit the ESC key or right click the mouse and select End Mode from the pop-up menu. Before placing more parts, we assign a footprint to the existing part and make copies of it. *To assign a footprint to a*

FIGURE 2-7
Placing the parts.

part, double click the part to display the Property Editor spreadsheet (see Figure 2-8). The cells may be shown vertically or horizontally. Find the PCB Footprint cell and enter res400. This is a basic through-hole, axial lead resistor footprint. Finding and selecting footprints is described in greater detail in Chapter 8. If the PCB Footprint variable is not shown, make sure that the <Current properties> option is selected in the Filter by: list. Once the footprint has been assigned, close the spreadsheet.

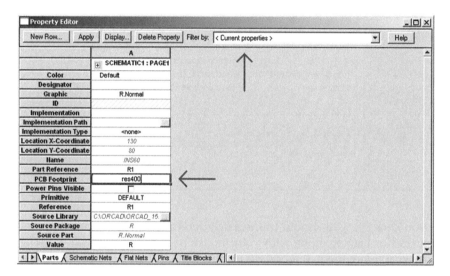

FIGURE 2-8
Assigning a footprint to a part with the Property Editor.

Next, duplicate the resistor. To do so, left click the resistor to select it and use standard Windows copy and paste procedures (Ctrl + C and then Ctrl + V, or Edit → Copy and then Edit → Paste from the menu) to place several copies of the resistor onto the schematic page.

WIRING (CONNECTING) THE PARTS

Place Wire
tool button

Next, connect the parts with wires. *To place wires,* hit the W key, select Place → Wire from the Place dropdown menu, or press the Place Wire tool button, ⅃. The cursor will turn into a crosshair. Place the cursor on a box at the end of one of the resistor's leads and left click to start a wire (see Figure 2-9).

Click on the end of another resistor lead to complete that wire. The crosshair will persist, so you can continue placing wires. Finish connecting wires to the resistors however you wish. Once you have finished connecting the circuit, press the ESC key or right click and select End Wire to stop the place wire cursor and get the pointer back. If you inadvertently click near a lead but not on it, the wire may appear to be connected but may not be (that is why it is important to have the grid enabled). If the connections are not made properly, you will have problems when attempting to generate a netlist. You will be able to

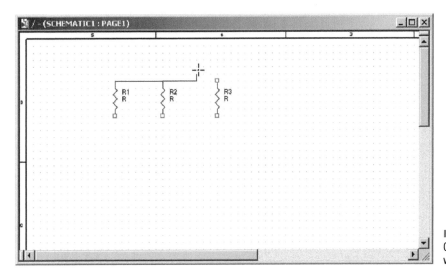

FIGURE 2-9
Connecting the parts with wires.

tell if a connection you made to a component was completed properly because the box at the end of the lead will disappear. At this point do not worry about power supplies or ground connections; this is just a "big picture" exercise to demonstrate the design flow process.

CREATING THE PCB EDITOR NETLIST IN CAPTURE

Once all the connections are complete, the next step is to create a netlist (a set of files that describes the circuit). Several types of netlists are possible, but you want *to generate a PCB Editor netlist*. Begin by making the Project Manager window active (instead of the Schematic Page window) and select the .dsn icon by left clicking it once. If the Schematic page is active, the Tools menu will not be available. Minimize the Schematic page if necessary to get to the Project Manager. As shown in Figure 2-10, select Tools → Create Netlist from the Tools menu. The Create Netlist dialog box will pop up as shown in Figure 2-10.

From the Create Netlist dialog box, select the PCB Editor tab. Select the Create or Update PCB Editor Board (Netrev) option. For the time being, use the default output board file name and path. Select the Create Open Board in OrCAD PCB Editor... option to automatically launch the PCB Editor applicaton. Click OK to start the process.

Capture will display a warning text box stating 'Directory "C:\...\projectname\allegro' specified as Netlist directory does not exist. Do you want to create it? Click Yes. Capture will display a warning text box stating Design Path/projectname.dsn will be saved prior to netlisting. Click OK.

You should then see the progress box shown in Figure 2-11 as the netlist is generated. Capture will then generate the netlist files, report the results in the Session log, and launch PCB Editor. If you had the View Output option checked, the three netlist files will also be displayed in Capture. Leave Capture open so

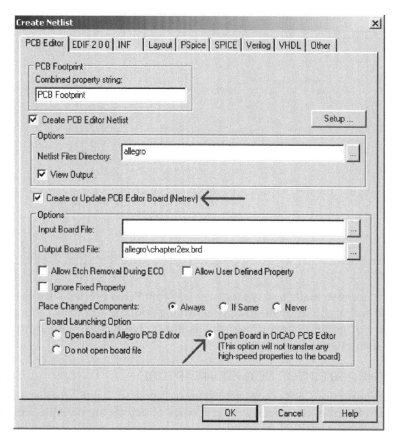

FIGURE 2-10
Creating a netlist for PCB Editor.

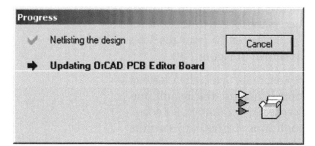

FIGURE 2-11
The netlist Progress dialog box.

that it and PCB Editor can communicate with each other if necessary. This will allow you to go back and review the circuit if you need to when you are working in PCB Editor.

DESIGNING THE PCB WITH PCB EDITOR

The PCB Editor Window

Once the PCB Editor application is running, you should end up in the board layout environment shown in Figure 2-12.

Figure 2-12 shows six areas of the work environment. These areas are discussed in detail in Chapter 3 and only briefly introduced here. The design window is where you work on your board. With the design window and the various tools controlled by the control panel, menus, and toolbars, you can control various aspects of components, traces, and planes. Examples of controllable aspects include whether items are visible or not, whether they are fixed or

FIGURE 2-12
The PCB Editor environment and initial view.

movable, component location and orientation, and trace and board characteristics, to name just a few. The command window displays messages and prompts and allows you to enter commands. The worldview window gives you a bird's-eye view of the location and size of the design widow relative to the actual PCB. The status window lets you know if the program is busy or idle and shows the coordinates of the crosshairs. The P and A buttons are related to the display and coordinates but are described later.

PCB editor begins with nothing in the design window and parts are not visible until they are placed, so we need to draw a board outline and place the parts.

CONTROLLING THE VIEW

The initial size of the design window is about 33 in. wide by 22 in. tall. To make a realistic board size for this simple example you want to zoom in. There are a couple of ways to do that. The zoom tools, , are Zoom to Points, Fit, In, Out, Previous, Selection, and Redraw, respectively. Click the Zoom to Points button, , place the cursor at about coordinate position 4600, 3600 (units in mils, i.e., thousanths of an inch) then left click and release. Move the cursor down and to the left until it is less than or close to coordinates 0, 0. As the mouse is moved, a rectangular box with crossbars will become visible. Left click and release again to finish the Zoom to Points command. The design window covers a work area that is a little more than 4 in. by 3 in., which is plenty large for the few components in this example.

Zoom tools

Zoom to Points button

FIGURE 2-13
New Board Outline dialog box.

FIGURE 2-14
The Options pane.

Drawing the Board Outline

To draw the board outline, select Setup → Outlines → Board Outline from the menu. The Board Outline dialog box will be displayed, as shown in Figure 2-13. With the dialog box still displayed, move your mouse over to the design area and left mouse click at point 0, 0 (or close to it). While watching the cursor coordinates on the status bar (lower right side of the design window), move the mouse up about 4 in. (4000 mils) and to the right about 3 in. (3000 mils) and left click. A dashed outline will be displayed, and the Board Outline dialog box changes to Edit mode. The Board Edge Clearance: determines how far from the edges of the board the place keep-in and route keep-in boundaries will be. The default value is 400 mils. For this design example, enter 200 mils in the text box and click OK.

When you click OK, another rectangle is automatically drawn inside the board outline. Actually there are two rectangles, one directly on top of the other. The two outlines are the route keep-in and package keep-in outlines. The route keep-in outline defines the edge of any plane layer and the boundary within which traces must stay. The package outline defines the area within which all components must reside.

If you hover your mouse over the inner rectangle, the data tips box will tell you which outline it is. You can turn that layer off to view the other rectangle. If the data tip says Package/Keepin (or Route Keepin), move your mouse over to the Options tab (see Figure 2-14) to display the pane, and select Package Keepin class from the list. Left click the colored box next to the All subclass. The box will turn black and the place keep-in outline will disappear. You should then see the other outline, and the data tip should tell you it is the Route Keepin/All outline (or the Package Keepin). You can turn this one off by selecting Route Keepin class from the Options tab as described previously.

You can select and view the different boundaries by selecting them from the class list in the Options pane or the Visibility pane in the control panel area at the right of the design window (see Figure 2-14). The class and subclass for each outline so far is Board Geometry & Outline, Route Keepin and All, and Package Keepin and All, respectively. Classes and subclasses are discussed in greater detail in Chapters 3 and 8 and the PCB Design Examples.

Placing Parts

Parts can be placed manually or automatically. Multiple automatic placement modes are available, but we wait until the PCB Design Examples to use them. *To place parts manually,* select Place → Manually... from the menu; click the Manual Place button, , on the toolbar; or select Place → Manually... from the menu bar. The Placement dialog box will be displayed, as shown in Figure 2-15. Select the Placement List tab if it is not already selected. Click the Components by refdes box; this will automatically select all the components. Leave the Placement dialog box displayed (i.e., do not dismiss it by clicking the OK or Cancel button). Move the mouse cursor over to the design window; a component will be attached to it. Left click in the design window to place the part. The next part in the list will automatically be attached to the mouse cursor. The box next to the part that was just placed will be unchecked (it is no longer in the queue). Once all the parts are placed, dismiss the Placement dialog box by clicking OK.

Manual Place button

FIGURE 2-15
Manually placing components.

Moving and Rotating Parts

To move parts, select the Move button, , click and release the left mouse button to select a part. Move the part to the desired location then click and release the left mouse button to place the part. You can continue moving other parts or right click and select Done from the pop-up menu to deactivate the move tool. *To rotate a part,* use the move tool as just described to select the part, then right click and select Rotate from the pop-up menu. The component will pivot around pin 1 as you move the mouse around. Left click to stop rotating and left click again to place it, then right click and select Done from the pop-up menu.

Move button

FIGURE 2-16
Parts of a component footprint.

Figure 2-16 shows the component footprints as well as the silk screen and assembly details. The parts of the footprint are

Text:

- a. Assembly top (Comp Value)
- b. Silkscreen top (Comp Value)
- c. Assembly Top (Ref Des)
- d. Silkscreen top (Ref Des)
- e. Assembly top (Tolerance)
- f. Silkscreen top (Tolerance)
- g. Assembly top (User Part)
- h. Silkscreen top (User Part)

Objects:

1. Place_Bound_Top (Package Geometry)
2. Pin_Number (Package Geometry)
3. Assembly_Top (Package Geometry)
4. Silkscreen_Top (Package Geometry)

Color button

You can change the visibility of the various parts of the footprint (and PCB as a whole) using the Color and Visibility dialog box. Click the Color button, ▦, or select Display → Color/Visibility from the menu. *To control the visibility of text objects,* select Components from the Group list, as shown in Figure 2-17. *To control the visibility of package objects,* select Package Geometry from the Group list as

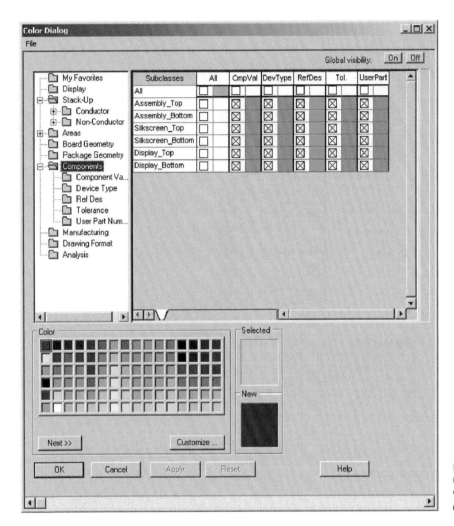

FIGURE 2-17
Controlling the color and visibility of components objects.

shown in Figure 2-18. *To control the visibility of padstacks,* select Stackup from the Group list. You control the visibility of the various parts by checking or unchecking the boxes then clicking Apply. You can customize the color of any text, etch, or detail object by selecting a color from the palate then clicking the color square of the item you want to customize. More will be said about footprint composition and construction in Chapter 8.

Routing the Board

USING THE AUTOROUTER

The next step is to route the board. *To route the board using the autorouter,* select Route → Route Automatic… from the menu; the Automatic Router dialog box

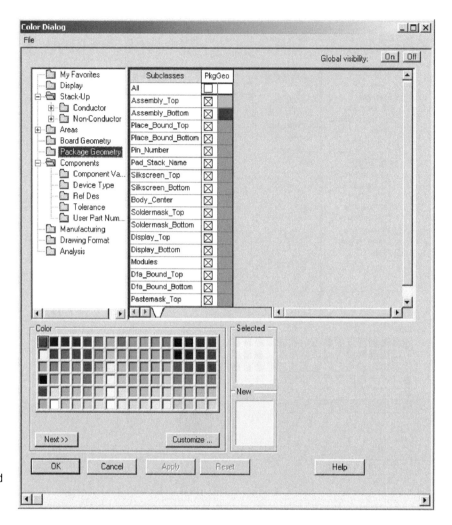

FIGURE 2-18
Controlling the color and
visibility of geometry
objects.

shown in Figure 2-19 will be displayed. The autorouter can be run in three modes. The different options will be demonstrated in the PCB Design Examples. For the time being, leave everything as is and click the Route button.

The autorouter will begin routing the board and a status box will be displayed (Figure 2-20). When the routing is complete the status box will be dismissed. Figure 2-21 shows how the board might look after routing.

In complex designs, you need to know how to use the manual routing tools to preroute critical traces and clean up after the autorouter. The manual routing tools are described next.

MANUAL ROUTING

In Chapter 3 and the PCB Design Examples, we take an extensive look at manually routing traces, so the manual routing is only briefly introduced here. First,

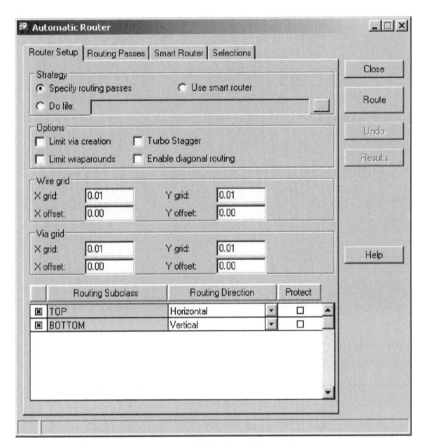

FIGURE 2-19
Starting the autorouter.

FIGURE 2-20
Automatic Router
Progress dialog box.

unroute (rip up) the traces routed by the autorouter. *To rip up a trace,* select the Delete button, ✖, and left click the trace you want to rip up. The trace will be highlighted. To complete the rip-up, click the trace again, select another trace, or right click and select Done from the pop-up menu. Ripping up multiple (or all) traces from a list is demonstrated in the PCB Design Examples.

To begin manual routing, click the Add Connect button, ⌐, on the toolbar. Left click a rat's nest to begin routing. Left click again to place a vertex. Clicking on

✖

Delete button

Add Connect button

FIGURE 2-21
Example of a routed board.

a pin ends the route, or you can right click and select Finish to end the route. To stop routing, right click and select Done from the pop-up menu.

Note: If the route tool does not seem to be working, display the Find pane and make sure that Ratsnest box is checked.

To add a via, select a net with the Add Connect tool, place a vertex, right click, and select Add Via from the pop-up menu. A via will be placed at the vertex and routing continues on the alternate layer listed in the Options pane.

Slide button

If you want *to push (move) a trace* without rerouting it, use the Slide button, ▣. Left click the trace, move your mouse pointer to the desired location, and left click again to place the trace.

By using the control panel's Options pane, you can change routing properties, such as the active layer (Act), alternate layer (Alt), the trace width, and line type. *To select a different routing layer*, use the Act dropdown list to select the desired layer. When you are finished routing, right click and select Done from the pop-up menu.

These are just a few of the basic tools you need to know about. In the PCB Design Examples, you will see how to use more of the tools and control routing properties, such as setting minimum and maximum trace widths and setting up plane layers.

PERFORMING A DESIGN RULE CHECK

After you have completed routing your board, you should check for errors. *To check for errors*, run the design rule checker (DRC) by selecting Tools → Update

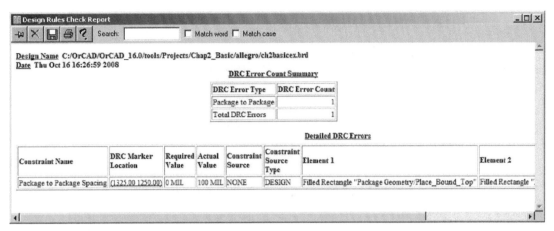

FIGURE 2-22
An example DRC report.

DRC from the menu. A message will be displayed in the console window (at the bottom of the screen) as to whether there were errors. If there are errors, the message in the console window does not go into detail about what they are, just that they exist. If you want to see a description of the errors, run a DRC report. *To run a DRC report,* select Tools → Quick Reports → Design Rules Check Report from the menu. An example of a DRC Report is shown in Figure 2-22. The report shows what would be reported if R2 and R3 were too close to each other (not the circuit in Figure 2-21).

Creating Artwork for Manufacturing

At this stage, PCB Editor has generated a design file that fully describes your board. This file is optimized for viewing, editing, and saving on your computer, but it is not in the format that many PCB manufacturers use for fabricating boards. The most common type of file system used in PCB manufacturing is the Gerber file system. PCB Editor has the capability of translating its .brd file to Gerber files. In PCB Editor, this is called *manufacturing* the design. PCB Editor allows considerable control over the manufacturing. As a result, several steps are required to generate all of the manufacturing files, especially with more complicated designs. Setting up the manufacturing process and generating the Gerber files is described in detail in Chapter 10. A quick overview is given here.

To create artwork for your design, select Manufacture → Artwork... from the menu or click the Artwork button, , on the toolbar. The Artwork Control Form shown in Figure 2-23 is used to create the artwork. Artwork files and NC drill and router file creation are described in detail in Chapter 10.

Artwork button

Congratulations, you have routed your first PCB using OrCAD PCB Editor! The objectives of this chapter were to demonstrate the basic steps of designing a

FIGURE 2-23
Artwork Control Form
dialog box.

circuit schematic and using PCB Editor to design a PCB. The following is a summary of the process:

1. Start Capture and set up a PCB project using the PC Board wizard.
2. Make a circuit schematic using OrCAD Capture.
3. Use Capture to generate a PCB Editor netlist and automatically start PCB Editor.
4. Make a board outline.
5. Position the parts within the board outline.
6. Route the board.
7. Generate manufacturing artwork.

In the next chapter, we cover the design flow in greater detail and learn more about the PCB Editor tool set; and in Chapters 8 and 9 and the PCB Design Examples, we learn more about footprint design, PCB layer design, and the steps to manufacturing the board design.

CHAPTER 3

Project Structures and the PCB Editor Tool Set

This chapter explains what you did when making the simple design in Chapter 2 and why. It also introduces and describes the PCB Editor tool set in greater detail, so that you will be well equipped to lay out more complicated boards in the PCB Design Examples in Chapter 9.

PROJECT SETUP AND SCHEMATIC ENTRY DETAILS

Capture Projects Explained

When you set up your project by following the File → New → Project menu path, you had five options from which to choose: a project, a design, a library, a VHDL file, or a Text file. The options we are most interested in are projects and libraries, and we look at those in great detail throughout the book. VHDL files are used in field programmable gate array projects and are not discussed here. A Text file is simply a text file (for making project notes, for example).

After you selected the New Project option to begin setting up your project, four more options were available to you in the New Project dialog box: Analog or Mixed-Signal A/D, PC Board Wizard, Programmable Logic Wizard, or Schematic. A flow diagram of these options and suboptions is shown in Figure 3-1 (see also Figure 2-2). Analog or Mixed-Signal A/D is used to simulate analog and/or digital circuits using PSpice. PSpice is used to develop and test models (Chapter 7), perform circuit simulations (Chapter 7 and the PCB Design Examples), and simulate transmission lines (PCB Design Examples). For now, we work mostly with the second option, the PC Board Wizard, while we focus on designing PCBs. The next option, Programmable Logic Wizard, is for working with programmable devices and is not discussed in this text. A Schematic is basically just a design file with only a schematic and a parts cache. The thick green line in Figure 3-1 shows the path you followed in Chapter 2, that is, File → New → Project → PC Board Wizard → No Simulation.

There are four types of OrCAD libraries. As shown in Figure 3-2, these four libraries have three file extensions. There are two types of OLB libraries: an LIB library and a symbols library, which contains .dra files as well as some

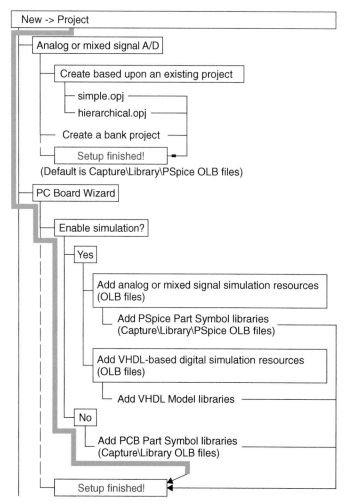

FIGURE 3-1
New Project design flow
and options.

other files that are explained in detail in Chapter 8. Inside the capture folder is a library folder, which contains one of the types of OLB files, and a folder called *pspice*, which contains the other types of OLB files. The OLB files located directly in the Capture library folder contain simple schematic part symbols and are the ones we used in Chapter 2. The libraries located in the pspice folder contain parts with schematic symbols, too, but the parts also contain PSpice templates, which are links to specific PSpice models. The PSpice models to which templates point are located in the main OrCAD pspice folder (shown at the bottom of Figure 3-2). Individual models are grouped into various PSpice library files, which contain the .LIB extension. The share folder contains the footprint models (called *symbols*), which are files with the .dra extension. Most parts in Capture are capable of having a PSpice template or a footprint assigned to them, but only some parts have them preassigned. It may seem

Address C:\OrCAD\OrCAD_16.0

Folders   ✕

⊟ 📁 OrCAD_16.0
   📁 bin
  ⊞ 📁 doc
  ⊞ 📁 share
   ⊟ 📁 pcb
    ⊟ 📁 pcb_lib
     📁 devices
    ⊞ 📁 symbols
  ⊟ 📁 tools
   📁 bin
   ⊟ 📁 capture
    ⊟ 📁 library
     ⊞ 📁 pspice
     📁 UserLib
    ⊞ 📁 templates
   ⊟ 📁 pspice
    📁 library
    📁 Templates
    📁 UserLib

PCB Editor Footprint Library
(*.dra)

Schematic Capture Library
(*.OLB)
(no simulation capabilities)

Schematic Capture Library
(*.OLB)
(with simulation capabilities)

PSpice Model Library
(*.LIB)
(PSpice Templates)

FIGURE 3-2
OrCAD libraries (extraneous files not shown).

backward, but only parts from the PSpice part library (located in the Capture/library folder) have preassigned templates and footprints, and the footprints are OrCAD Layout footprints not PCB Editor footprints. So, in Chapter 2, when you worked with the PCB Project Wizard, you selected libraries from the Capture/library folder, which had parts with schematic symbols but had no PSpice simulation capabilities or footprints assigned to them.

Once you clicked Finish in the PCB Project Wizard box, Capture opened up the Project Manager window shown in Figure 3-3. Behind the scenes, Capture generated two files: an OrCAD project file (name.OPJ) and a design file (name.DSN). These files will be in the directory you chose when you set up the project.

As you look at the Project Manager window, you can see that a project contains three folders labeled Design Resources, Outputs, and Referenced Projects. Initially, the Outputs and Referenced Projects folders are empty. When a netlist is created, netlist files are placed in the Outputs folder. The Design Resources folder contains one design (represented by the icon) and a Library folder. A project can have only one design, but a design can have several subfolders, which in turn may contain several different items. The Library folder contains links to the libraries used by your design. We discuss library management in Chapter 7. The design contains at least one Schematic folder (the root folder) and a Design Cache folder. A design can contain multiple Schematic folders, and each Schematic folder can contain multiple schematic pages. The Design Cache folder contains a record of each part you used in your design. If you modify one of the parts on a Schematic page, Capture makes a copy of it (leaving the original part in the library unchanged) and adds a record of the modified part to the design

FIGURE 3-3
Project
Manager
window.

cache. A design with one Schematic folder and one or more Schematic pages connected together by off page connectors is called a *flat design*. A design with more than one Schematic folder and one or more Schematic pages per folder or that contains hierarchical blocks is called a *hierarchical design*. Hierarchical designs are not discussed here, since they are not used with PCB design projects. For more information on project details, see Chapter 2 of the *Capture User's Guide*, under "Starting a New Project" (cap_ug.pdf in the OrCAD_16.0/ DOC folder).

Capture Part Libraries Explained

Once the project is set up, the next step is to open the Schematic page (if it is not already open) and begin placing parts from the Place dropdown menu. When the Place Part dialog box opens (see Figure 3-4), it shows a list of libraries (in the Libraries window) and a list of the parts (in the Part List window) within a library. If you select a part within one of the libraries, you can see what the part looks like in the preview window, as pointed out in Figure 3-4(a). The libraries listed in the Libraries window are ones that were added by the PC Board Wizard when you set up the current project and libraries that were added to the list during previous projects. You can add other libraries to the list by clicking the Add Library... button. Likewise, you can remove a library from the list by selecting it and clicking on the Remove Library button. The libraries that are listed may be from the Capture or the PSpice library. It is not obvious

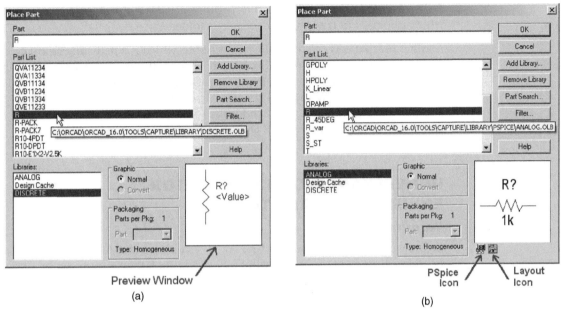

FIGURE 3-4
Place Part from Capture or Capture/PSpice library: (a) Capture parts, (b) PSpice parts.

which library it is from just by looking at the name (compare Figures 3-4(a) and 3-4(b)). However, if you have Windows ToolTips turned on and hold your mouse over the library name, the ToolTips text box will show the path of the library (see Figure 3-4(a)). From the path, you can tell which type of library it is. Another important note is that, if a PSpice or Layout library (not a PCB Editor library) is associated with the Capture part, you will see one or both of the icons below the part preview window, as shown in Figure 3-4(b). If no icons are present, it means that it is just a Capture schematic part that has no footprint or PSpice template (models) assigned to it.

As you place parts in your design, Capture keeps track of the parts and stores the information in a database file generated the moment you place your first part. This file has a .DBK extension. After you finish your design and tell Capture to make the PCB Editor netlist, Capture generates three more files that describe which parts were used and how they were connected. The files are pstxnet.dat (the netlist file), pstxprt.dat (the reference designator and device type file), and pstchip.dat (a device definition file also used for pin swapping, etc.); and they are located in the Outputs folder. These files are also used when back-annotating information from PCB Editor to Capture. More on back-annotation is discussed in the PCB Design Examples. For the most part, these files are used behind the scenes, and you need not know much about them for what we cover in this text. If you want to know more about them, please see *Allegro®*

PCB Editor Users Guide: Transferring Logic Design Data (algrologic.pdf, p. 18) which is located in the OrCAD documents folder.

Another file that gets created is the netrev.lst file, which is a netlisting report generated when the PCB Editor netlist files are created. If there is a problem during the netlisting process, this is where the errors and warnings are documented. If you have trouble trying to create a netlist, open this file with Microsoft Word or another text application and search for the word *error* or *warning*. An example of a netlist error is given in the PCB Design Examples.

UNDERSTANDING THE PCB EDITOR ENVIRONMENT AND TOOL SET

Terminology

Before we start talking about the tool set, we examine a couple terms that should be defined up front, so the descriptions of the tools make more sense. These important terms follow:

- **Classes** and **subclasses**—another way of thinking about layers and objects on layers. From the PCB perspective, there are layers: routing layers, plane layers, soldermask layers, and the like. But being able to group or separate things into classes gives you more control over the design. Generally speaking, a class is a broad grouping of things and a subclass is a more specific thing that can be grouped into a class. For example, Etch (defined later) is a class, and Top and Bottom Etch are subclasses of Etch. Some subclasses are distinct but have the same name. They are distinct because they belong to different classes. An example of this situation is the Silkscreen subclass, one of which belongs to the Board Geometry class (lines or text belonging to the board), another belongs to the Component Reference (text only, belonging to components only) class, and yet another belongs to the Manufacturing class (a grouping of all types of the other two subclasses). By having separate subclasses of silkscreen (or soldermask, say), you can control the color and visibility of each type of silkscreen object. As you go through the design examples, this will become very clear.
- **Clines**—where *cline* is short for connection line. A cline is a routed copper trace that makes electrical connections to pins. Clines are also called *etch*. Unrouted connection lines are called *net*s or *rat's nest*. Lines that make up graphic objects are just lines.
- **Etch**—objects on routing and plane layers that become copper patterns during the manufacturing process. Etch objects are on subclasses that belong to the Etch class and can be traces (clines), areas (rectangles), or text. Etch objects can be connected to nets or may stand alone.
- **Elements**—a general term for things that you put into your board design. You can have text elements, graphic elements (e.g., silkscreen outlines), and components (footprint symbols).

- **Frectangle**—a filled rectangle. A frectangle is drawn using one of the Shape tools as described later. Frectangles can have special properties, and not all rectangles are frectangles. Some of the differences between plain rectangles and frectangles are described later. There are also dynamic copper shapes, which can be rectangles but are different from frectangles. Frectangles and dynamic copper objects are demonstrated in the design examples.
- **Symbols**—elements you can place in your design that can be made up of one or more other elements. For example, a component footprint is called a *symbol* (specifically a footprint symbol), which can comprise padstacks, outline details, and text elements. Other types of symbols include flash symbols, mechanical symbols, and drawing symbols, to name a few. Symbols are used extensively in Chapter 8 and the PCB Design Examples.
- **Functions**—a part of an integrated circuit (e.g., a logic gate).

PCB Editor Windows and Tools

At this point in Chapter 2, PCB Editor was launched, the board design (*name.* brd) was opened with all default settings, and a blank work area was presented to you. From this point you made a board outline, placed the parts on your board, autorouted the board, and produced the artwork for it. To be fluent at these tasks, you need to know how to use the PCB Editor tools. We now take a tour of the PCB Editor environment and tool set.

Note: From the time you start your design in Capture to the time the artwork is produced in PCB Editor, nearly 30 files can be generated, which together fully describe your design. If you save more than one project in the same folder, it can become very cluttered. It is a good idea to set up a "MyProjects" folder that contains subfolders for each project.

The Design Window

The Design window is the working environment for a board design (see Figure 2-12). From the Design window, you have access to the tools you need to handle parts, route traces, and perform back annotations (design updates from PCB Editor to Capture). Hundreds of tasks can be executed from the Design window menus. Since they are covered in detail in the *Allegro® User's Guide*, a detailed discussion of the menus is not given here, but the key menu options are discussed in the Examples as the need arises. The toolbar is discussed next.

The Toolbar Groups

By default, two toolbars are displayed. You can move them wherever you want or turn them off. *To add tool bars*, select View → Customize Toolbar... from the menu. This section does not describe the tools in great detail but serves as an introduction only. The use of the tools are demonstrated in the design examples and described further in *Allegro® PCB Editor User Guide: Getting Started with Physical Design* (algrostart.pdf).

APPLICATION MODE GROUP

PCB Editor works in two application modes: [a] Generaledit and [b] Etchedit. The differences between the two modes are subtle, and most of the time you need not consciously choose between one or the other. The biggest difference between the two modes is which items are available and checked by default in the Find pane and which ones of those items are higher up in selection priority. When in General Edit mode, the highest level hierarchical type elements (items at the top of the list) enabled by the Find filter are highlighted by default and become more easily selectable with your mouse. But in Etch Edit mode, the lowest level hierarchical type elements enabled by the Find filter (items at the bottom of the list) become more easily selectable. You can execute most commands in either mode, but you will find that some commands are easier to perform in one mode than the other and you may have to wrestle with the Find and Options panes settings to do so. You can determine which application mode you are in by looking at the application box in the status bar (described later). See *Allegro® PCB Editor User Guide: Getting Started with Physical Design* (algrostart.pdf) for further information.

EDIT GROUP

The edit group is made up of the following tools: [a] Move, [b] Copy, [c] Delete, [d] Undo/Redo, and [e] Fix/Unfix. These tools can be used on graphical objects, text, and etch. The Delete tool deletes (erases) objects, but it is used to rip up traces (it does not delete nets) and unplace parts. The Fix/Unfix tool is used to prevent objects (e.g., shapes and text) from being moved or deleted, and it prevents routed traces (clines) from being ripped up.

VIEW GROUP

a. Unrats All/Rats All—controls the visibility of all unrouted nets (also know as *rat's nest*). You can also control the visibility of individual nets using the Rats Off property with the Constraint Manager (described later) or by selecting Unrats All then selecting Display → Show Rats → Net in concert with the Find filter tab to then display specific nets.

b. Zoom—allows you to zoom to a specific area (Zoom points), the entire design (Zoom fit), in or out, to a previous view, or to a selection (by drawing a box across a part and some nets, you use Zoom to view every-thing that was highlighted).

c. Redraw—refreshes the display.

SETUP GROUP

The following tools constitute the setup group:

a. Grid Toggle—allows you to turn the grids on or off. The displayed grid res-olution depends on the settings in the Define Grid dialog box (from Setup menu) and the type of layer (class) that is active. Classes related to etch lay-ers use the Etch grid resolution and non-etch-related classes use the Non Etch resolution. If the grid resolution is too dense for a given Zoom level, PCB Editor automatically adjusts the grid density view but not the functionality.

b. Color Pallet—allows you to control the color and visibility of classes and subclasses. The Color dialog box is described in greater detail later.

c. Shadow Toggle—controls the intensity of visible layers. Shadow mode makes highlighted items stand out more clearly.

d. Stack-up—launches the Layer Cross Section dialog box, where you define your board's stack-up as described later.

e. Constraint Manager—launches the Constraint Manager, where you set autorouter specifications (e.g., trace width and spacing) and design rule constraints. The Constraint Manager is discussed further later.

f. Keep-in Router.

g. Keep-in Package—automatically selects the correct classes and drawing tools so that you can quickly draw these areas in your board design.

SHAPES GROUP

These are drawing tools and are different from the Add (line and rectangle) group objects described later. They are used to make shapes that interact with the DRC tool and are used to create dynamic shapes that adjust automatically (e.g., plane layers that heal and you can plow through) when interacting with other Etch objects. You can also make static shapes with them, which are used for keep-in/-out areas and the like. Shapes can be filled or unfilled, static or dynamic. Some classes, such as the Package Keep-in class, automatically know what type of shape it needs (unfilled), so even if you unknowingly choose a static filled rectangle, it will automatically become unfilled. Many other classes do not. For example, you can place a filled rectangle on the GND plane, but it may not connect to pins properly. Copper areas on negative plane layers should be Dynamic shapes.

a. Shape Tools—The following tools are used to make the shapes: Shape Add (polygon), Shape Add Rect(tangle), and Shape Add Circle.
b. Shape Select—used to edit existing shapes (stretch shapes, add or move vertices, etc.).
c. Shape Void tools—are used to remove areas of shapes. For example, if you want to create an opening in the plane layers because a mounting hole is inserted in the design and has a route keep-out area, then you can use a void shape to make the opening.
d. Edit Shape Boundary—is used to add or remove areas to a shape at the shape's boundaries. It is similar to using the Merge and Void commands but is easier to use and can be used only on the edge of a shape.
e. Island Delete—deletes pieces of planes or shapes that become detached from the main shape due to spacing requirements around groups of pins. These floating pieces can become EMI issues and should be removed.

Note: When making changes that affect dynamic shapes some changes may not take effect immediately. To redraw shapes, select Display → Status → Update shapes from the menu.

MANUFACTURE GROUP

a. Artwork—opens the Artwork Control Form dialog box, where you set up the classes and subclasses for which you want manufacturing data generated.
b. NC Drill Param(eters)—opens the NC Parameters dialog box, where you can set up drill format (Enhanced Excellon format, etc.).
c. NC Drill Legend—is used to set up then display the drill legend (drill chart). Drill symbols in parts on your design then also become visible.
d. ODB—exports database information into a Valor ODB++ database, which contains all CAD/EDA, assembly, and manufacturing data. ODB is not enabled in basic OrCAD PCB Editor.

DISPLAY GROUP

a. Show Element—displays a text box that provides information and properties about the object you select (nets, shapes, components, etc.). Use it in concert with the Find filter to locate objects or nets without actually having to click on them.
b. Show Constraints—is similar to Show Element but reports only constraints.
c. Show Measure—measures the distance between two pick points and displays the measurement in a pop-up text box.

d. Highlight/Dehighlight—marks objects to make them easier to see. Also used in conjunction with intertool communication between Capture and PCB Editor.

e. Highlight Sov—checks for cline segments (traces) routed over voids (not enabled in basic OrCAD PCB Editor).

f. Waive DRC—used to override DRCs that you know about and do not need or want to fix.

MISCELLANEOUS GROUP

a. Report—opens the Reports dialog box so that you can pick a report to view (e.g., DRC, bill of materials). Selecting this button is the same as selecting Tools → Reports from the menu.

b. DRC update—updates the DRC status. It does the same thing as selecting Display → Status → Update DRC from the menu. You can then use Display → Status to see how many errors there are, or you can use the Report button to show the list of errors if there are any.

PLACE GROUP

a. Place Manual—displays the Placement dialog box so that you can place parts or other objects from the list.

b. Place Manual–H—does the same thing as Place Manual but does not automatically show the Placement dialog box. To show the dialog box, right click and select Show from the pop-up menu.

ROUTE GROUP

a. Net Schedule—sets priorities on net routing. It is not enabled in basic OrCAD PCB Editor.

b. Add Connect—is used to manually route traces (produce clines).

c. Slide—is used to move or stretch a routed trace. It automatically uses autorouter algorithms to prevent most (but not all) DRCs.

d. Delay Tune—is used to route differential pairs. It is not enabled in basic OrCAD PCB Editor.

e. Custom Smooth—is used to reroute routed traces to make the routing more efficient and remove unnecessary corners and route distance. It performs the same basic function as Gloss but is used to work on one trace at a time. Select this tool then the trace you want smoothed.

f. Vertex—is used to add a vertex to a line or cline (use shape select to add or move vertices on shapes).

g. Create Fanout—is used with Setup → Design Parameters → Route → Create Fanout Parameters to define how fan-outs should be done. Select this tool then click on a pin or component to make it happen.

h. Spread between Voids—is not enabled in basic OrCAD PCB Editor.

ADD GROUP

These objects are graphical in nature only. Use these tools for making graphics and text on silkscreen and assembly layers. The tools are

a. Add Line.
b. Add Rectangle.
c. Add Text.
b. Edit Text.

Note: Edit text is used to change the text characters. If you need to change the text properties (size and spacing etc.), you need to use the Options pane and Setup → Design Parameters → Text → Setup Text Sizes.

Control Panel with Foldable Window Panes

The control panel is an area containing three tabs that show or hide collapsible window panes. The panes are shown in Figure 3-5 in their default condition (no tools or commands active). The panes are dynamic, so what is displayed depends on what tool is active or what type of object you selected at that moment. From these panes, you can control visibility and selectability of objects and what particular tools can do (extent of effect, etc.) when you use them. The panes are collapsible, so that they do not take up design window space, but you can pin them in place by toggling the stick pin icon. You can also close them altogether by clicking the X in the upper right corner. To display them select View → Window → Options/Find/Visibility from the menu bar. Each pane is described briefly. You see examples of them in action during the design examples.

VISIBILITY PANE

The Visibility pane is a shortened version of the Color dialog box (described later). It provides a handy way to turn on and off routing and plane layers (or specific elements on those layers). The colored boxes and check boxes are switches you can use to toggle on and off specific items or entire rows or columns. The colored box in the Options pane (described later) works the same

FIGURE 3-5
Control panel window panes.

way, and with it, you can control the visibility of more elements using the class and subclass lists.

FIND FILTER PANE

The Find pane acts as a selection filter for tools and commands. By selectively checking object boxes, you can restrict which objects will be selected when you perform mouse picks in your design. The order of the object types indicates a level of hierarchy, so if all objects are enabled and you attempt to perform a mouse pick in a congested area of your board, the top-down order indicates which object type will likely be picked. Unchecking an object prevents that type of object from being selected. Note: If you attempt to perform a command on an object and PCB Editor does not let you do it, check the Find pane; chances are, the box for that type of object has been unchecked.

You can further narrow searches and selections by using the Find By Name area. Select the type of thing you are looking for from the dropdown list then click the More... button. A dialog box will pop up, which will allow you to pick specific objects from a list by name or property.

OPTIONS PANE

The Options pane is very dynamic and you will use it often. Its appearance is determined by what tool you have selected or what command you are running. Figure 3-6 gives an example, where (left) shows what the Options pane looks like when the Delete tool is active and (right) shows what the Options pane looks like when the Add Connect tool is active. Along with the Visibility pane

FIGURE 3-6
Options pane with (left)
Delete tool and (right)
Add Connect tool active.

and the Color dialog box, you can turn on or off layers (classes) or parts of layers (subclasses) by toggling the colored squares. When you first start out learning PCB Editor, it is easy to forget about the Options pane, but you want to keep it in mind, as it gives you significant control over your tools. You might want to pin this one up until you get used to relying on it.

Command Window Pane

The Command window (Figure 3-7) provides you with information, gives instructions, and allows you to enter commands at the Command prompt. Most of the commands you often use are also located on the toolbar and in the menus, but the Command window can give you greater control of the tools (if you know the commands). The documentation folder in the OrCAD directory has manuals (called *command references*) for these commands. The manuals are listed alphabetically by the first letter of the command followed by coms. So, for example, if

```
last pick:  2200.00  1300.00
Loading mfg.cxt
NC Parameters file 'nc_param.txt' not found! ... loading defaults
Pick first element.
Pick element to slide.
Pick text location.
Command >
```

add text

FIGURE 3-7
Command window pane.

you are looking for instructions on how to use the Command window to perform a manual routing task (add connect) look in the acoms document folder (or use the Online Help tool and search for Add Connect).

WorldView Window Pane

The WorldView shown in Figure 3-8 is located in the lower right-hand corner of the design window and gives you a bird's-eye overview of large boards. The white square represents your monitor, and the green outline represents the PCB's outline. WorldView is interactive. If you use the highlight tool and select a part or net (whether by clicking on it or by selecting it from a list using the Find pane), the object will be displayed in the WorldView. If you right click inside of it, a pop-up menu is displayed that lets you change the size and location of the view in the design area.

FIGURE 3-8
WorldView window pane.

Status Bar

The status bar (shown in Figure 3-9) is located along the bottom of the design window, below the Command window and WorldView window. At the far left of the bar, the Command Status section lets you know if a command is active and running. If a tool has been selected (e.g., Move), the name or function of the tool is listed at the far left; and when a command is running, the colored box turns red. When the command has finished executing, the box turns green again.

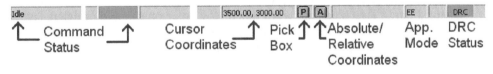

FIGURE 3-9
Status bar.

The x, y coordinates indicate the location of your cursor. The accuracy of the display is based on the design accuracy, which you can change in the Design Parameters dialog box (select Setup → Design Parameters → Design from the menu). The coordinates are absolute (based on the origin of the design) or relative (based on the last mouse pick). You can toggle between the two at any time using the A (or R) button, as described later.

The P button is interactive. Clicking it produces a Pick dialog box, which you can use to enter coordinate points in the work area using the keyboard rather than selecting a point with the mouse. This is useful for drawing outlines and so forth on large designs, so that you need not pan around the design trying to find and select a particular point. The x and y coordinates in the Pick box are entered with space between them, not a comma.

The A (or R) button determines whether the coordinates are absolute or relative. Absolute coordinates show the cursor position relative to the design origin. Relative coordinates show the cursor x and y distance relative to the last pick point you made (whether a tool was active or not).

The Application Mode text box informs you of whether PCB Editor is in Etch Edit (EE) mode or General Edit (GE) mode (described previously). The DRC Status box lets you know at a glance if the design rule check is up to date and if any errors exist by the color of the box. If the box is any color other than green (see Figure 3-8), then you need to update the DRC and use a DRC report to locate any errors if they exist.

Color and Visibility Dialog Box

Color button

The Color dialog box (Figure 3-10) is used to define custom colors for classes and subclasses and allows you to control the visibility of specific objects belonging to those classes (Xs indicate the object is visible, open boxes indicate they are invisible). The Color dialog box can be displayed by selecting the Color button, ▦, on the toolbar. As mentioned previously, the Color dialog box performs some of the same functions as the colored and check boxes in the Visibility and Options panes, but as the figure shows, you have much greater control of objects and layers.

Layout Cross Section (Layer Stack-Up) Dialog Box

Xsection button

The Layout Cross Section dialog box, shown in Figure 3-11, is where you define your board's layer stack-up. From this dialog box, you can add or delete layers, define their physical and electrical properties (if you want to), and define positive and negative properties to routing and plane layers, respectively. The Layout Cross Section dialog box is displayed using the Xsection button, ▦. Examples of setting up layer stack-ups are given in the PCB Design Examples.

Constraint Manager

Cmgr button

The Constraint Manager, shown in Figure 3-12, is where you set routing and component placement rules for your board design. The Constraint Manager can be launched using the Cmgr button, ▦, by selecting Setup → Constraints... from the menu, or by typing cmgr at the > prompt in the Command window.

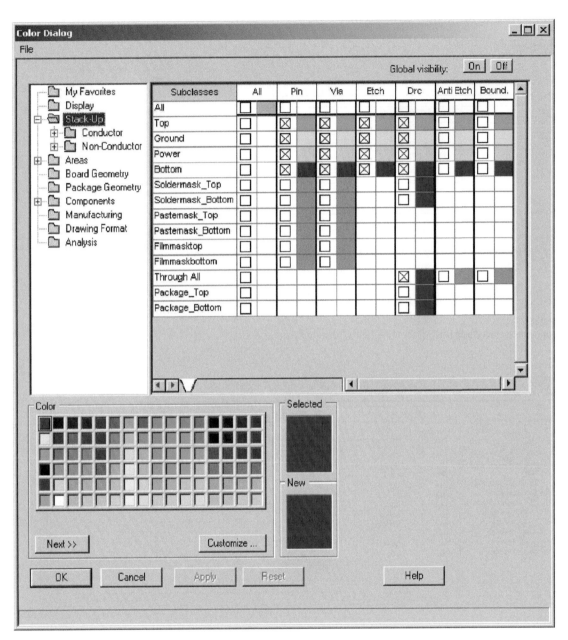

FIGURE 3-10
Color dialog box.

The Constraint Manager has four tabs and each tab has several views (indicated by folders and icons). From these tabs and views, you have very precise control over routing characteristics for every net (e.g., trace width and spacing, what vias are used, routing enabled, locking traces, etc.), every component, and shapes.

FIGURE 3-11
Layout Cross Section dialog box.

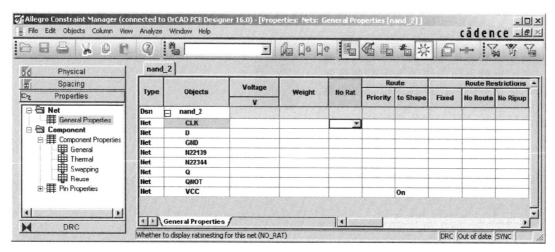

FIGURE 3-12
Constraint Manager dialog box.

Note: The Constraint Manager takes a moment to load, but if it is taking a long time to load, check the Command window. This tool will not load if a command is still active, and the Command window (and status bar) will tell you if that is the case. If so, you need to stop the current command by right clicking in the work space and selecting Done from the pop-up menu.

PADSTACK DESIGNER

The Padstack Designer is used to define and modify padstacks (see Figure 3-13). The padstacks can be ones from the PCB Editor library or ones in a PCB design only. You use Padstack Designer both during PCB layout work and separately during footprint development. Padstack Designer can be launched from within PCB Editor (from the Tools menu) or in stand-alone mode from the Windows Start menu. Many examples of using the Padstack Designer are presented in Chapter 8 (footprint design examples) and the PCB Design Examples.

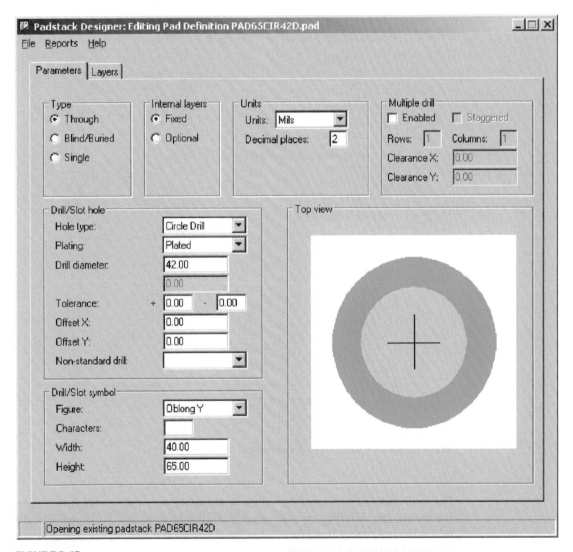

FIGURE 3-13
Padstack Designer.

Manufacturing Artwork and Drill Files

Once the board is laid out and routed, the artwork must be generated. The information in the board design is separated into specific data files (Gerber files), which are used by the board manufacturer to make the different parts of your board. The Artwork Control Form (Figure 3-14) is used to specify all the different types of layers for which Gerber files will be created and the format of the files. You can add as many layers as necessary to fully define your board design. Chapter 10 describes how to set up the Artwork Control Form based on a board layout example.

FIGURE 3-14
Artwork Control Form dialog box (Film Control tab).

The drill files are generated by selecting Manufacture → NC → NC Drill... from the menu. The drill files are generated from drill settings that you specify in the NC Parameters dialog box (shown in Figure 3-15). The use of this dialog box is discussed further in Chapter 9.

FIGURE 3-15
NC (drill) Parameters dialog box.

The number and types of files generated by the Artwork Control Form depends on the complexity of your board and the requirements of your board manufacturer. Most of the artwork files correspond to the layers with which you work in the layout environment, but some of the files do not. For example, files are generated for etch and silkscreen layers and drill files, all of which depend on the output format you chose. The two most common types of output formats are RS-274D and RS-274X (also known as *Excellon*). The Excellon format is typically used, but your board manufacturer should tell you which specific files it needs and which format it prefers. Artwork instructions are set up from the Manufacture → Artwork menu, and we will look at specific settings in Chapter 9.

Understanding the Documentation Files

Some of the terminology used in the PCB fabrication business is left over from the early PCB manufacturing days. *Drill tapes* and *apertures* are such terms. Nowadays, a drill tape is just an electronic file that describes drill holes and sizes just like any of the other files describes its data. However, originally the drill tape was actually a role of paper or Mylar with holes punched into it that described hole sizes and locations for early computer numerical control (CNC) machines, but it is still sometimes called a *drill tape*. Here is another example. In Chapter 1, the current technology of photoplotting and laser direct imaging was discussed. In contrast, the older technology used a xenon flash lamp and a shutter to expose photosensitive film or glass plates. The shutter controlled the exposure and an aperture controlled the size of the exposed area. Any shape or drawing could be made by having apertures of different sizes and shapes. Light would be shone through the selected aperture and moved in the $\pm x$ direction, and the film was moved in the $\pm y$ direction. Opening and closing the shutter in one area (to make a pad) was called a *flash*, while holding the shutter open and moving the light source or film in the x and y directions (to make a trace) was called a *draw*. The technology today is more advanced, but the concept is the same, and the same terminology is used. Figure 3-16 shows a list of

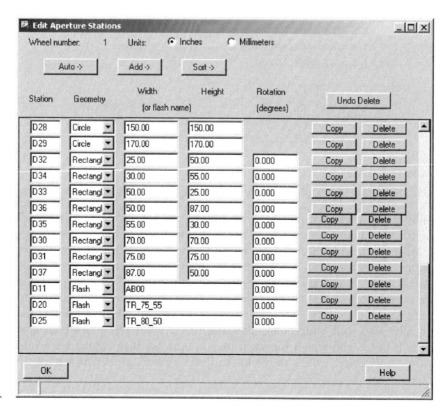

FIGURE 3-16
Edit Aperture dialog box.

the apertures the Artwork Control Form uses, where you will notice D codes and flash geometries (circles, rectangles, and flash symbols). A Gerber D code is a just a chronological number that specifies the size and shape of the aperture used on a given layer. Flash symbols (e.g., D code D20 in the figure) are used to produce thermal reliefs, which are used to connect plated through holes to copper planes. The procedure for drawing thermal flash symbols is described in Chapter 8 and the use of them is described in the PCB Design Examples.

You should now be familiar with the PCB design flow and PCB Editor tool set. In the following chapters, we look at how to assign footprints to parts from within Capture, how to make your own Capture parts, how to make new footprints with PCB Editor, and how to set up the PCB Editor's Constraint Manager to route complex boards.

Introduction to Industry Standards

In the previous chapters we looked at how to use the OrCAD tools to design circuit schematics in Capture, route PCBs with PCB Editor, and postprocess the design for fabricating the board. We have not, as of yet, taken a look at how to design the PCB itself. Chapters 4 through 6 provide an introduction to PCB design. Chapter 4 introduces industry standards related to PCB design.

Not every PCB that is manufactured makes it into service. The ratio of the number of PCBs that enter service to the number of PCBs manufactured is called *yield* (in percent). The higher the yield the better because failed boards cost time and money and produce waste. There are several failure points that can be addressed to increase yield. To have high yield we need three things. First the board has to be manufacturable, second it has to perform properly (signal integrity and quality), and third it has to be reliable (it has to work for the full length of its expected life span).

Being manufacturable means two things. The bare board has to be able to be *fabricated* given standard fabrication allowances (SFA) and the board also has to be able to be *assembled*; that is, parts need to be able to be attached to the board with proper solder joints without damaging the parts or the board.

Performance refers to both mechanical and electrical considerations. Mechanically, the board must physically fit into its enclosure and it must be able to handle its environment with respect to ambient temperature, vibration, and humidity. Electrical performance refers to whether it meets the operational design constraints (power, I/O, etc.), is immune to outside interference, and does not cause interference to neighboring equipment.

Being reliable means it meets the above considerations over the expected life of the device. If it is designed correctly it should not fail before the expected end of life unless the user exceeds specified operational design criteria. If the user stays within the operational guidelines and the unit still fails, then there was a bare board defect (a manufacturing defect that was not caused by your design), or there was a circuit or PCB design defect that resulted in a poor assembly process, which resulted in damage to components or board, or the circuit was operating too close to limits, which stressed the system, and failed early due to accelerated stress aging. If reliability problems exist then a failure analysis is conducted and the board is redesigned.

INTRODUCTION TO THE STANDARDS ORGANIZATIONS

When you begin a new PCB design you may be asking how big and what shape should the board outline be, where should the parts be placed and in what order, what kind of layer stack-up should be used, how wide and far apart should the traces be routed, and what grounding and shielding techniques should be used? Is there a "right" way to do it, and who says so?

There are several standards related to PCB design. The organizations below set standards that may be guides, rules for certification, or even laws. To cover all aspects of these standards would fill an entire book by itself. Some of the standards organizations that you may have heard of or will hear of are listed below with a brief description of who they are and what they do. The discussion in this chapter is limited to the basic standards for PCB design. A listing of applicable design standards is presented in Appendix A.

Institute for Printed Circuits (IPC—Association Connecting Electronics Industries)

The IPC is a global trade association consisting of more than 2300 member companies. It is an organization made up of contributors from industry and includes designers, board manufacturers, assembly companies, suppliers, and original equipment manufacturers. Contributing members bring lessons learned and known good practices to the table, and they document and disseminate the knowledge base through industry-accepted standards. Over the past several years the IPC standards have replaced many of the military standards (MIL-STD) and are sources that you should be familiar with. A list of IPC standards is provided in Appendix A. You can also visit the IPC Web site (www.ipc.org), where you can download a free land pattern viewer.

Electronic Industries Alliance (EIA)

The EIA is a national trade organization comprising over 1000 U.S. manufacturers and high-tech associations and companies. Its primary focus is promoting the market development and competitiveness of the U.S. high-tech industry in the global economy through domestic and international policy efforts. It has influence on design standards set by contributing groups, which include the following:

- CEA—the Consumer Electronics Association
- ECA—the Electronic Components, Assemblies, and Materials Association
- GEIA—the Government Electronics and Information Technology Association

- JEDEC—the JEDEC Solid State Technology Association
- TIA—the Telecommunications Industry Association
- EIF—the Electronic Industries Foundation
- ISA—the Internet Security Alliance

Joint Electron Device Engineering Council (JEDEC)

The JEDEC (now also known as JEDEC Solid State Technology Association) is the semiconductor engineering standardization body of the EIA. It is an association of several hundred organizations that represents all areas of the electronics industry. Its primary focus with regard to PCB design is in standardizing discrete and integrated circuit semiconductor devices and packages. You need to know the package specifications in order to design footprints for your PCB. You can access many of the standards online at www.jedec.org. A list of package specifications is provided in Appendix B.

International Engineering Consortium (IEC)

The IEC is a nonprofit organization that works with business and educational communities. It conducts research and reports findings in publications and conferences to address industry opportunities and challenges to aid industry and academia. You can find out more about the IEC (including a free online educational program) at www.iec.org.

Military Standards

MIL-STD are maintained by Defense Supply Center Columbus (DSCC), which is a field activity of the Defense Logistics Agency, whose purpose is to provide logistics and contract management support to the U.S. armed forces. The Department of Defense develops and procures an incredible amount of material and engineering services through private contractors. MIL-STD set and communicate standards on how things are to be designed, built, and tested in a controlled, known, and acceptable manner so that all who bid on contracts know exactly what is expected of them, so that they can be successful and competitive. In recent years specialized MIL-STD have been replaced by commercial standards such as IPC standards for PCB design and manufacturing and many other fields of engineering and manufacturing (ASME, etc.). Some of the MIL-STD are becoming obsolete, but new commercial standards are based on and update the old MIL-STD. For example, the newer IPC-2221A PCB standard originated from MIL-STD-275. You can obtain many of the old MIL-STD for free at the DSCC Web site, www.dscc.dla.mil/Programs/MilSpec/DocSearch.asp.

American National Standards Institute (ANSI)

The ANSI is a private, nonprofit organization that administers and coordinates voluntary consensus standardization and conformity assessment in the United States. Organizations may become accredited by ANSI, which signifies that their procedures meet ANSI's requirements for due process. There are a couple of PCB and electronic design standards produced through a joint effort between ANSI and other standards organizations (see IEEE below). You can find out more about ANSI at www.ansi.org.

Institute of Electrical and Electronics Engineers (IEEE)

The IEEE is a developer of technology standards that are designed to build consensus in an open-based process with input from interested parties. IEEE is a central source of standardization in fields such as telecommunications and power generation.

Examples of IEEE standards related to schematic design and PCB layout include IEEE/ANSI 315-1975, Graphic Symbols for Electrical and Electronics Diagrams, and IEEE-1445-1998, IEEE Standard for Digital Test Interchange Format, respectively. Visit IEEE at www.ieee.org.

CLASSES AND TYPES OF PCBS

The design approach for a PCB depends on many factors including its intended end use, design and fabrication complexity, acceptable fabrication allowances, and type of component and attachment technology. Standard classifications have been established to aid designers, fabricators, and consumers in communicating with each other on these issues. The classifications include performance class, producibility level, and type of construction.

Performance Classes

PCBs can fall into any of three end-use performance classes. Throughout many of the IPC standards (IPC-7351, Section 1.3; IPC-D-330, Section 1.1.42.6; IPC-CM-770E, Section 1.2.1) material performance and tolerance levels are determined by the class rating. Performance classes are based on things like allowed variation in copper-plating thickness, feature location tolerance, and hole diameter tolerance (plated and unplated), to name a few. The three classes are as follows:

- Class 1, General Electronic Products, includes general consumer products like televisions, electronic games, and personal computers that are not expected to have extended service lives and are not likely to be subjected to extensive test or repairability requirements.
- Class 2, Dedicated-Service Electronic Products, includes commercial and military products that have specific functions such as communications, instrumentation, and sensor systems, from which high performance is expected over a longer period of time. Since these items usually have a higher cost they are usually repairable and must meet stricter testing requirements.
- Class 3, High-Reliability Electronic Products, includes commercial and military equipment that has to be highly reliable under a wide range of environmental conditions. Examples include critical medical equipment and weapons systems. They typically have more stringent test specifications and possess greater environmental robustness and reworkability.

Producibility Levels

Producibility levels are described in detail here. The levels are not a set of explicit requirements but a way of describing how complex a design is and the precision required to produce the particular features of a PCB or PCB assembly. Smaller features (trace widths, etc.) require stricter tolerances, which increases the design complexity. The IPC standards (IPC-7351, Section 1.3.1; IPC-CM-770E, Section 1.2.2; IPC-D-330, Section 1.1.42.6) provide several tables that assist the designer in determining the complexity as it relates to SFAs. For example, issues such as tolerances for interconnecting lands and conductor width tolerances are described in the standards. The three producibility levels are

- Level A, general design—preferred complexity.
- Level B, moderate design—standard complexity.
- Level C, high design—reduced producibility complexity.

Fabrication Types and Assembly Subclasses

PCB fabrication types (IPC-CM-770E, Section 1.2.3) are indicated by a number; the higher the number, the greater the sophistication required to make the board. Issues that are related to the fabrication type are the number of copper layers (e.g., single layer, two layer, or multilayer) and the types of vias used to connect the layers, etc. The six fabrication types defined by IPC are

- Type 1, single-sided printed board.
- Type 2, double-sided printed board.
- Type 3, multilayer printed board without blind or buried vias.
- Type 4, multilayer printed board with blind and/or buried vias.
- Type 5, multilayer metal-core printed board without blind or buried vias.
- Type 6, multilayer metal-core printed board with blind and/or buried vias.

Each PCB type can be further defined by an assembly subclass, which describes how components will be attached to the board (IPC-CM-770E, Section 1.2.2). The subclasses are as follows:

- Subclass A, through-hole devices (THD) only.
- Subclass B, surface-mounted devices (SMD) only.
- Subclass C, mixed THD and SMD (simple).
- Subclass X, complex THD/SMD, fine pitch, ball grid array (BGA) packages.
- Subclass Y, complex THD/SMD, ultrafine pitch, chip-scale packaging.
- Subclass Z, complex THD/SMD, fine pitch, flip-chip packaging.

Typically more sophisticated types and subclasses require stricter tolerances (producibility levels). Higher performance class boards are made more reliable by using stricter producibility levels and lower (easier) fabrication types and assembly classes.

IPC Land Pattern Density Levels

The density levels are used to gauge PCB footprint (land pattern) designs with regard to how densely a board can be populated and with regard to the difficulty of trace routing and fabrication. More will discussed in Chapter 5 as to how this relates to footprint design in PCB Editor. The three IPC land pattern density levels (IPC-7351, Section 1.4) are

- Density Level A, most land protrusion (largest courtyard and least density).
- Density Level B, nominal land protrusion (median courtyard and median density).
- Density Level C, least land protrusion (smallest courtyard and highest density).

INTRODUCTION TO STANDARD FABRICATION ALLOWANCES

No manufacturing process is perfect and it is therefore subject to tolerance limitations. PCB manufacturing is no exception. Design tolerances include drill-hole location and diameter, copper plating and etching, and solder-mask resolution, to name a few. Manufacturing tolerance becomes increasingly important as the number of layers increases and line widths and spacing decrease. The tolerance errors can add up at each manufacturing step and result in a scrapped board.

It is important to be aware of manufacturing limitations so that you stay within the boundaries of the manufacturer's capabilities. Industry standards exist to set minimum performance and process guidelines. Individual manufacturer's also have their own capabilities, and you need to be aware of them as well. Just because your design meets certain minimum industry standards does not mean that every manufacturer has the ability to manufacture or assemble the PCB as designed. The following discussion covers the major design issues to look out for and references to the appropriate standards.

Registration Tolerances

As described in Chapter 1 (see also Coombs 2001, p. 17.1.4), many steps and design files are required to fabricate a multilayer PCB. The design parameters in each step have to line up with the next, or misregistration can occur, which can result in manufacturing defects and a nonoperational board. One of the PCB's most vulnerable spots is the plated through hole because it requires accurate alignment of many layers performed over several manufacturing steps.

Breakout and Annular Ring Control

Figure 4-1 shows how fabrication allowances result in a final hole tolerance. Figure 4-1(a) shows the ideal hole, which has a specified diameter and location. Figure 4-1(b) shows the uncertainty of the final hole diameter due to drilling

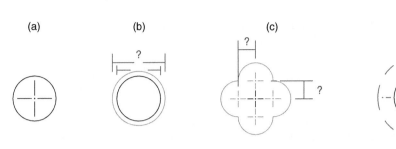

(a)   (b)   (c)   (d)

FIGURE 4-1
Example of fabrication tolerance and dimensional uncertainty.

tolerances. Figure 4-1(c) shows the uncertainty of the hole location. Figure 4-1(d) shows the desired hole compared to the possible hole after considering the combined uncertainties. Feature dimensions on each layer of a PCB have uncertainty, which, when combined, can result in a bad board if allowances are not designed into the board.

In addition to hole tolerances, there are also tolerances on trace/land location and size due to plating thickness and variation in etch rate and consistency. The combination of these uncertainties can result in problems such as loss of annular ring control and subsequent land breakout as shown in Figure 4-2 (Coombs 2001, p. 42.2). Plated through holes can often function with breakout, but reliability is greatly reduced. By knowing limitations of fabrication processes and following design guides you can greatly reduce the occurrence of defects and increase yield.

(a)   (b)

PCB DIMENSIONS AND TOLERANCES

The following sections briefly introduce design limitations that you should be aware of.

FIGURE 4-2
Breakout of plated through hole (PTH) due to misalignment: (a) ideal PTH, (b) PTH breakout.

Standard Panel Sizes

There are 16 industry standard board panel sizes (per ANSI/IPC-D-322 and IPC-2221A, Fig. 1, p. 28). The boards are identified by a letter and a number, which represent the x and y dimensions. Sizes range from AI to D4 as shown in Table 4-1. An example of a size C2 panel is listed in the table, where C2 is 7.1 × 6.7 in.

Table 4-1	Standard Copper Clad Panel Sizes			
	Number			
Letter	**1**	**2**	**3**	**4**
A	2.4 × 3.2	2.4 × 6.7	2.4 × 10.2	2.4 × 13.8
B	4.7 × 3.2	4.7 × 6.7	4.7 × 10.2	4.7 × 13.8
C	7.1 × 3.2	7.1 × 6.7	7.1 × 10.2	7.1 × 13.8
D	9.5 × 3.2	9.5 × 6.7	9.5 × 10.2	9.5 × 13.8

Sizes are given in inches.

Being aware of standard panel sizes may help reduce costs since smaller PCB designs can be panelized onto one large panel. If you have flexibility in specifying the size of your board, you can do so in a way that will allow you to maximize the number of boards on one panel. This helps reduce cost by minimizing the number of parts being handled and the amount of waste generated. This is not always an option as PCB size is often driven by design constraints that are out of your control. Small boards may have to be panelized, though, if they are to be assembled and soldered by automated processes since automated machinery has minimum size limitations (and maximums as well) on the size of PCBs that they can process.

Tooling Area Allowances and Effective Panel Usage

When manufacturers make PCBs they need to have an area around the PCB outline to place tooling holes and certain manufacturing data marks. This area is called a *tooling area*. The required tooling area ranges from 0.375 to 1.5 in. (typically 1.0 in.) as measured from the edge of the board outline to the panel boundary (ANSI/IPC-D-322). On panelized designs the distance between board outlines is typically 0.1 to 0.5 in. The goal is to utilize as much of a panel as possible without having to go to the next larger size (IPC-D-330 Section 2, Table 2-6, p. 9).

Most of the time these considerations are handled by the board manufacturer and are transparent to the board designer, but being aware of the issues may allow you to optimize the board layout and reduce production costs.

Standard Finished PCB Thickness

As described in Chapter 1, a PCB is an assembly of one or more cores joined together by sheets of partially cured epoxy (called *prepreg*). By stacking combinations of various core thicknesses and sheets of prepreg, a wide variety of finished board thicknesses can be achieved. Table 4-2 lists the typical board

Table 4-2	**Typical Finished Board Thicknesses**	
Inches	**Mils**	**Millimeters**
0.020	20	0.51
0.030	30	0.76
0.040	40	1.02
0.062	62	1.6
0.093	93	2.4
0.125	125	3.2
0.250	250	6.4
0.500	500	12.7

thicknesses in the industry. The following sections discuss standard core, prepreg and copper thickness, and tolerances. Unless you are designing controlled impedance PCBs you may not be immediately concerned about the discussion of each of the thicknesses described below. The information is presented here for completeness and as a reference for the curious and those who are interested in designing controlled impedance PCBs and can be used when working with PCB Editor's Layer Cross Section editor, which is discussed in Chapter 9.

Core Thickness

Cores are made up of substrates (i.e., fully cured laminate epoxy), which are then plated with copper on one or both sides. Table 4-3 shows typical laminate epoxy thicknesses without copper cladding or foil (see also Coombs 2001, Table 5-5, p. 5-12, and IPC-4101, Table 3-7). Copper (foil and plated cladding) thicknesses are described below.

Table 4-3 Typical Laminate Core Thicknesses (without Copper)			
Mils		**Millimeters**	
Range	**Avg**	**Range**	**Avg**
0.98–4.69	2.8	0.025–0.119	0.072
4.72–6.46	5.6	0.120–0.164	0.142
6.50–11.8	9.1	0.165–0.299	0.232
11.8–19.6	15.7	0.300–0.499	0.400
19.7–30.9	25.3	0.500–0.785	0.643
30.9–40.9	35.9	0.786–1.039	0.913
40.9–65.9	53.4	1.040–1.674	1.357
65.9–101	83.4	1.675–2.564	2.120
101–141	121	2.565–3.579	3.072
141–250	195	3.580–6.350	4.965

Prepreg Thickness

The prepreg that is used to join the plated cores during PCB stack-up comes in sheets of various thicknesses. Table 4-4 shows the various prepreg types and their sheet thicknesses before curing (Coombs 2001, Table 6-3, p. 6.8, and Table 10-2, p. 10.4). Board manufacturers stack up combinations of sheets to obtain the desired board thickness. The actual thickness of a sheet once it is in a board and cured depends on whether it is between plane layers or signal layers, because signal layers tend to sink into the prepreg, which results in a thinner end thickness. The dielectric constant of the prepreg also varies by manufacturer and is discussed in Chapter 6.

Table 4-4	Standard Prepreg Thicknesses	
	Thickness range	
Prepreg type	**(mils)**	**(mm)**
106	1.5–2.3	0.038–0.058
1080	2.3–3.0	0.058–0.076
2313	3.5–4.0	0.089–0.102
2116	4.5–5.3	0.114–0.135
2165	5.0–6.8	0.127–0.173
2157	5.8–6.5	0.147–0.165
7628	7.0–7.8	0.178–0.198

Copper Thickness for PTHs and Vias

Electroless copper plating is used to plate holes to make PTHs and vias. While drilled holes are being plated some of the external surfaces are also plated. The thickness of the plating is typically 20 to 100 microinches both in the holes and in the surfaces depending on the board manufacturer's processes (IPC-2221A, Table 4-3; Coombs 2001, p. 28.8).

Later in the fabrication process other plating processes are used to "finish" the board. After finishing processes are completed, the external surfaces of a PCB can be much thicker, but the wall thickness of a PTH is usually 1 mil or less. Most of the time it does not enter into the drill size calculations unless the hole is very small (most finished hole sizes are 8 mils or larger), but the information is presented here for completeness. Table 4-5 shows the minimum PTH wall thickness as presented in the literature (IPC-2221A, Table 4-3).

Table 4-5	Minimum Electroless Plating Thicknesses (Surfaces and Holes)			
	Classes 1 and 2		**Class 3**	
Minimum thickness	**(mils)**	**(mm)**	**(mils)**	**(mm)**
Average	0.79	0.020	0.98	0.025
Thin	0.71	0.018	0.79	0.020

Reference: IPC-2221A, Table 4-2 (partial data).

Copper Cladding/Foil Thickness

When you order your PCB the manufacturer will need to know how thick you want the copper. The thickness of the copper depends on how much current the trace will be required to carry and the required impedance of the traces (for controlled-impedance PCBs). The thickness also plays a role in how narrow the traces can be because thicker copper takes longer to etch and can result in variations in trace width and etchback effects as described below. As mentioned above, various finishing processes (etching and plating) alter the final cladding or foil thickness. The processes are described in detail in the IPC standards and Coombs and are not described here. Table 4-6 shows the copper thicknesses by weight and gauge as described in the literature (Coombs 2001, Tables 5-4, 5-11; lPC-D-330, Section 2, Table 2-16; IPC-4101, Table 1-2, p. 3). The values listed here are for reference only.

Table 4-6	Nominal and Finished Copper Thickness by Weight and Gauge (±10%)					
Area wt (oz/ft^2)	Nominal thickness		Internal minimum finished thickness		External minimum finished thickness	
	(mils)	(mm)	(mils)	(mm)	(mils)	(mm)
0.148 ($^1/_8$)	0.20	0.005	0.12	0.0031	0.91	0.0231
0.25 ($^1/_4$)	0.34	0.009	0.24	0.0062	1.03	0.0262
0.35($^3/_8$)	0.47	0.012	0.37	0.0093	1.15	0.0293
0.50 ($^1/_2$)	0.68	0.017	0.45	0.0114	1.32	0.0334
0.75 ($^3/_4$)	1.01	0.026	0.76	0.0193	1.62	0.0410
1	1.35	0.034	0.98	0.0249	1.89	0.0479
2	2.70	0.069	2.19	0.0557	3.10	0.0787
3	4.05	0.103	3.41	0.0866	4.32	0.110
4	5.40	0.137	4.63	0.118	5.49	0.139
5	6.75	0.171	5.92	0.150	6.32	0.160
6	8.10	0.206	7.13	0.181	7.28	0.185
7	9.45	0.240	8.35	0.212	8.22	0.209
10	13.5	0.343	12.0	0.305	10.9	0.277
14	18.9	0.480	16.9	0.428	14.3	0.364

COPPER TRACE AND ETCHING TOLERANCES

When copper is etched (instead of milled) the edge of the copper trace is neither a completely smooth nor a vertical wall. The roughness (called the *edge definition*) occurs because of mask resolution limitations, nonuniformity of the acid circulation, gas bubbles during etching, etc. The wall will have a slight angle to it because, as the acid begins to work its way into the exposed copper, a sidewall begins to form, which also is attacked by the acid. As Figure 4-3 shows, the copper near the etch resist begins to be removed under the mask.

Copper trace · w' · Etch resist
Etchback
FR4 · W

FIGURE 4-3
Result on finished
trace width due to
etchback.

This effect is called *etchback* or *undercutting*. If the etching process is stopped as soon as the last bit of copper cladding or foil is removed from the surface of the board, the trace width at the bottom will be the initial size of the mask width (which is defined by the Gerber files). For this reason the wider value (W, not w', in Figure 4-3) is used for design calculations when trace width dimensions are required.

Width tolerances are specified in Table 10-3 of IPC-2221A (p. 78) and range from ±4.0 mils to ±0.6 mil for 1½ oz copper depending on reducibility level and plating. The designer should be aware of the width variations when calculating trace widths for controlled impedances and for current handling ability. For general design considerations traces should be made as wide as practical. Per IPC-2221A the minimum trace width and spacing is 3.9 mils. Individual board manufacturers may have their own etching and spacing tolerances. Typical minimum trace widths are 4 to 8 mils. It is a good idea to call and ask what their capabilities are or check their Web sites.

STANDARD HOLE DIMENSIONS

Holes are drilled into PCBs by various techniques including twist bits, router bits, lasers, and plasmas. The capabilities with regard to placement and size accuracy and the speed can vary considerably. But the board designer needs to know what size of drill hole to specify in a padstack. Standard drill bit sizes are specified in ANSI standard B19.11M. However, not all board manufacturers carry every size bit. When you specify a particular hole size manufacturers have different ways of "adjusting" it to fit their capabilities. Some may round (up or down) to the nearest drill size available or they may always round up to the next largest drill bit to make sure that the hole is never too small. This affects annular ring width, which can lead to breakout as described earlier. IPC-D-330 (Section 2, Table 2-4) lists minimum plated through hole sizes by board thickness and class.

When laminate materials are drilled they become soft due to frictional heating. The softened laminate then smears during the drilling process, which coats the surface of the copper and can prohibit good plating (Coombs 2001, p. 10.7). To solve this problem the laminate inside the hole is etched back after being drilled to desmear the hole. Etchback enhances plating of PTHs (Coombs 2001, p. 48.30), but excessive etchback can cause partial delamination and internal shorts when combined with misregistration and may make the hole larger than requested. This is one reason why adequate clearance is required between the plated through hole and the ground plane (Coombs 2001, p. 17.9).

After the hole is drilled and desmeared it is plated. Depending on the manufacturer's plating methods and processes, the plating can add as much as 1 mil of thickness on all surfaces, which means that the diameter of the hole could be as much as 2 mils smaller than the specified drill diameter. The padstack drill size in PCB Editor is the actual drill size, not the finished hole size. In most cases the variance in available drill bit size and plating is not a problem. But if your holes are very small and/or the lead diameters are very close to the size of the drill hole, the hole variations can cause a lead to not fit into its through hole. Padstack calculations that account for plating widths and tolerances are described in the next chapter. Another problem that can occur is that, if the hole size is very small compared to the thickness of the board, the plating process may not be satisfactory. The "proper" hole diameter to PCB thickness is called the *aspect ratio*.

Aspect ratio (AR) is the ratio of the plated hole diameter to the PCB thickness. It is recommended that the AR be between 3:1 and 5:1 for Level A boards (IPC-2221A, Table 5-1, p. 27; also IPC-CM-770E; and IPC-D-330, Table 6-30), with 3:1 being a good target (Coombs 2001, p. 42.3).

The AR for Level B boards is 6:1 to 8:1, and Level C boards may be as high as 9:1 or higher. Check with your board manufacturer to know what its capabilities are, because if the aspect ratio is too high, plating problems can occur inside the hole, which can lead to open circuits from incomplete plating and barrel cracking.

Padstack design, which includes all of the issues described above, is covered in detail in Chapters 5 and 8.

SOLDERMASK TOLERANCE

Due to photolithography misregistration and swell of the soldermasks, lands can be partially covered. Soldermasks that are patterned on solder-plated lands may be damaged when the solder reflows during soldering operations and can adversely affect solderability. This may be especially troublesome on very small parts (SOT23). To reduce these risks, the soldermask openings are often larger than the lands (oversized). There are two basic categories of soldermask materials: liquid screen printed and photoimageable masks. The recommended oversizing for liquid-screen-printed coatings is 16 to 20 mils, and the recommended oversize of photoimageable masks is 0 to 5 mils (IPC-2221A, 4.5.1.2, p. 24). Many of the PCB Editor footprints have clearances of between 0 and 10 mils. Many fabricators will adjust the soldermask as necessary for their processes. However, some do not and expect (or assume) that you will do it. Check with your board house to find out if it requires oversizing and, if so, how much and who it expects to do it.

REFERENCES

ANSI B19.11M.

ANSI/IPC-D-322, Guidelines for Selecting Printed Wiring Board Sizes Using Standard Panel Sizes. Lincolnwood, IL: Institute for Interconnecting and Packaging Electronic Circuits.

Coombs, C. F., Jr. (Ed.) 2001. *Coombs' Printed Circuit Handbook*. 5th ed. New York: McGraw-Hill.

IEEE-1445-1998. IEEE Standard for Digital Testing Interchange Format.

IEEE/ANSI 315-1975. Graphic Symbols for Electrical and Electronics Diagrams.

IPC-2221A. May 2003. Generic Standard on Printed Board Design. Northbrook, IL: IPC/Association Connecting Electronic Industries (easy reading, ~100 pages, lots of pictures).

IPC-4101. June 2002. Specification for Base Materials for Rigid and Multilayer Printed Boards. Northbrook, IL: IPC-Association Connecting Electronics Industries.

IPC-7351. February 2005. Generic Requirements for Surface Mount Design and Land Pattern Standard. Northbrook, IL: IPC/Association Connecting Electronic Industries (easy reading, ~74 pages, lots of pictures).

IPC-D-330.

IPC-CM-770E. January 2004. Guidelines for Printed Board Component Mounting. Northbrook, IL: IPC/Association Connecting Electronic Industries.

SUGGESTED READING

Highly suggested reading for new PCB designers is given below. A more thorough list is given in Appendix B.

1. MIL-STD-275, Printed Wiring for Electronic Equipment (similar to and superseded by IPC-2221A) (*free download*).
2. MIL-HDBK-198, Selection and Use of Capacitors (*free download*).
3. MIL-HDBK-199, Selection and Use of Resistors (*free download*).

OTHER ITEMS OF INTEREST

1. ASME B18, Series Standards for Mounting Hardware (includes tables for standard screw sizes and recommended drill hole sizes, etc.).
2. MIL-HDBK-5961A, List of Standard Semiconductor Devices (*free download*).
3. MIL-HDBK-1861A, Selection and Use of Electrical and Electronic Assemblies, Boards, Cards, and Associated Hardware (*free download*).
4. 47CFR15, FCC Rules on Radio Frequency Devices (including unintentional radiators) (*free download*).

CHAPTER 5

Introduction to Design for Manufacturing

INTRODUCTION TO PCB ASSEMBLY AND SOLDERING PROCESSES

For a PCB to be manufacturable the bare board has to be able to be fabricated within standard fabrication allowances (SFAs), and the board has to be able to be assembled given the different component technologies. As discussed in the previous chapters, there are many steps to designing and fabricating a PCB. When a design is complete and submitted to a board house, the house must be able to perform the manufacturing steps that the design calls for. Whether a PCB is manufacturable or not really begins with and includes parts creation and schematic entry in Capture, padstack and footprint design, parts placement and trace routing, and artwork production in PCB Editor.

But it does not end there. Once the board has been fabricated, it is of little use without the functional parts. Those parts need to be able to be attached to the board without them or the board being damaged. Parts attachment encompasses positioning the parts (which depends on proper parts placement and orientation) and making reliable solder joints between the component's leads and the board's mounting pads (which depends on good padstack and footprint design). This chapter provides the information necessary for padstack and footprint design and component placement for the design of manufacturable PCBs.

Assembly Processes

Printed circuit boards may be manually assembled or assembled by automated machinery. Assembly processes depend on the class of component technology (Classes A through Z as described earlier) and the number of boards to be assembled at one time. Some companies may both fabricate and assemble boards under one roof, while some companies may specialize in PCB fabrication only and others in PCB assembly only. The method of assembly plays a role in how you lay out your PCB because of clearance and orientation issues and soldering processes. A brief summary of assembly processes is given here along with component placement, orientation, and spacing considerations.

MANUAL ASSEMBLY PROCESSES

Manual assembly is typically used for prototype and low-volume work and in post-automated assembly for odd-form components. Both surface-mount technology (SMT) and through-hole technology (THT) components may be assembled manually. In low-volume work an assembly line of several assemblers may be used, in which each person is responsible for attaching specific components. The assembly processes may be interrupted several times to functionally test sections of the PCB as it is assembled. Manual assembly may involve both manual placement and soldering or a mixture of manual placement and automated soldering (described below).

Manual assembly can be tedious work. Consistent component placement and orientation can aid manual assemblers. For example, orienting polarized components (capacitors and diodes) in the same direction and orienting integrated circuits (ICs) so that pin 1 on all ICs is located in the same direction can significantly reduce assembly defects and increase yield.

AUTOMATED ASSEMBLY PROCESSES (PICK AND PLACE)

Automated insertion processes (pick and place) exist for both surface-mounted devices and through-hole devices which include both radial- and axial-leaded devices. Automated machines are programmed to extract parts from reels or bins and place the components on the PCB in the correct location and orientation. Through-hole devices can typically be populated at rates from 20,000 components per hour (CPH) to 40,000 CPH (Coombs 2001, Section 41.2.5, p. 41.10). Data for programming automated placement machinery can be supplied by output from PCB Editor and other CAM programs.

Through-hole devices are usually packaged as roles or strips of components, which are taped together by their leads. Through-hole components are usually placed only on the top side of the board so that the leads can be wave soldered and the components themselves are not exposed to molten solder. Soldering processes are described below. The typical automated process steps for through-hole devices is insertion of dual inline packages first, then axial-leaded devices, then radial-leaded devices, and finally odd-form devices. After the components have been inserted, the board is most often wave soldered (wave soldering is described below), but can be reflow soldered (intrusive reflow; see Coombs 2001, p. 43.10), also described below.

Surface-mounted devices are commonly packaged in tubes, matrix trays, tape and reel, and bulk. Surface-mounted devices may be mounted on one or both sides of a PCB. When attached to the top side only, solder paste is screen printed onto the PCB's solder pads. The parts are then placed onto the board by the pick-and-place equipment with the component lead terminations set into the paste, temporarily holding the parts in place. The board is then run through a reflow oven, which melts and then cools the solder, thereby attaching the part to the board. Surface-mounted devices can typically be populated at rates from 10,000 to 100,000 CPH (Coombs 2001, Section 41.3.2, p. 41.15).

When surface-mounted devices are placed on both sides of a PCB or when a board contains both surface and through-hole devices, a sequential, reflow/ wave soldering process is employed. First the top-side surface-mounted devices are attached to the board using the solder paste and reflow process described below. Next, through-hole devices are inserted from the top and held in place by clinching (bending) the leads on the bottom, by gluing the part to the top, or in some cases by friction between the lead and the hole. The board is flipped over and adhesive dots are applied to the bottom of the board by an automated dispenser. The bottom-mounted surface-mounted devices (SMDs) are then positioned manually or by pick-and-place machines onto the glue dots. The adhesive holds the bottom-side surface-mounted devices in place until the solder joint has been completed. The assembly is run through an oven to cure the adhesive. The board, with through-hole components on the top and SMDs on the top and bottom, is then run through a wave-soldering station, which solders the through-leads and the bottom-mounted SMDs. The previously soldered top-mounted SMDs remain soldered on the top.

When a PCB has only SMDs but has them on both sides, a two-step reflow soldering process is sometimes used. The top-mounted SMDs are attached to the board first using a high-temperature solder paste, and then the board is run through a high temperature reflow oven. With the top-side components securely in place, the board is flipped over and the bottom-mounted SMDs are attached with the lower temperature paste and reflow process as described below.

Soldering Processes

Soldering is used both to attach components physically to the PCB and to provide electrical conductivity between the component's leads and the PCB traces. For the soldering process to be successful an intermetallic compound, or alloy, must be formed between the solder and the base material (the leads and traces). To protect the solder joint areas from oxidation, contact areas on new PCBs receive a surface finish by being dipped in a solder bath and hot-air solder leveled or are plated by some other plating process such as electroless nickel or palladium (Coombs 2001, p. 32.1). Just prior to or during soldering, the surfaces to be joined are cleaned (deoxidized) with flux so that the solder can flow over and wet the surfaces.

There are two general soldering methods: mass soldering (which includes wave, oven reflow, vapor phase reflow, and conduction reflow) and directed energy (which includes hot gas, hot bar, laser, iron, and pinpoint torch) (see Coombs 2001, p. 43.10; IPC-CM-770, p. 34). Only manual, wave soldering, and oven-reflow soldering are discussed here and only briefly.

MANUAL SOLDERING

Manual soldering is used for a wide variety of applications from complete PCB assembly to simple repair work and touch-up. There are several types of manual soldering tools available, including but not limited to hot-air pencils, soldering irons, and induction coils. Other than slower soldering speed, the biggest drawbacks to manual soldering are the increased risk of electrostatic discharge during handling and thermal gradients caused by localized heating of the board and parts. Parts placed on the board that will be manually placed and soldered require no special layout consideration as far as spacing and orientation other than the basics described below. However, it is helpful to the assembler if parts placement affords room to work and similar parts are aligned and oriented in a consistent manner as described above.

WAVE SOLDERING

During wave soldering the board is held by its edges on a conveyor, fluxed, and preheated as shown in Figures 5-1 and 5-2. The conveyor moves the board past a standing wave of molten solder so that only the bottom side of the board is exposed to the solder. Wave soldering can be used for both through-hole devices (THDs) and SMDs, but reflow is preferred for SMDs. The through-hole components are placed on the top with the leads protruding out the bottom of board, which is prefluxed. As the conveyer moves the PCB into the solder wave, solder wicks up the barrel and creates fillets on the top and the bottom. SMDs are glued to the board, fluxed, and run through the wave. Very small components or large tantalum caps can be problematic with wave soldering. The small parts cause problems during the gluing process because in some cases the glue dot is larger than the component and the glue can ooze over onto the solder pads. Large SMDs cause problems because of thermal stresses that can lead to component cracking.

FIGURE 5-1
Wave soldering (side view): DIP, dual inline package; PTH, plated through hole.

When surface-mounted devices are wave soldered (i.e., mounted to the bottom side of the board), the designer needs to know (or specify) the direction the board will travel through the wave. The components should be placed on the

Board traveling into wave

FIGURE 5-2
Wave soldering (rear view).

board as shown in Figure 5-3 so that smaller parts are not shadowed by larger parts, which can cause poor solder joints on the smaller parts, and so that solder bridging does not occur across the leads.

Solder bridging can occur because of the fine lead pitch of some SMDs. As each pin leaves the solder wave, it tends to draw some of the solder from the pin beside it as the molten solder attempts to reduce its surface tension (for the

FIGURE 5-3
SMD component orientation for wave soldering.

orientation shown in Figure 5-3). This causes a problem at the last two rows of pins as there are no pins that follow to draw the excess solder away from them, which can result in the last two pins on a side being bridged by the excess solder. One method to minimize the bridge is to place solder thieves on the trailing edges of SMD ICs as shown in Figure 5-3. The solder thieves are extra pads placed after the last pads that pull the excess solder away from the pads. Solder thieves can also be made by simply making the last pads a little larger and extending farther back than the typical pads on that device.

If SMD parts are on top of the board and are reflow soldered prior to wave soldering, fan-out vias need to be located away from lands to prevent solder migration away from the SMD lands and down into the via. This can occur as heat conducts from the solder wave up via barrels and re-reflows the solder, which then draws the solder down the via by capillary action. The suggested spacing is ≥20 mils (IPC-7351, Section 3.4.6.2) between the edge of the via and the edge of the pad. A default spacing can be set in PCB Editor using the Fanout tab in the Automatic Router Parameters dialog box.

Some boards may be too large or too small to be wave soldered. Large boards sag as they are heated unless special holding fixtures are used. Very small boards may need to be panelized with breakaways (tab routes or V scores) so that they can be handled by automated equipment without having to make specialized board holders.

REFLOW SOLDERING

There are various types of reflow soldering, but the discussion here is limited to oven-type reflow soldering. Reflow soldering is most often used for surface-mount devices, but through-hole components can also be soldered this way (called *pin-in-paste* or *intrusive reflow soldering*). A schematic diagram of a reflow oven is shown in Figure 5-4.

Solder paste (which contains solder and flux in a viscous paste) is applied to a bare board with a stencil. The components are placed into position (usually by pick-and-place machines) so that the leads sit in the solder paste. The assembly

FIGURE 5-4
Reflow soldering oven (side view).

is placed on a conveyor and run through a temperature-controlled oven that heats the boards and parts evenly. As the assembly heats up the flux is activated and the solder reflows (melts). Surface tension of the molten solder tends to self-align the parts. However, if parts are not in proper alignment and thermal gradients do not melt the solder on all pads at the same time, the component may stand up on one end (called *tombstoning*). Once the assembly has reached the proper temperature and the solder has reflowed, the assembly exits the oven and is cooled to solidify the solder.

Solderable components need to be selected for both wave and reflow soldering. It is also important not to place all of the thermally massive parts in the same area of the board, as this can cause thermal gradients across the board and poor solder joint formation in the cooler areas (Coombs 2001, p. 43.12; for board design, see Coombs 2001, p. 43.25).

COMPONENT PLACEMENT AND ORIENTATION GUIDE

A PCB assembly consists of the bare board, the attached components, and connectors. The board design can have a significant impact on how easily components can be placed on and attached to the board and how reliable the end product is.

The board topology (class and level), the component technology (SMD or through hole), and the soldering method used to attach the parts (reflow vs. wave soldering) play significant roles in how parts should be placed and spaced on the board.

Component placement and orientation depend on the type of component (THT or SMT), the assembly method (manual, wave soldered, or reflow), and the electrical performance requirements (electrical discussed below). These by themselves do not make the board function better but they do make it easier to assemble, inspect, test, and troubleshoot. General guidelines are

1. Components should be placed so that they are neat and organized with uniform spacing and alignment.
2. Components should be oriented such that component edges are parallel to the board edges (with the exception noted below for wave soldering).
3. If a board is machine soldered, through-hole components should be mounted on a single side that is opposite the solder whenever possible.
4. When placing components on both sides of the board and when mixing through-hole and SMD technologies, keep in mind that multiple assembly phases may be required to place all of the components, which increases cost and potential failure points and makes rework more difficult.
5. Do not put plastic leaded chip carriers (PLCCs) or large tantalum capacitors on the bottom side of board (i.e., do not wave solder them), as they can easily crack due to thermal stresses.
6. A 100-mil (2.5-mm) grid should be used if possible, but a 20-mil (0.5-mm) grid or even 2-mil (0.05-mm) grid can be used if required for component leads that are not on standard grids (IPC-2221A, Section

5.4.2, p. 31 and Section 8.1.2, p. 56). When a metric grid is used, a grid of 2.54 mm (0.100 in.), 1.27 mm (0.05 in.), 0.64 mm (0.025 in.), or 0.50 mm (0.020 in.) should be used (IPC-7351, Section 3.4.1.4, p. 20).

7. A 0.100-in. (2.54-mm) grid should be used for boards that will undergo bed-of-nails testing.

8. Polarized capacitors and diodes should all be oriented consistently throughout the board for ease of inspection and testing.

9. When machine-vision-assisted assembly processes are used, add fiducials (global and local) to aid in component placement.

10. If the design allows, connectors should be placed on the short side of the board when using automated soldering processes.

11. Allow adequate space along board edges during component placement for handling the board and to accommodate mounting hardware.

12. Components that weigh more than 5.0 g per lead should be mechanically supported if the board will experience vibration (IPC-2221A, Section 5.2.7, p. 29).

13. Thermal management during soldering processes and circuit operation should be considered throughout the design process.

14. Electrical considerations usually have priority over mechanical considerations when there is a conflict between the two, unless it will result in a mechanical failure of the board.

15. For mixed-signal (analog/digital) PCBs, components should be segregated to minimize the effect of switching noise on analog circuits. High-power circuits should also be segregated from low-power and low-noise circuits.

Component Spacing for Through-Hole Devices

The following tables provide minimum recommended spacing guidelines for discrete and IC through-hole devices.

DISCRETE THDS

For discrete THDs, see Tables 5-1 and 5-2.

INTEGRATED CIRCUIT THROUGH-HOLE DEVICES

For integrated circuit THDs, see Table 5-3.

MIXED DISCRETE AND IC THROUGH-HOLE DEVICES

For mixed discrete and IC THDs, see Table 5-4.

HOLES AND JUMPER WIRES

For holes and jumper wires, see Table 5-5.

Component Spacing for Surface-Mounted Devices

The following tables provide minimum recommended spacing guidelines for discrete and IC surface-mount devices.

Table 5-1 Minimum Recommended Spacing for Discrete, Axial THDs

Parameter	Mils	Millimeters	PCB Editor default
Side to PCB edge	$a = 75$	$a = 1.9$	Depends on pad to track space setting. No DRC error occurs as long as place outline does not cross the package keep-in outline.
End to PCB edge	$b = 90$	$b = 2.29$	
End to end	100	2.54	DRC error occurs if place outlines overlap.
Side to side	100	2.54	DRC error occurs if place outlines overlap.
When body diameters are <100 mils (2.54 mm)			
Side to side	$a = 70 + \frac{1}{2}D_1$	$a = 1.78 + \frac{1}{2}D_1$	DRC error occurs if place outlines overlap.
When $D_2 > D_1 > 100$ mils	$b = 10 + \frac{1}{2}D_1 + \frac{1}{2}D_2$ (a and b ≥ 100 mils minimum)	$b = 0.25 + \frac{1}{2}D_1 + \frac{1}{2}D_2$ (a and b ≥ 2.54 mm minimum)	
Side to end	$95 + \frac{1}{2}D_1$	$2.41 + \frac{1}{2}D_1$	DRC error occurs if place outlines overlap.
When one or more body diameters are >100 mils			

Reference: IPC-2221A, Fig. 7-1. p. 51.

Table 5-2	Minimum Recommended Spacing for Discrete, Radial THDs	
Parameter	**Mils**	**PCB Editor default**
PCB edge	$r = \frac{1}{2}$ the diameter of the device or $\frac{1}{2}$ the height, whichever is greater, *and* $r \geq 60$ mils (1.52 mm).	No DRC error occurs as long as place outline does not cross the package keep-in outline.
Others parts	$r = \frac{1}{2}$ the diameter of the device or $\frac{1}{2}$ the height, whichever is greater.	DRC error occurs if place outlines overlap.

Table 5-3	Minimum Recommended Spacing for Through-Hole Mounted ICs		
Parameter	**Mils**	**Millimeters**	**PCB Editor default**
Side to PCB edge	$a = 100$	$a = 2.54$	No DRC error occurs as long as place outline does not violate the package keep-in outline.
End to edge	$b = 75$	$b = 1.91$	
End to end	200	5.08	No DRC error occurs as long as place outlines do not overlap. Could violate IPC standard and not cause DRC error.
Side to side	100	2.54	

Table 5-4   Minimum Recommended Spacing between Through-Hole Discretes and ICs

Parameter	Mils	Millimeters	PCB Editor default
a ($D_1 > 100$)	$115 + \frac{1}{2}D_1$	$2.91 + \frac{1}{2}D_1$	No set rule. DRC error occurs if place outlines overlap.
b ($D < 100$)	200	5.08	
c ($D < 100$)	100	2.54	
d ($D_1 > 100$)	$40 + \frac{1}{2}D_1$	$1.02 + \frac{1}{2}D_1$	

Table 5-5   Minimum Recommended Spacing for Holes and Jumper Wires

Parameter	Mils	Millimeters	PCB Editor default
Hole to hole (Plated or nonplated)	(a) So as not to violate pad spacing rules. (b) So that residual laminate is >20 mils (0.5 mm) between holes.		DRC error occurs if pad-to-pad spacing rules are violated.
Jumper wires (any direction)	100	2.54	DRC error occurs if pad-to-pad spacing rules are violated.

DISCRETE SMDS

See Table 5-6 for discrete SMDs.

Table 5-6	Minimum Recommended Spacing for Discrete SMDs			
Parameter		**Mils**	**Millimeters**	**PCB Editor DRC**
Side to PCB edge and/or end to PCB edge		60	1.5	DRC error occurs if space from pad to edge of board outline is less than pad-to-track spacing rule or if place outline crosses package keep-in outline.
End to end and/or side to side		*Size 0603 or larger*		Spacing determined by relationship of place outline to pads and body. DRC error occurs if place outlines overlap.
		20	0.50	
Pad to via		*Smaller than 0603*		DRC error occurs if distance between edge of pads is less
		12	0.30	
		20	0.50	

References: IPC-2221A, p. 56; IPC-7351. Tables 3-3 to 3-9, Figs. 3-14 and 8-35.

INTEGRATED-CIRCUIT SMDS

For IC SMDs, see Table 5-7.

MIXED DISCRETE AND IC SMDS

Use the greater of any of the preceding spacing rules for the components involved.

Mixed THD and SMD Spacing Requirements

Use the greater of any of the preceding spacing rules for the components involved (usually the THT spacing).

FOOTPRINT AND PADSTACK DESIGN FOR PCB MANUFACTURABILITY

Many footprints are included with the PCB Editor software, but you will need to make your own at some point. Chapter 8 describes the PCB Editor tools that

Table 5-7	Minimum Recommended Spacing for IC SMDs			
Parameter		**Mils**	**Millimeters**	**PCB Editor default**
Component side to PCB edge and/or end to PCB edge		60	1.5	DRC error occurs if space from pad to edge of board outline is less than pad to track spacing rule or if place outline crosses package keep-in outline.
End to end (body)		20	0.50	Spacing determined by relationship of place outline to pads and body. DRC error occurs if place outlines overlap.
Side to side (pad to pad)				

References: IPC-2221A, p. 56; IPC 7351, Tables 3-3 to 3-9.

are used to design the footprints, and the following describes the design considerations and industry standards related to designing footprints. Footprint design for SMDs and THDs are significantly different, but both require consideration of PCB manufacturability and assembly. A footprint (land pattern) in PCB Editor consists of padstacks, silk-screen elements, and a place boundary outline (see Chapter 8 for specifics).

There are no specific component spacing constraints in PCB Editor as there are route spacing constraints. The footprint design (the size of the place boundary outline) determines the spacing and therefore the maximum board density possible (without incurring DRC errors). Per the IPC standards there are several spacing recommendations, which are dependent on the board classification and package types. PCB Editor's footprints do not necessarily fall within specific IPC design standards. So if you need to produce a Level C board, you will need to modify or make new footprints that fall within the specific guidelines. Or if you want footprints that specifically meet Level A requirements to increase yield and reliability, then you may also need to design special footprints or modify existing ones. The following discusses some of the design issues and provides references to the industry standards.

Land Patterns for Surface-Mounted Devices

When you are doing a board layout and need a part that is not in the PCB Editor footprint library, there are several things you can do to construct one. If PCB Editor has a footprint that is similar to the one you need, but it has the wrong number of pins, you can save a copy of the existing part with a new name and use it as a pattern to add pins and resize the place outline and silk screen as necessary. If PCB Editor has no footprint that you can use as a pattern, the next step is to check with the manufacturer's data sheet to see if it has a suggested land pattern. If it does not provide the necessary information, then you may be able to find the design parameters from the IPC Land Pattern Viewer. If all else fails you will need to design one from scratch using the package information from the component's data sheet or one of the JEDEC package standards.

Component data sheets and the JEDEC package standards provide information about the package dimensions, but not the PCB land pattern, typically. The IPC standards provide guidance on the land pattern dimensions relative to the lead dimensions but does so in a way that requires additional calculations to construct the padstacks and footprints in PCB Editor. When designing footprints in PCB Editor you need to know the size of the padstack and the distances between the centers of the pads. The IPC standards typically provide spacing information relative to the edges of the pads.

So for convenience, a way of translating the JEDEC package dimensions directly into PCB Editor footprint dimensions while maintaining compliance with IPC standards is needed.

When you look up a part's data sheet, the dimensions are usually with regard to the package. An example of an eight-pin small outline integrated circuit (SOIC) is shown in Figure 5-5. The part manufacturer may also provide the JEDEC specification. Figure 5-6 illustrates an example of JEDEC standard MS-012 for an SOIC package. Figure 5-6 is reprinted with permission of JEDEC.

From the manufacturer's information and/or the standards, we need to be able to determine the padstack width and height and the spacing between the pads in both the x and the y directions as shown in Figure 5-7, in order to create a new footprint in PCB Editor.

FIGURE 5-5
Data sheet package dimensions (typical convention).

We begin by looking at the design of the padstack followed by footprint design using the established padstacks.

FIGURE 5-6
JEDEC package dimensions (typical convention).

Symbol	Common dimensions			Note
	MIN	NOM	MAX	
b	0.31	–	0.51	7.8
b1	0.28	–	0.48	7.8
E	6.00 BSC			
e	1.27 BSC			
L	0.40	–	1.27	

FIGURE 5-7
Footprint dimensions
(typical convention).

SMD PADSTACK DESIGN

A good padstack promotes the best possible solder joint between a component termination (lead) and the PCB. The padstack must allow for component dimensional variations, PCB fabrication tolerances, placement tolerances, and solder fillet specifications. Through-hole devices are relatively large and therefore more forgiving of these tolerances, but SMDs are typically much smaller and are therefore more sensitive to manufacturing and placement variations. IPC-7351 (which superseded IPC-SM-780/2) is the standard for surface-mount land pattern design for both padstack and footprint design.

As shown in Figure 5-8 the solder pad (padstack) needs to be larger than the component lead to allow a proper solder joint. The required pad oversize is defined in IPC-7351 (pp. 8–14), where the term J_T defines the distance from the end of the pad to the toe of the lead, J_H defines the distance from the end of the pad to the heel of the lead, and J_S defines the distance from the sides of the pad to the sides of the lead.

FIGURE 5-8
SMD padstack requirements: (a) side view, (b) toe view.

The values for J_T, J_H, and J_S depend on the type of component and the desired density level (A through C); nominal values are provided in Tables 5-8, 5-9, and 5-10.

Table 5-8	Nominal Toe Solder Fillet Values (J_T) by Package Type	
	Nominal density	
Package type	**Mils**	**Millimeters**
Gull wing (SOG)	14	0.35
J lead (SOJ)	14	0.35
Chip components (0603 and larger)	14	0.35
Chip components (smaller than 0603)	4	0.10
Small outline (SO)	12	0.30
Tantalum capacitors	6	0.15
Leadless chip carrier	22	0.55
MELF	16	0.40
Butt joints	31	0.80

To design a padstack in PCB Editor we need to know the width of the padstack, W_P, and the height of the padstack, H_P, to be able to fill in the required values in the Padstack Designer (Figure 5-9).

Table 5-9 Nominal Heel Solder Fillet Values (J_H) by Package Type		
	Nominal density	
Package type	**Mils**	**Millimeters**
Gull wing (SOG)	14	0.35
J lead (SOJ)	−8	−0.20
Small outline (SO)	0	0.00
Chip components (all)	−2	−0.05
Tantalum capacitors	20	0.50
MELF	4	0.10
Leadless chip carrier	6	0.15
Butt joints	31	0.80

Table 5-10 Nominal Side Solder Fillet Values (J_S) by Package Type		
	Nominal density	
Package type	**Mils**	**Millimeters**
Gull wing (SOG) (pitch greater than 0.625 mm)	1	0.03
J lead (SOJ)	1	0.03
Gull wing (SOG) (pitch less than 0.625 mm)	−1	−0.02
Chip components (0603 and larger)	0	0.00
Chip components (smaller than 0603)	0	0.00
MELF	2	0.05
Small outline (SO)	−2	−0.04
Tantalum capacitors	−2	−0.05
Leadless chip carrier	−2	−0.05
Butt joints	8	0.20

Using the parameters from the component data sheet or the JEDEC standard, Eqs. (5.1) and (5.2) can be used to determine maximum sizes for W_P and H_P.

$$W_{P(\text{MAX})} = E_{\text{MIN}} - (E_{\text{MAX}} - 2L_{\text{MIN}}) + 2J_{\text{T}} + 2J_{\text{H}}$$
$$+ \sqrt[2]{(E_{TOL(\Delta)})^2 + F^2 + P^2}, \tag{5.1}$$

FIGURE 5-9
Pad design using Padstack Designer.

where $W_{P(MAX)}$ is the maximum pad width (see Figures 5-7 and 5-9); $E_{(MIN\ and\ MAX)}$ is the distance between the ends of the leads per the JEDEC dimensions (Figure 5-6); $E_{TOL(\Delta)}$ is the tolerance of E per the JEDEC dimensions or calculated from $(E_{MAX} - E_{MIN})$ (Figure 5-6); L is the length of the lead that will be soldered to the pad per the JEDEC dimensions (Figure 5-6); J_T and J_H are solder fillet allowances as described in Tables 5-8 and 5-9, which are derived from IPC-7351; F is the PCB fabrication tolerance (IPC-2221A, Table 10-3, 0.1 mm or 4 mils typically); and P is the placement tolerance of pick-and-place machines (depends on the machine, 0.15 mm or 6 mils is typical); and

$$H_{P(MAX)} = b_{MIN} + 2J_S + \sqrt{(b_{TOL(\Delta)})^2 + F^2 + P^2} \qquad (5.2)$$

where $H_{P(MAX)}$ is the maximum pad height (see Figures 5-7 and 5-9), b_{MIN} is the minimum lead width, J_S is the solder fillet allowances as described in Table 5-10, $b_{TOL(\Delta)}$ is the tolerance of b per the JEDEC dimensions or calculated from $(b_{MAX} - b1_{MIN})$ (Figure 5-6), and F and P are as described for Eq. (5.1).

These equations were derived from various tables in IPC-7351, but the actual standard also includes rounding factors, which are not included in Eqs. (5.1) and (5.2). Tables 5-8, 5-9, and 5-10 were also derived from various tables in IPC-7351, but the source standard includes additional data for greater and lesser density levels, while only the nominal density levels are included here. Please see the source for full dimensional considerations or download the IPC Land Pattern Viewer from the IPC Web site (a demo version is available for free).

Once the pad is designed (i.e., you have calculated the width and height), use the procedures given in Chapter 8 to create the pad with Padstack Designer.

SMD FOOTPRINT DESIGN

Once the padstacks are designed, they need to be correctly located to complete the footprint design. An example of how IPC-7351 defines land pattern parameters is shown in Figure 5-10 for an eight-pin SOIC. The component outline defines the outermost boundary of the IC including both the edge of the package and the end of the leads. Next a courtyard is defined around the part that includes the body and the basic land pattern. The courtyard is adjusted outward from the part outline to protrude into and consume PCB real estate. The amount of protrusion (called *courtyard excess*) determines the minimum separation between components during placement. The greater the courtyard excess, the less densely the PCB can be populated. Density levels are categorized by Levels A through C, as listed in Table 5-11 for various SMD packages (from IPC-7351, Tables 3-2 through 3-14).

Pad

Component outline

Courtyard outline

Excess

Protruded courtyard

FIGURE 5-10
IPC-7351 land pattern description of an eight-pin SOIC.

PCB Editor does not use the courtyard concept used by the IPC-7351 standard for its SMD footprints. Figure 5-11 shows the eight-pin SOIC place boundary outline in PCB Editor overlaid onto the IPC protruded courtyard outline. Notice that the PCB Editor place boundary outline does not include the excess space at the ends of the body or around the leads.

For most of the PCB Editor footprints included with the software library, the courtyard excess is nonexistent. Also in PCB Editor's Constraint Manager, there is no package spacing constraint that you can set. The DRC looks for violations only where place outlines cross. So satisfying density level requirements requires that you either modify the footprint definitions or set the place grid conservatively and manually check each of the components individually using Tables 5-6 and 5-7 as guides.

Table 5-11	Courtyard Excess (Protrusion) by Density Level for Various SMDs					
	Density Level A		Density Level B		Density Level C	
Package type	Mils	Millimeters	Mils	Millimeters	Mils	Millimeters
Gull wing (SOG)	20	0.50	10	0.25	4	0.10
J lead (SOJ)	20	0.50	10	0.25	4	0.10
Small outline (SO)	20	0.50	10	0.25	4	0.10
Chip components (0603 and larger)	20	0.50	10	0.25	4	0.10
Tantalum capacitors	20	0.50	10	0.25	4	0.10
MELF	20	0.50	10	0.25	4	0.10
Leadless chip	20	0.50	10	0.25	4	0.10
Chip components (smaller than 0603)	8	0.20	6	0.15	4	0.10
Butt joints	59	1.50	31	0.80	8	0.20

Standardized footprint design for through-hole devices is perhaps not as widely known as the SMD standards. This may be partly because of the greater variety of package styles of the through-hole devices and partly because their typically larger size makes them less sensitive to many of the design and manufacturing issues related to SMDs. As a result land pattern standards or guides for through-hole devices are harder to come by for many devices and may not exist for others. In this section we look at how to design footprints and padstacks for through-hole devices.

Land Patterns for Through-Hole Devices

FOOTPRINT DESIGN FOR THROUGH-HOLE DEVICES

Through-hole devices fall generally into one of two categories: axial leaded or radial leaded. An example of a radial-leaded capacitor is shown in Figure 5-12. The footprint design for this type of device is determined strictly by the construction of the device. Clearly the padstacks have to be located where the leads extend from the body. Radial-leaded devices include pin grid arrays

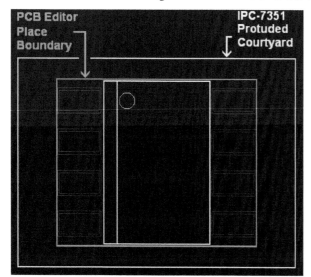

FIGURE 5-11
Comparison of PCB Editor place boundary to IPC courtyard.

and many discrete transistor devices such as TO220 and T092 packages. The only variable is the padstack design with regard to the lead diameters. Padstack design is described below.

(a)

(b)

Place outline

Component outline
(Silkscreen)

FIGURE 5-12
Radial-leaded through-hole device: (a) axial-leaded capacitor, (b) PCB Editor axial footprint.

Footprint designs for components with axial leads are highly variable compared to the radial-leaded devices, but the IPC standards (IPC-CM-770, Section 11.1.8, p. 67) and other sources in literature provide general guidance on footprint design. Figure 5-13 illustrates the design parameters for an axial-leaded component (in this case carbon resistor). The location of the padstacks depends on the length of the body and the location of the lead bends (lead

FIGURE 5-13
Generic footprint design parameters for axial leaded components.

form). The minimum required distance between the padstacks is calculated using Eq. (5.3),

$$L_P = L_B + 2(R + L_A), \tag{5.3}$$

where L_P is the pad spacing (center to center), L_B is the length of the body, R is the bend radius allowance (as read from Table 5-12), and L_A (as read from Table 5-12) is the length of the lead extension from the end of the body to the beginning of the bend. Note that R and L_A are dependent on the diameter of the lead, D_L, and that L_E is the sum of L_A and R. Padstack calculations are described in the next section. Once this minimum padstack spacing is calculated, the padstacks should be placed on the closest standard grid location. Standard grid spacing is described under Component Placement and Orientation Guide. In most cases a 100-mil grid is used. The total length of both leads should not exceed 1 in. (25 mm) unless the component is mechanically supported.

Table 5-12	**Bend Radius and Lead Extension Allowances**	
D_L (mils)	R (mils)	L_A (mils)
$D_L < 31$	$1.0 \times D_L$	31
$31 \leq D_L \leq 47$	$1.5 \times D_L$	D_L
$D_L > 47$	$2.0 \times D_L$	D_L

PADSTACK DESIGN FOR THROUGH-HOLE DEVICES

The biggest vulnerability of PCB fabrication is the plated through hole because of registration errors, aspect ratio, plating, etc. The likely failure points are breakout (annular ring control) due to misalignment and opens due to thermal stress and cycling during soldering processes (assembly and rework). But if we follow the standards and fabricator's design guides, then we will usually have little trouble.

In designing the PTHs for through-hole components, we need to consider the lead-diameter/drill-hole relationship, the PCB-thickness/drill-hole relationship (aspect ratio), the width of the copper ring around the hole (annular ring), the clearance of plane layers from the PTH, and the capabilities of the board manufacturer (see Figures 4-1 and 4-2). Both Coombs and the IPC standards are consulted here to define a process for designing PTHs in *PCB Editor*.

HOLE-TO-LEAD RATIO

The size of the plated through hole should be large enough so that the lead can easily slip into the hole, but it should not be so large that it prevents capillary action during soldering operations. There are two methods for determining the required hole size relative to the diameter of the lead that will be soldered into it. The first method is derived from a combination of IPC-2221A (Section 9.1.3, p. 74) and Coombs (Section 42.2.1, p. 42.3). This method takes into account the plating thickness (either assumed or known) and allows the designer to define a tolerance factor. This method produces a clearance between the lead and the PTH that varies with the diameter of the lead. The drill diameter is calculated using Eq. (5.4),

$$D_H = (D_L + 2T_P) \times k, \tag{5.4}$$

where D_H is the diameter of the drill hole in Padstack Designer, D_L is the diameter of the lead (per the component's data sheet or by measurement), T_P is the thickness of the plating inside the hole (if not known, use $T_P = 1$ mil), and k is a user-defined tolerance factor, for which $1.05 < kÐ \leq 3.0$ (1.5 is recommended).

As an example, if the diameter of a component lead, D_L, is 32 mils, then the drill hole should be $D_H = (32 + 2 \times 1) \times 1.5 = 51$ mils. With the exception of the variable k, this method is explicit and easy to use. However, as lead diameters become larger, the clearance between the lead and the hole becomes wider, and at some point the clearance may become too large for proper capillary action to take place. The point at which this occurs is not well documented in the literature.

The second method is to use a lookup table in which the drill-hole size is dependent on the lead diameter and the desired producibility level (A–C) as shown in Table 5-13 (derived from IPC-2222, Table 9-3, p. 20). So as an example using this method, if we use the same 32-mil lead diameter and the data sheet specifies a 10% tolerance on the lead diameter, then the maximum lead diameter is 35.2 mils and the minimum is 28.8 mils. And if we want a producibility level of A, then the hole diameter would be between 36.2 (35.2 + 10) and 56.8 mils (28.8 + 28).

Table 5-13 Hole-to-Lead Size Relationship by Producibility Level

Hole size	Level A		Level B		Level C	
	Mils	Millimeters	Mils	Millimeters	Mils	Millimeters
Minimum = max lead diameter +	10	0.25	8	0.20	6	0.15
Maximum = min lead diameter +	28	0.70	28	0.70	24	0.60

If you search the Internet, you can find other algorithms for calculating the "proper" hole-to-lead ratio. Which method you choose is up to you. The two methods here were derived from the IPC standards and other sources, but they are not hard and fast requirements.

PTH LAND DIMENSION (ANNULAR RING WIDTH)

Once the size of the drill hole has been determined, the next step is to determine the diameter of the pad (also called the *land*). The difference between the pad diameter and the drill-hole diameter is called the *annular ring*. This is the copper area where the solder joint is made with the component lead. The larger the pad diameter is, the larger the solder joint will be—to a point. Too large a pad does not necessarily make a better solder joint and only increases the amount of heat required to make the solder joint. Too small a pad can cause a weak solder joint, and the pad can be easily damaged by heat or mechanical stress and lift off from the board. The pad diameter recommended by IPC (IPC-2221A, p. 73) is calculated using Eq. (5.5),

$$D_P = a + 2b + c, \tag{5.5}$$

FIGURE 5-14
PTH design parameters.

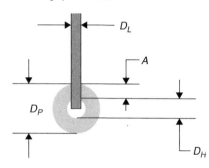

where D_P is the pad diameter (see Figure 5-14), a is the finished hole size ($a = D_H - 2T_P$ from Figure 5-14), b is the minimum annular ring requirements from Table 5-14 (IPC-2221A, Table 9-2), and c is the standard fabrications allowances from Table 5-15 (IPC-2221A, Table 9-1).

As an example, if we use the 32-mil lead from the example above (using a hole size of 50 mils) and we are calculating the size of an external pad (TOP or BOTTOM in Padstack Designer) and we want a Level A producibility level, then the pad size would be $(50 - 2 \times 1) + (2 \times 2) + 16 = 68$ mils.

Table 5-14	Annular Ring Requirements for Internal and External Rings	
	Internal	**External**
Mils	1	2
Millimeters	0.025	0.05

Table 5-15	SFAs for PTH Design		
	Level A	**Level B**	**Level C**
Mils	16	10	8
Millimeters	0.40	0.25	0.20

CLEARANCE BETWEEN PLANE LAYERS AND PTHS

Plated through holes and vias typically have lands on all layers even when there is no connection to it on that layer; these are called *nonfunctional lands*. IPC-2222 (Section 9.1.4, p. 18) states that nonfunctional lands should be used on internal layers when possible, but nonfunctional lands are not required on every layer if the board is over 10 layers thick, and they are not required on plane layers. When a via or plated through hole passes through a plane layer, a minimum amount of space is required between the plane and the plated hole or its nonfunctional land to allow for fabrication allowances, to minimize wicking of plating solution into the laminate from the through hole (resulting in internal shorts to internal plane layer), and to meet voltage withstanding (insulation) requirements.

IPC-2221A (Section 6.3.1, p. 42) states that the trace-to-trace spacing requirements (see also Table 6-8 in Chapter 6 of this text) apply to the clearance between the plane edge and the PTH or its land. Furthermore, per IPC-2222 (p. 17, Fig. 9-1), the minimum clearance between the plane edge and the edge of internal nonfunctional lands or plated holes is to be a minimum of 10 mils (0.25 mm), that is, the clearance diameter is to be 20 mils (0.51 mm) larger than the drill diameter.

Normally the padstacks in PCB Editor do not contain nonfunctional lands on plane layers, so the clearance diameter should be 20 mils (0.51 mm) larger than the drill diameter (i.e., the edge of the plane should be 10 mils from the edge of the plated through-hole barrel). Padstacks that are not designed this way may cause a problem during fabrication or operation of the board.

If the land (pad) is designed per Eq. (5.5) above and the minimum clearance between the hole and the plane is used, the clearance diameter will typically be several mils larger than the lands. If the lands are larger than the clearance diameter, however, there will be an overlap of the lands and the plane layers as shown in Figure 5-15, even if the clearance (dimension d) meets IPC standards. This should be avoided as there will be capacitive coupling between the pad and the plane, which can alter the characteristic impedance of the trace and could cause cross-talk problems at high frequencies. Characteristic impedance is described in greater detail in Chapter 6.

Another thing to pay attention to is that the clearance diameter should be no larger than necessary, as large slots in the ground (return) plane can result if several pins are close together (as in a board or socket connector), which can cause problems with high-speed digital and high-frequency analog circuits when signal lines pass between the pins. This forces return signals on the ground planes to go around the open slot in the ground plane and increase

FIGURE 5-15
PTH or via with oversized lands overlapping plane.

the loop inductance of the circuit. We will look at loop inductance and ground planes in Chapter 6.

SOLDERMASK AND SOLDER PASTE DIMENSIONS

The soldermask opening is typically *larger* than the pad (land) diameter in Padstack Designer. Board manufacturers sometimes require a particular over-sized opening (5 mils is not uncommon). If they do not automatically adjust the size for you, then you will need to modify the soldermask opening before you create the PCB artwork.

The solder paste dimensions are usually the same size as the external lands (TOP and BOTTOM layers). Most of the PCB Editor footprints that come with the default library do not have a solder paste defined. If you will need the solder paste definitions make sure that they are properly specified for your specific assembly needs.

REFERENCES

Coombs, C. F., Jr. (Ed.) 2001. *Coombs' Printed Circuit Handbook.* 5th ed. New York: McGraw-Hill.

IPC-2221A. May 2003. Generic Standard on Printed Board Design. Northbrook, IL: IPC/Association Connecting Electronic Industries.

IPC-2222. February 1998. Sectional Design Standard for Rigid Organic Printed Boards. Northbrook, IL: IPC/Association Connecting Electronic Industries.

IPC-7351. February 2005. Generic Requirements for Surface Mount Design and Land Pattern Standard. Northbrook, IL: IPC/Association Connecting Electronic Industries.

IPC-CM-770. January 2004. Guidelines for Printed Board Component Mounting. Northbrook, IL: IPC/Association Connecting Electronic Industries.

IPC-SM-780/2. July 1998. Component Packaging and Interconnecting with Emphasis on Surface Mounting. Northbrook, IL: IPC/Association Connecting Electronic Industries.

CHAPTER 6

PCB Design for Signal Integrity

Desirable electrical characteristics of a circuit and its PCB include low noise, low distortion, low cross talk, and low radiated emissions, to name a few. The purpose of this chapter is to introduce the issues that cause PCB performance problems and how to route the PCB to minimize them and maximize signal integrity.

CIRCUIT DESIGN ISSUES NOT RELATED TO PCB LAYOUT

Circuit design constraints are primarily the responsibility of the circuit design engineer and will not be covered in detail here, but some of the issues will be mentioned briefly since the symptoms of poor circuit design can be confused with PCB design problems.

Noise

Noise generally refers to any signal that interferes with or degrades a signal of interest. It is often used with an adjective for problems, such as phase noise, switching noise, cross-talk noise, and reflection noise. In this text we will limit the term *noise* to mean random or pseudo-random, natural signals, which are generally not a result of the PCB design. Functional problems such as cross talk or ringing (which are PCB-related problems) will be named as such. From this perspective there are two basic categories of noise: background noise and intrinsic component noise. These noise problems are generally addressed by the circuit designer, not the PCB designer, but are discussed here briefly for completeness.

BACKGROUND NOISE

Background noise is an uncontrolled signal that originates from the system or environment your board is working in. For example, if your circuit is an audio amplifier that is supposed to amplify a speaker's voice as he or she speaks into a microphone, but a crowd of people is talking around the speaker or a jet plane flies overhead, both the speaker's voice and the background sounds will be amplified and the signal would be considered noisy or said to suffer from a low signal-to-noise ratio. There is nothing you can do about it from the PCB design perspective. Sensors may also be noisy because of their sensitivity, but that is also a circuit design issue and needs to be handled long before the PCB is laid out.

INTRINSIC NOISE

There are four basic types of intrinsic noise: thermal noise, shot noise, contact noise, and popcorn noise. Thermal noise (also known as *Johnson noise*) is due to the motion of electrons in a conducting material. It is present in any material that exhibits a resistance to current flow and is a function of temperature. It is white noise (is constant over frequency) and is prominent in resistors and semiconductor devices. Shot noise is also white noise and is due to potential barriers and is also prevalent in semiconductor devices.

Contact noise (also called *excess noise in resistors* and $1/f$ *noise*) is due to imperfect connections at contact junctions or interfaces. It is not constant over frequency and can be fairly large at low frequencies. Your best defense against this type of noise is good-quality connectors and good solder joints.

Popcorn noise (also called *burst noise*) is typically proportional to $1/f^2$ and is worse in high-impedance circuits. It is caused by manufacturing defects in semiconductors and ICs.

Distortion

Distortion is an issue more related to analog circuitry because of the nature of continuous signals. In analog circuitry all voltages between the power supply rails may be of significance. Digital signals are not continuous: they are either HI or LO and usually nothing in between matters. As long as voltage levels meet threshold specifications, there is no ambiguity and therefore no quality issues. Ringing on the rising and falling edges of a square wave might be considered distortion, but that is handled differently, as described below. Distortion of a sinusoidal signal (which normally has a single spike on a frequency spectrum) begins to occur in amplifiers as the sine waves either are clipped or experience a phase reversal. Op amps have amplitude limits imposed by the power supplies, their drive capabilities, and their frequency response. If the amplitude of a sinusoidal output signal (as determined by the input signal times the gain) exceeds the output capability of the op amp, then the output signal will be clipped off and begin to resemble a square wave. Square waves are composed of many sine waves, which are primarily odd harmonics of the fundamental frequency of the square wave. The dominant harmonic is typically the third one, so as a sine wave begins to clip the onset of third harmonic distortion is observed.

If the input signal exceeds the op amp's input limits (as imposed by the power supply rails), the output signal will also be distorted. Some amplifiers simply clip the signal (causing third harmonic distortion), while other op amps experience phase inversion, which also causes harmonic distortion.

These problems are caused by the circuit design and component selection and are not the fault of the PCB design. These effects are mentioned because, if you are not used to them or do not know about them, they can be confused with PCB layout problems. Along with harmonic distortion ringing will produce unwanted frequency components, which can be seen with a spectrum analyzer and may be confused with other forms of distortion or noise. Ringing is caused by reflections, which in turn are caused by impedance mismatches on PCB traces, which *is* a function of the PCB design.

Frequency Response

Both analog and digital circuits have frequency limits. In digital circuitry, if frequency limits are exceeded the signal level may rise and fall before a gate has a chance to switch states. This may give the appearance that the signal is attenuated or that the receiving gate is "not seeing" the signal. This too is a circuit design problem and not a PCB problem. In circuit design we need to make sure that the components selected are within design constraints.

When signals exceed the frequency limits of analog circuitry, the output signal will also be attenuated, and distortion will result if the sine wave begins to look like a triangle wave at the output of the frequency-limited component. This is a function of the amplifier's slew rate, $-3\,\text{dB}$ BW, and gain bandwidth product. Again these issues need to be handled at the circuit design level, well before the PCB design stage.

ISSUES RELATED TO PCB LAYOUT

Electromagnetic Interference and Cross Talk

There are three goals in designing PCBs for electrical performance and signal integrity: (1) The PCB should be immune from interference from other systems, (2) it should not produce emissions that cause problems for other systems, and (3) it should demonstrate the desired signal quality. A common factor relating these three issues is electromagnetic waves. As Figure 6-1 shows, "noise" can be introduced into your PCB from outside sources, and it can produce noise that is radiated to other systems and to itself.

When electromagnetic waves get into your system, this is referred to as *electromagnetic interference* (EMI). On the flip side, your

FIGURE 6-1
The enemy—electromagnetic interference.

PCB can be the source of EMI and cause problems for other systems. The ability for systems to "play nice together" is referred to as *electromagnetic compatibility* (EMC). The FCC has established rules for many types of systems regarding EMI and EMC, which, depending on your application, your PCB may have to abide by. Properly laying out your PCB can greatly reduce EMI and improve EMC. In this section we take a look at how to minimize EMI and its effects. There are many good books available that address these issues in greater detail. The material in this chapter is not intended to duplicate those works but to provide an overview of the issues and provide insight on how to design PCBs with PCB Editor with regard to signal integrity issues.

The method by which systems and circuits can "reach out and touch" another circuit is inductive and capacitive coupling of electromagnetic fields.

In the 1820s Faraday and Henry showed that an electric current could be produced in a conductor by changing the current in another, nearby conductor (Serway 1992, p. 806). And years later Maxwell showed that changing electric fields also produce magnetic fields. These fields are the source of many woes in PCB design. We begin by looking at magnetic fields and inductive coupling.

Magnetic Fields and Inductive Coupling

As shown in Figure 6-2 a magnetic field vector, **B**, is developed around a conductor when current flows through the conductor, into the "X" end and out of the "Dot" end of the conductor. The right-hand rule (from Ampere's law) is used to determine the direction of the field: If the thumb of the right hand points in the direction of conventional current flow (movement of positive charges), then the magnetic field curls in the direction of the fingers. This is defined mathematically by a cross product called the *Biot-Savart law*. By applying some calculus, which will not be shown here, an equation can be derived for the scalar (nonvector) magnitude, B, of the magnetic field vector, **B**, near the conductor.

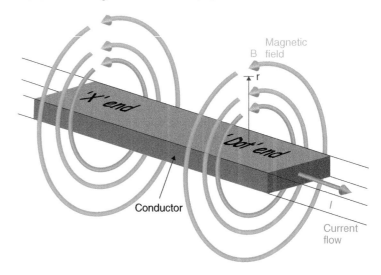

FIGURE 6-2
Magnetic field caused by a current-carrying conductor.

The magnitude of the magnetic field a distance, r, from a long straight conductor (Serway 1992, p. 838) is given by Eq. (6.1),

$$B = \frac{\mu_0 I}{2\pi r^2},\tag{6.1}$$

where B is the magnitude of the magnetic field in Wb/m^2, μ_0 is the permeability of free space ($\mu_0 = 4\pi \times 10^{-7} Wb/A \times m$), I is the current in amps (A), and r is the distance from the conductor.

The total magnetic field within or through an area is called magnetic flux, Φ, which has units of Wb and is described by Eq. (6.2) (Serway 1992, p. 849; $Wb/m^2 \times m^2 = Wb$),

$$\Phi = BA \cos(\theta),\tag{6.2}$$

where B is the magnetic field magnitude per unit area (Wb/m^2), A is the area intersected by the magnetic field (m^2), and θ is the angle between B and A.

Magnetic flux expands or contracts in proportion to changes in current flow. As the flux expands or contracts around the conductor, we see from Faraday's Law of Induction, given in Eq. (6.3) (Serway 1992, p. 877), that a voltage is induced into the conductor. This is known as *self-induced electromotive force* (emf):

$$E = -\frac{d\Phi}{dt}.\tag{6.3}$$

The minus sign in Eq. (6.3) is a result of Lenz's law, which states that the emf induced into the conductor produces a current in the conductor that creates a magnetic flux that will oppose the changing magnetic flux. This effect is called *self-inductance*. The self-inductance tends to limit how fast the current can change in a conductor. This is what makes an inductor have inductance and oppose AC currents in analog circuits and fast rise times in digital circuits. The magnitude of the inductance, L, is shown in Eq. (6.4):

$$L = \frac{N\Phi_m}{I}.\tag{6.4}$$

The inductance is directly proportional to N, the number of turns on a coil ($N = 1$ for a PCB trace and its return path), and the magnetic flux and is inversely proportional to the current, I.

If a secondary conductor is near a primary conductor, which is carrying current as shown in Figure 6-3, some of the flux will be felt by the secondary conductor. If the current through the primary conductor changes with respect to time (as in the case of a rising edge of a digital signal or an AC signal) the magnetic field (and therefore the magnetic flux) also changes with respect to time as it

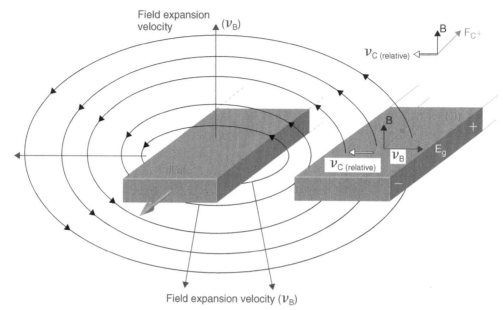

FIGURE 6-3
Voltage induced into adjacent trace by varying magnetic fields.

increases in strength and expands outward (or decreases in strength and contracts inward) from (or toward) the primary conductor.

When the flux that is impinging on the secondary conductor changes with respect to time, as it expands or contracts, we see again from Faraday's Law of Induction in Eq. (6.3) that a voltage, E_g, is induced into the secondary conductor. Using the right-hand rule for electric generators, the direction of the induced current can be determined. To use the right-hand rule point the thumb in the direction of the motion of the conductor (relative to the expanding/shrinking flux—ν_C), point the first finger in the direction of the B field, and the middle finger will point in the direction of the force applied to the positive charges $(\mathbf{F}_c^+)$ and therefore the positive current flow. The resultant voltage and current are shown in Figure 6-3. Notice that the current in the secondary conductor flows in the direction opposite that of the current in the primary conductor as a result of the induced voltage. This produces a secondary magnetic field which opposes and partially cancels the primary magnetic field (by Lenz's law).

If the induced current flow in the secondary conductor is changing with respect to time (which it is because the primary conductor is causing it to) and it is in close proximity to the first conductor (which it is), then the secondary conductor will also induce an emf back into the primary conductor. The magnetic flux that goes back and forth between the two conductors is called *mutual inductance*.

The emf (voltage) generated into the primary conductor (by the secondary conductor) will be in a direction that aids the original current flow in the primary conductor as the secondary flux tries to oppose the primary flux. If the original flux is partially canceled, then the self-inductance is also partially canceled and the changing current in the primary conductor is not as limited (i.e., it sees less inductance).

It would seem that unlimited current could flow in the primary inductor since the secondary conductor aids it, but the secondary inductor also experiences its own self-inductance and is therefore limited. Also the amount of mutual inductance between the two conductors is limited by how much of the flux couples the two conductors. This is similar to the coupling coefficient in transformers and in both circumstances is never 100%.

When a trace on a PCB induces a voltage into an adjacent signal trace we call that *cross talk*, which is bad because it generates noise in the adjacent signal trace. But if the second conductor is the PCB's ground plane, that is good because it reduces the trace's inductance and therefore the overall *loop inductance* of the circuit. For the time being we will stick with using the term *return plane*, rather than *ground plane*, for reasons described later.

Loop Inductance

In Figure 6-3 only segments of the traces were shown. Of course for current to flow through the circuit, a closed path must exist, as shown in Figure 6-4. Any conductor in the circuit that carries current will produce a magnetic field as indicated by the circular arrows.

An equation for inductance is given in Eq. (6.5) (Serway 1992, p. 905),

$$L = \mu_0 n^2 A\ell, \qquad (6.5)$$

FIGURE 6-4
Loop inductance of a closed circuit.

where n is the number of turns (1), $A\ell$ is the volume that the circuit occupies, and μ_0 is the relative permeability of the material in which the circuit exists. For most materials used in PCBs, $\mu_0 = 1$. So the inductance is a function of the volume of the inductor and the number of turns of the conductor around the space. Therefore, inductance is dependent on the circuit geometry, by which a smaller volume results in a smaller circuit or loop inductance.

If we look at the closed loop circuit shown in Figure 6-4 with respect to its volume, we can see that, if the circuit is physically large and makes a large loop, it will have what is referred to as high loop inductance.

If on the other hand we arrange the circuit as shown in Figure 6-5, we can see from Eq. (6.5) that the loop inductance will be less since there is less volume. If you notice the direction of the arrows (the magnetic fields) you can see that the source current and the return current magnetic fields oppose each other, thereby reducing the flux and inductance.

FIGURE 6-5
Closed loop circuit with
low loop inductance.

Figure 6-6(a) shows the resultant magnetic field of two conductors in close proximity where the currents are in the same direction. As indicated the magnetic fields circulate in the same direction and aid each other. This is the case for an inductor in which the turns are wound in the same direction and build up an overall strong magnetic field.

Figure 6-6(b) shows the resultant magnetic field when the currents of the two conductors are in opposite directions. The magnetic fields oppose each other and result in partial flux cancellation. The amount of cancellation depends on the amount of mutual inductance, which in turn depends in part on the distance between the conductors.

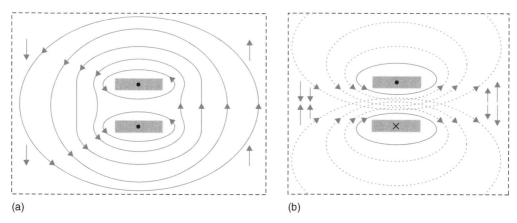

(a)   (b)

FIGURE 6-6
Aiding and opposing magnetic fields: (a) aiding fields, (b) opposing fields.

The situations in Figures 6-4, 6-5, 6-6(a), and 6-6(b) are steady-state DC circuits, but if we apply the concept in Figure 6-3 and Eq. (6.3) for time-varying currents, then we can see that, if the return trace on a circuit board is in close proximity to the signal trace, then the inductance of the traces is reduced; that is, the loop inductance of the closed-loop circuit is reduced. If the loop inductance is reduced in an AC circuit, then by Eq. (6.6) it can be seen that there will be less inductive reactance (X_L), less voltage drop, and less cross talk (fewer EMI problems):

$$X_L = 2\pi f L \tag{6.6}$$

To maintain a small X_L (and low loop inductance), we need to have the return path as wide as possible (low self-inductance) and as close as possible to the signal path wherever possible (maximum coupling and small cross-sectional areas). The easiest way to do this is by using a plane layer as the return path.

The return plane has historically (and most often inappropriately) been called the *ground plane*, but it is being referred to more often as an *image plane* or a *return plane*. In PCB design a return plane has low inductance (and therefore low self-inductance) and it is everywhere the signal trace is and therefore allows for maximum coupling between the signal trace and the plane for any and all widths of the signal trace.

From this discussion then we can say that one of the most important functions of the return (image) plane is to reduce loop inductance. Reducing loop inductance (and the magnetic fields related to it) provides a low-impedance return path for power and signal lines and reduces unwanted cross talk to nearby conductors. It should also be stated that cross talk between unrelated conductors is also reduced by keeping them farther apart (i.e., *r* is large).

Electric Fields and Capacitive Coupling

We saw in the previous paragraphs that keeping signal and power lines close to their return paths provides flux cancellation and reduces loop inductance, which is beneficial in all respects. But what happens with the electric fields under these circumstances and what is the effect on the circuit? Figure 6-7(a) shows electric field lines for a single charged conductor, which can represent a signal trace that is a long way away from its return path. Figure 6-7(b) shows electric field lines between two oppositely charged conductors, which can represent a signal or power line close to its return path. As shown, the solitary conductor radiates electric field lines in all directions, while the coupled conductors contain (or at least concentrate) the electric field between them.

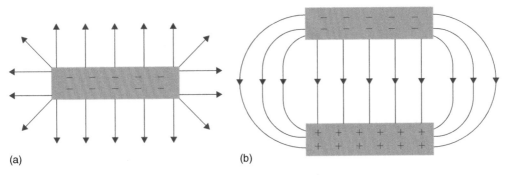

(a)   (b)

FIGURE 6-7
Electric fields on conductors: (a) a single charged conductor, (b) a field between oppositely charged conductors.

The Ampere–Maxwell law (Serway 1992, p. 851) states that "magnetic fields are produced both by conduction currents and by changing electric fields." So, to minimize cross talk, it would seem to be in our best interest not to allow traces to radiate electric fields in an uncontrolled manner but to keep the signal (and power) traces close to their return paths. This is true for both magnetic and electric fields.

But what happens to the capacitance of the traces relative to the return plane when we do this? The equation of a parallel plate capacitor (in farads) in air is given in Eq. (6.7) (Serway 1992, p. 712),

$$C = \frac{\varepsilon_0 A}{d},\tag{6.7}$$

where C is capacitance (in farads), ε_0 is permittivity of free space, A is the area common to the parallel plates, and d is the distance between the plates. As indicated, as the plates become closer together or as the area becomes larger, the capacitance increases. This also holds for the capacitance between a trace and the return plane, although as we will see later the equation is slightly different and the units are in F/in. to make it easier to calculate capacitance for various trace lengths.

From Eq. (6.8) we can see that as the capacitance, C, increases the capacitive reactance decreases:

$$X_C = \frac{1}{2\pi f C}.\tag{6.8}$$

By combining Eqs. (6.7) and (6.8) as shown in Eq. (6.9), we can see that, by keeping traces and power planes close to their return planes (small d), the capacitive reactance between the trace and the return plane is reduced (coupling is increased):

$$X_C = \frac{d}{2\pi f \varepsilon_0 A}.\tag{6.9}$$

And by keeping unrelated signal traces farther apart (large d), the reactance between the traces is higher and the coupling (cross talk) is reduced. With both magnetic and electric fields, the wider the return path (the areas of the conductor), the better the coupling, and the closer the signal or power conductor is to the return plane, the better the coupling.

GROUND PLANES AND GROUND BOUNCE

What Ground Is and What It Is Not

In the previous discussion, the term *return plane* was used instead of the term *ground plane*. Unless the return plane is physically connected to the earth by some means, it really has nothing to do with "ground," The ground symbol, ⟂, has long been used on schematics (in academia and in practice) to indicate a connection to the point to which all closed-circuit currents must return. An example is shown in Figure 6-8. This gives the impression that ground,

Ground Symbol

somehow, is an omnipresent, unfaltering current sink and equipotential reference. Equipotential means that the voltage is the same everywhere, regardless of how much current is flowing through it. This is a myth.

Although the depiction of the ground connection shown in Figure 6-8 is convenient to use on the schematic, in reality there has to be a physical, real-world connection. And, just so you know, the official ground symbols per IEEE Std 315-1975 (ANSI Y32.2-1975) are given in Table 6-1. The IEEE disclaims any responsibility or liability resulting from the placement and use in the described manner. However, it is quite common that the ground and return symbols are used otherwise. Since the ground concept has been around a long time, it is unlikely that its use will change overnight. The important thing is that it is clear what the symbol means. The standard states that the symbols can be given supplementary information (such as names) on the schematic to specifically annotate the symbol's purpose or function.

FIGURE 6-8
Typical depiction of "ground."

Table 6-1	IEEE/ANSI Standard Ground Symbols	
Symbol	**Name**	**Purpose/usage**
	Earth GND	A direct connection to the earth.
		A direct connection to a vehicle's or an airplane's frame that serves the same function as earth ground.
	Noiseless GND	Used to indicate a low or noiseless earth ground.
	Safety GND	Used to indicate a ground connection that serves a safety function against electric shock.
	Chassis GND	A connection to a chassis, or frame, or similar connection of a *printed circuit board* and may be completely different from earth ground.
	Return	Used to indicate common return connections.

Symbols and definitions in Table 6-1 reprinted with permission from IEEE Std. 315-1975, Copyright 2003, by IEEE. All rights reserved.

There are two basic ground and source connection schemes, parallel and series connections, as shown in Figure 6-9. A parallel ground system is shown in Figure 6-9(a). A parallel ground system is also called a *separate ground system*, since the current flow in each branch is supplied by and returns to the source

FIGURE 6-9
Typical signal and return connection schemes: (a) parallel connected, (b) series connected.

through completely separate paths. A series-connected ground system is shown in Figure 6-9(b). The series-connected ground system is also referred to as a *common ground system* or *daisy chain*, since the current flows in the two branches share a common path.

At the schematic level the circuits in Figure 6-9 are identical; mathematically the circuits are also identical, that is, $I_T = I_1 + I_2$ and $V_{R1} = V_{R2} = V_S$. Furthermore, as indicated in Figure 6-9(b), you could connect R_1 to the common return path at either point (1) or point (2) without changing the meaning of the circuit description in any way. However, on the PCB, the two circuits shown in Figure 6-9 can be significantly different. There may even be significant differences between points (1) and (2) in Figure 6-9(b).

It was said earlier that any conductor that carries current will produce a magnetic field. Even if there is very tight coupling between the signal conductor and its return conductor, some inductance will always exist because the coupling is not 100.0% complete; that is, there is less than perfect mutual inductance from the primary trace to the secondary trace (the return plane) and from the secondary trace back to the primary trace.

With this understanding we can see from Figure 6-10 that there is an "unseen" schematic within the PCB. As a result, although the relationship $I_T = I_1 + I_2$ is

FIGURE 6-10
The actual circuit—the hidden schematic.

still true, it should be realized that $V_{R1} = V_{R2} = V_S$ is no longer true because there are voltage drops along the shared and individual impedances between the source and each of the loads. Furthermore points (1) and (2) and any other points along the return path are not equipotential. The ideal, equipotential ground plane does not exist in practice.

Common impedance is particularly troublesome when high-power or noisy signals share common return paths ($Z_{COMMON1}$ and $Z_{COMMON2}$) with low-power signals. High-speed digital signals operating near low-level analog signals are an example.

Clearly then, it would seem that the best return system is the parallel system shown in Figure 6-9(a). However, a potential problem arises when we try to implement this approach on the PCB. Figure 6-11 shows the routing scenarios in which Figure 6-11(a) is the parallel connection and Figure 6-11(b) is the series connection. As described above and shown in Figure 6-11(a), the current paths do not interfere with each other in the parallel connection since

(a)   (b)

FIGURE 6-11
Signal and return connection schemes in PCB Editor: (a) parallel (separate) connected, (b) series (common) connected.

their paths are completely separate, and since each signal path is directly above its return path, there is tight coupling between the signal and the return and therefore the inductance is minimized. The problem is that it would become incredibly cumbersome to route a PCB using this approach with even a few additional components and worse on a high-density, multilayer PCB. If components are moved, movement of the signal and return paths would have to be carefully coordinated, resulting in many opportunities for routing "errors" that could be difficult to detect.

The series connection in Figure 6-11(b) is obviously much easier to route, but it loses the benefit of the separate signal and return paths. Even with the signal and return paths tightly coupled, the common paths (impedances) could be problematic for circuits operating at high frequency or with fast rise times.

Ground (Return) Planes

Above, we said that we want the return path to be as wide as possible and as "everywhere" as possible, which, when taken to the extreme, causes the return path to become a plane. But since (at first glance) it appears that a plane is potentially a common return path (a common impedance), the question arises as to whether this is really the best solution to overcome the inconvenience of the routing problem. If we reroute the circuit of Figure 6-8 as shown in Figure 6-12, in which the return path is the GND plane in PCB Editor (where the thermal reliefs indicate a connection to the GND plane), we can analyze the situation. If the signal path is relatively close to the return path, the return signal will automatically flow through the GND plane directly below the signal trace. The reason this happens is that by doing so the loop inductance of the circuit is minimized. As is commonly known in DC circuits, current follows the path of least resistance. Perhaps not as commonly known, AC currents will follow the path of least impedance and, particularly on PCBs, the path of least inductance. The only way for that to happen is if the return current travels directly under the signal trace on its way back to the source. This is true no matter what kind of crazy path the signal trace makes, as long as no discontinuities exist in the return plane.

So (in Figure 6-12) even if R_2 was directly between R_1 and the PCB connector, the return paths would not be common as long as the signal traces to R_1 and R_2 did not overlap (which can happen if the signal traces were routed on different layers) or become "too" close. We will look at the appropriate trace separation (what *too close* means) in the routing section discussed later.

FIGURE 6-12
Pseudo-common return path using a "ground" plane.

Ground Bounce and Rail Collapse

In a typical PCB the power distribution system contains one or more power and ground planes. The power and ground planes are like very wide traces (have little inductance) and are usually adjacent to each other (high capacitance). This is exactly what we want for the power distribution system. However, despite these advantages, a problem occurs in high-speed digital systems when gates switch from one state to another. The problem in general is known as *switching noise*.

Figure 6-13 shows a representative CMOS logic gate. The capacitor, C_L, represents all of the capacitances related to construction of the CMOS transistors $Q1$ and $Q2$. When the gate switches from one state to another, C_L has to charge (or discharge) before the gate can reach its steady-state value. For example, at a logic state of 0, $Q1$ is off and $Q2$ is on, the output (and VC_L) is at V_{SS} (0 V). When the gate tries to switch to a high state, logic level 1, $Q1$ turns on, $Q2$ turns off, and C_L begins to charge to V_{DD}. During this transition the gate consumes significant power because for a brief moment $Q1$ and $Q2$ are both partially on. A short circuit exists from V_{DD} to V_{SS} through $Q1$ and $Q2$ and through C_L (which has low impedance while it charges). Because a high current results (even if only briefly) the voltage at the V_{DD} pin tends to drop until the switching is complete and C_L is fully charged. A similar thing happens when the gate tries to change from a logic 1 to a logic 0 state, except that as C_L tries to discharge through $Q2$ (which is turning on as $Q1$ turns off) the voltage at the V_{SS} pin tends to rise until the switching is complete and C_L is fully discharged.

FIGURE 6-13
A CMOS logic gate.

Because the power and ground planes are not superconductors, there is a drop in voltage between the supply pins of the gate and where power is connected to the PCB (same for the return plane). Remember that there is always some amount of resistance and inductance—even on the so-called ground plane. This is shown in Figure 6-14, in which we see the supply voltages across the PCB while the gate is switching. The drop in the positive rail is called *rail collapse* and the rise in ground potential is called *ground bounce*. Note that, since there is really nothing magical about the so-called ground plane, the term *rail collapse* can refer to the ground rail rising as well as the supply rail dropping.

The solid line in Figure 6-14 shows that a relatively linear drop in positive rail voltage (or rise in ground potential) occurs along the distance from the connector

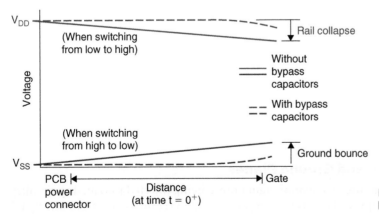

FIGURE 6-14
Illustration of ground bounce and rail collapse.

to the switching gate when none of the ICs on the PCB have bypass capacitors. The worst voltage drop occurs at the gate itself, but any gates located between the switching gate and the connector will "feel" the voltage drop as well.

The dashed lines in Figure 6-14 indicate the voltage drop (or rise in ground potential) when all of the ICs have bypass capacitors. The capacitors act as current reservoirs and help hold up the positive rail voltage and keep down the return (ground) rail voltage except at close proximity to the gate that is switching (although it is minimized there as well).

The primary purpose of bypass capacitors in digital circuits is to promote a stable PCB power distribution system and prevent rail collapse and ground bounce—that is, to keep switching noise off from the rails. Conversely, the purpose of analog bypass capacitors is to keep any power supply noise that does exist from getting to the analog circuitry; that is, the bypass capacitors act as low-pass filters and short out power supply transients (noise) before they get to the amplifiers.

Since analog circuits (particularly amplifiers) are usually designed to operate strictly between (and a safe distance from) the rails, they rarely cause rail collapse but are usually the victims of it. The purpose of amplifier circuits is to amplify small signals, which are often in the millivolt or microvolt range; thus a very quiet circuit environment is highly desirable. Since digital switching noise can be as much as 100 mV or more, making analog and digital systems work together on the same PCB can be a significant challenge.

Even when a PCB's power distribution system is well designed (low inductance planes and lots of bypass caps) switching noise can be a significant problem for analog circuits (and other digital circuits for that matter) as the switching currents surge across the PCB planes. This is particularly true if the analog circuitry is between the PCB connector and the noise digital circuitry. Any voltage drops along the path will be seen as noise by the analog system.

It was said earlier (see Figure 6-12) that, as long as traces did not overlap or if they were not too close, the return currents would tend to stay directly under the signal traces and cross talk would be minimized. However, in very high-frequency analog or high-speed digital systems, return currents (whether signal or power return) may deviate from the ideal path because of imperfections in the PCB's copper plating and variations in the laminate materials. As a result, high-frequency analog and high-speed digital return currents may actually spread across a return plane "looking" for the path of least inductance. This occurs particularly when a signal leaves one layer and goes to another through a via and the return currents do not have an easy path from the one return plane to another that is closer to the new signal layer.

Split Power and Ground Planes

The solution to the problem of digital noise being injected into analog circuitry through the supply planes is to segregate the analog components from the digital

(a)

(b)

(c)

(d)

FIGURE 6-15
Different types of power/ground planes: (a) continuous plane; (b) split plane; (c) moated plane; (d) isolated, continuous planes.

ones and eliminate common return paths. Segregating the components is straightforward; the components are physically placed in different places on the board. Eliminating common return paths can be accomplished by splitting the ground and power planes into separate areas. The various planes are shown in Figure 6-15. A typical plane is one continuous sheet (an entire layer dedicated to a single power or ground connection) as shown in Figure 6-15(a). But it is possible and advantageous to break up the plane into sections or to create completely separate planes for the digital and analog areas. Figure 6-15(b) shows a split plane that provides isolated areas on a single layer while providing an electrically common reference point. This configuration is common where power electronics are placed over the plane area that is close to the board connectors, and analog and digital electronics are placed over their respective return planes. This allows all circuits on the board to be referenced to a common ground but forces the specific return currents to stay within their own areas. This is demonstrated in the PCB Design Examples.

Return planes can also be completely separate areas by using moats as in Figure 6-15(c) or distinct continuous planes as shown in Figure 6-15(d). Moated planes are sometimes used as local ground or reference planes for high-speed clocks or small sections of a circuit that require their own regulated supply or ground potential, and isolated planes are used where parts of the system do not share a common ground reference or power supply system.

Care must be exercised when using split or multiple isolated ground and power planes. Even if the planes are separated physically, noise can be capacitively coupled from one plane to the next as shown in Figures 6-16(a) and 6-16(b). To minimize noise coupling between analog and digital planes, split planes on different layers should be prevented from overlapping each other, as shown in Figure 6-16(c), or should be separated with a shield plane, as shown in Figure 6-16(d). This is demonstrated in Example 3 in the PCB Design Examples.

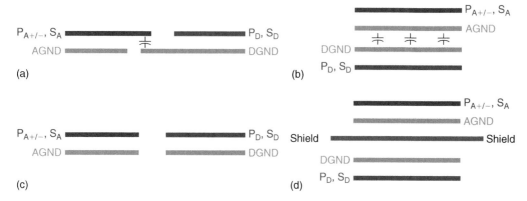

FIGURE 6-16
Power and ground plane stack-up scenarios: (a) coupling between overlapped planes, (b) coupling between parallel planes, (c) nonoverlapping split planes, (d) shielded isolated planes. Key: P, power; S, signal; A, analog; D, digital.

On rare occasions the analog and digital return or reference planes of a PCB may be on different layers (and not overlapping or separated by a shield plane) but must be referenced to a common point (for example, when working with analog-to-digital and digital-to-analog converters). So the question becomes how to keep them physically separated but electrically connected. The easiest way is by using the isolated planes (Figure 6-15(c)) and then connecting the planes at a point using a plated through hole as shown in Figure 6-17(a) or a shorting bar. Moated planes (Figure 6-15(c)) can also be connected using the shorting bar. Both of these methods can be used in PCB Editor and are demonstrated in the PCB Design Examples.

PCB ELECTRICAL CHARACTERISTICS

Characteristic Impedance

It was stated earlier that to minimize cross talk we want to minimize trace (loop) inductance and maximize the capacitance to the return plane. What does this do to the characteristic impedance, Z_0, of a trace?

Perhaps the first thing to do is to look at what characteristic impedance really is. For example RG58 is a coaxial cable that is often used as a shielded transmission

(a)   (b)

FIGURE 6-17
Methods of shorting together separate plane layers: (a) a via as a short, (b) a copper trace (strip) as a short.

line in 50-Ω systems. Actually, RG58 is about 52 Ω, not 50 Ω. But even so, what does that mean? If you use an ohmmeter to measure the resistance from the center conductor to the shield you will see that it is neither 52 nor 50 Ω. So how is its characteristic impedance 52 Ω?

Figure 6-18 shows a model of a transmission line, which consists of series inductors and parallel capacitors. This is called the *lumped-element model*, which assumes that the series resistance is negligibly small and that the transmission line is infinitely long (or at least long enough to watch what happens). Each *LC* "lump" represents a finite section of the transmission line, and the sum total of the elements is representative of the total inductance and capacitance of the transmission line.

FIGURE 6-18
A lumped-element transmission line model.

We begin the analysis with all of the capacitors discharged and all currents at zero. At time $t = 0$s the switch is shut, which applies the source voltage, V_S, to the transmission line through the source resistance R_S as shown in Figure 6-19. Initially C_1 acts as a short circuit so $I = V_S/R_S$. Current, I, begins to charge capacitor C_1, and a return current will also flow out of the bottom of C_1 back to the source (note that this is a *displacement* current as postulated by Maxwell

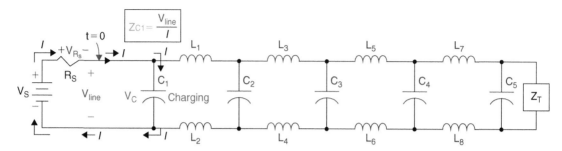

FIGURE 6-19
A signal applied to the transmission line.

rather than a *conduction* current as defined by Ampere). The instantaneous impedance is $Z_{C1} = V_{line}/I$.

As C_1 charges (no longer acting like a short), current begins to flow into L_1. Each inductor pair (L_1 and L_2, L_3 and L_4, etc.) is mutually coupled, so the magnetic field of L_1 induces the return current in L_2.

As current flows past L_1, C_2 begins charging positively on the top side; and as L_2 forces return current to flow back to the source (due to mutual inductance), C_2 begins charging negatively on the bottom side (relative to its top lead).

At some point C_1 becomes fully charged to a value of $V_{C1} = V_{line} = V_S - I \times R_S$ and then the displacement current no longer flows through C_1, so the instantaneous impedance at C_1 is $Z_{C1} = \infty$ and $Z_{C2} = V_{line}/I$. Current continues down the line, charging up each capacitor in turn to a value of $V_{Cn} = V_{line}$.

As each capacitor along the way is charging, the instantaneous impedance across the line is $Z_{Cn} = V_{line}/I_{Cn}$ as shown in Figure 6-20. As each capacitor becomes fully charged, its impedance goes to infinity because the displacement current through it goes to zero. As seen from the source (V_S) the impedance of the line is $Z_{line} = V_{line}/I = V_{line}/I_{Cn}$ and is dynamic since it travels along the line. Furthermore, the impedance farther down the line is unknown.

FIGURE 6-20
The instantaneous impedance propagates along the transmission line.

The speed at which the instantaneous impedance travels along the line is dependent on the inductance and the capacitance of each section. It was said above it is desirable to have as little loop inductance as possible (which will never be zero) and as much capacitance as possible (which will never be infinite). Thus there will always be finite inductive reactance (X_1) and capacitive reactance (X_C) during any transient. However, the capacitors that are charged have nothing to do with the impedance (since they look like open circuits) and the inductors that have steady-state current flowing through them have nothing to do with the impedance (since they look like shorts). The capacitors and inductors farther down the line have nothing to do with the impedance either, since they do not see any action until the capacitors and inductors before them have approached a steady-state condition. Until the voltage (V_{line}) reaches the load, Z_T, the source actually has no idea the load, Z_T, even exists; neither does it know how many sections of L and C there are until all of the previous sections have reached steady state. If the impedance of each section is the same all along the line, then we call the instantaneous impedance the characteristic impedance of the transmission line and give it the special symbol Z_0.

Before we consider what happens to the current flow and line voltage in Figure 6-20 once all of the capacitors are charged and the line voltage and current reach Z_T, we need to take a closer look at the behavior of the transmission line. From the above discussion we see that it takes a finite amount of time for the applied voltage (minus the voltage drop, V_{RS}) to propagate down the line, and, as the applied voltage propagates, it essentially behaves as a wave. In fact the effects described here are due to wave properties and not directly due to electrons flowing (at least not like we normally think of them). The key to understanding Z_0 (and reflections and ringing, as we will see shortly) is in understanding how and at what speed the waves travel.

If you ask an average person how fast electricity travels, you will usually get the answer that it travels at the speed of light. Except in one particular case, that answer is not correct. If we think of electricity as flowing electrons, then electricity actually travels at only about 1 cm/s (Bogatin 2004, p. 211), pretty slow really. This seems counterintuitive since when we turn on a light switch the lights come on seemingly immediately, as if the "electricity" traveled at the speed of light from the switch to the light bulb. But what does travel at (almost) the speed of light is the electromagnetic wave that is launched into the wiring by the switch closing.

Figure 6-21 can be used to explain the difference between the speed of electrons and the electromagnetic wave velocity. The figure shows a copper tube, which contains marbles that are separated by small springs. If an additional marble (No. 5) is shoved into the tube, marble 4 is shoved further into the tube, compressing the spring between it and marble 3. Note that in this early stage marbles 2 and 1 have no idea what is going on yet. As No. 5 is shoved into No. 4's place the rest of the marbles must "do the wave" to make room for it. Eventually all of the marbles have slid over by one marble space and marble 1 pops out the other end.

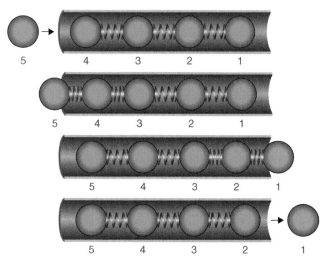

FIGURE 6-21
Wave velocity vs. particle velocity.

Notice now that all of the marbles have moved a distance of only one marble space, but the effect of this movement (a wave) is felt at the end of the tube in about the same amount of time. The speed of the wave is determined for the most part by the value of the spring constants and partly by the momentum of the marbles.

So in a transmission line the electrons travel slowly, but the electromagnetic (EM) waves travel fast. The speed of the EM wave is determined by how quickly the magnetic fields in the inductors and the electric fields in the capacitors can be built up or dissipated, which is influenced by the material properties and geometry of the PCB through which the wave travels.

The velocity of an EM wave through a medium is described by Eq. (6.10),

$$v_{EM} = \frac{1}{\sqrt{\varepsilon_0 \varepsilon_r \mu_0 \mu_r}}, \tag{6.10}$$

where v_{EM} is the velocity of the EM wave in a given material, ε_0 is the permittivity of free space (8.89×10^{-12} F/m), ε_r is the relative permittivity (dielectric constant) of the material (a unitless constant relative to ε_0), μ_0 is the permeability of free space ($4\Pi \times 10^{-7}$ H/m), and μ_r is the relative permeability of the material (a unitless constant relative to μ_0).

You may recall that the speed of light, c (a special EM wave), in free space is

$$c = \frac{1}{\sqrt{\varepsilon_0 \mu_0}}. \tag{6.11}$$

So we can rewrite Eq. (6.10) as

$$v_{EM} = c \times \frac{1}{\sqrt{\varepsilon_r \mu_r}}. \tag{6.12}$$

As stated, the terms ε_r (relative permittivity) and μ_r (relative permeability) are unitless. Furthermore μ_r is equal to 1 in free space and in most polymers (including FR4 laminate), so we can further simplify Eq. (6.12) as shown in Eq. (6.13):

$$v_{EM} = c \times \frac{1}{\sqrt{\varepsilon_r}}. \tag{6.13}$$

From Eq. (6.13) we see that the velocity of an EM wave (which comprises both electric and magnetic fields) in a PCB varies inversely with the relative permittivity, ε_r.

Relating this observation with Eq. (6.7), we can state (without rigorous proof) that the capacitance of a transmission line is determined by the geometry of the transmission line and the relative dielectric constant (ε_r) within the transmission line. And the inductance of a transmission line (specifically the loop inductance) is determined by the geometry, but μ_r falls out since it is equal to 1 (see Eq. [6.5] and Figures 6-4 and 6-5).

In practice, calculating the characteristic impedance, capacitance, and inductance can be fairly complex, depending on the geometry of the circuit, but fortunately that has been done for us for the most common transmission line configurations. The equations are shown in Tables 6-2, 6-3, 6-4, and 6-5. The PCB designer has full control over the trace width (w) and partial control over the trace thickness (t) by selecting the ounces per square foot but may have little or no control over the thickness of the laminate (h). These equations are solved for w and presented later in this chapter (in Tables 6-6 and 6-7).

Table 6-2	**Surface Microstrip Transmission Lines**		
Microstrip transmission lines		$Z_0(\Omega)$	C_0 (pF/in.)
Surface		$Z_0 = \dfrac{k}{\sqrt{\varepsilon_r + 1.41}} \ln\left(\dfrac{5.98h}{0.8w + t}\right)$	$C_0 = \dfrac{0.67(\varepsilon_r + 1.41)}{\ln\left(\dfrac{5.98h}{0.8w + t}\right)}$
Surface differential		$Z_{\text{diff}} = 2Z_0\left[1 - 0.48 \times e^{\left(-0.964\frac{d}{h}\right)}\right]$	
		Z_0 same as surface microstrip	

$L_0 = (Z_0^2 \times C_0)/12$ in nH/in, where $k = 87$ for $15 < w < 25$ mils and $k = 79$ for $5 < w < 15$ mils. Restrictions: $0.1 < w/h < 3.0$ and $1 < \varepsilon_r < 15$ (typically 4.0 to 4.5 for FR4).

Reflections

So the next question is, What happens when the voltage "wave front," V_{line}, reaches the termination impedance, Z_T? The answer is that it depends on what Z_T is.

Let's assume for a minute that Z_T is an open circuit. When the last capacitor, C_5, in Figure 6-18 is charged and V_{line} reaches Z_T (which equals infinity), then all capacitors are charged along the line (and their impedance equals infinity), so all current stops—or at least it would like to. But it cannot, because all of

Table 6-3	Embedded Microstrip Transmission Lines	
Microstrip transmission lines	**$Z_0(\Omega)$**	**C_0 (pF/in.)**

Embedded

$$Z_0 = \frac{k}{\sqrt{\varepsilon_r + 1.41}}\ln\left(\frac{5.98h}{0.8w + t}\right)\left(1 - \frac{h_1}{0.1}\right)$$

$$C_0 = \frac{1.41\varepsilon_{r,\text{eff}}}{\ln\left(\frac{5.98h}{0.8w + t}\right)}$$

OR

$$Z_0 = \frac{k}{\varepsilon_{r,\text{eff}} + 1.41}\ln\left(\frac{5.98h_2}{0.8w + t}\right)$$

$$\varepsilon_{r,\text{eff}} = \varepsilon_r(1 - e^{-1.55H/h_2})$$

Embedded edge coupled differential

$$Z_{\text{diff}} = 2Z_0\left(1 - 0.48 \times e^{\frac{-0.964}{2h-t}}\right)$$

$h = h_1 = h_2$
Z_0 same as embedded microstrip

$k = 87$. Restrictions: 0.1 $w/h < 3.0$, 1 $\varepsilon_r < 15$, and $Z_0 < Z_{\text{diff}} < 2Z_0$.

Table 6-4	Balanced Stripline	
Stripline transmission lines	**$z_0(\Omega)$**	**C_0 (pF/in.)**

Balanced (symmetric)

$$Z_0 = \frac{60}{\sqrt{\varepsilon_r}}\ln\left[\frac{1.9(H + h + t)}{0.8w + t}\right]$$

$$C_0 = \frac{1.41\varepsilon_r}{\ln\left(\frac{3.81h}{0.8w + t}\right)}$$

OR

$$Z_0 = \frac{60}{\sqrt{\varepsilon_r}}\ln\left[\frac{1.9(2h + t)}{0.8w + t}\right]$$

Differential (edge coupled)

$$Z_0 = 2Z_0\left(1 + 0.347 \times e^{\frac{-298}{2h+t}}\right)$$

Z_0 same as symmetric stripline

Restrictions: $w/(h - t) < 0.35$ and $w/h < 2.0$, $t/h < 0.25$, and $0.005 < w < 0.015$ in.

the inductors have current, I_{line}, through them and they will not allow I_{line} to stop instantly. As the magnetic fields of L_7 and L_8 begin to collapse to try to maintain their current (remember that they are mutually coupled and influence each other), they continue to shove current into C_5, raising its voltage a bit more (we will see later what *a bit more* means).

Table 6-5	Unbalanced Stripline		
Stripline transmission lines		$Z_0(\Omega)$	C_0 (pF/in.)
Unbalanced (asymmetric)	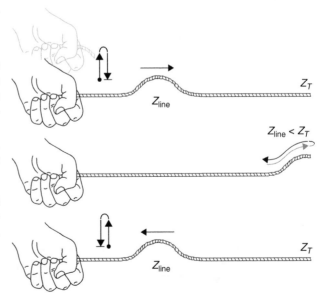	$Z_0 = \dfrac{80}{\sqrt{\varepsilon_r}} \ln\left[\dfrac{1.9(H + h + t)}{0.8w + t}\right]$ $\left[1 \dfrac{h}{4(H + h + t)}\right]$	$C_0 = \dfrac{2.82\varepsilon_r}{\ln\left[\dfrac{2(h - t)}{0.268w + 0.335t}\right]}$
Differential (broadside compled)		$Z_{\text{diff}} = \dfrac{82.2}{\sqrt{\varepsilon_r}} \ln\left(\dfrac{5.98H}{0.8w + t}\right)(1 - e^{-0.64})$	

Z_0 same as unbalanced stripline

References: IPC-2141, p. 12; IPC-2221A, p. 44; IPC-D-330, Section 2, Table 2-1 2; Brooks 2003, p. 203; Montrose 1999, pp. 171–177.

Note: Unless otherwise noted, $L_0 = \dfrac{Z_0^2 \times C_0}{1000}$ in nH/in.

The magnetic fields of each inductor pair (L_3 and L_4, L_1 and L_2, etc.) will collapse, one after the next, back toward the source and raise the voltage of its nearest capacitor, all the way down the line. This new voltage front (V_{line} + a bit more) propagates back from Z_T toward the source with the collapsing magnetic fields until all of the magnetic fields have collapsed and all of the capacitors have this new charge on them. An analogy of a reflection from a high impedance termination is shown in Figure 6-22, which shows a person launching a wave into a rope. If the rope experiences little or no friction, the wave will propagate down the rope unattenuated. If the end of the rope is loose (a high impedance), the wave will be reflected back toward the person, who will feel an identical wave returned.

In the example above the reflected wave has the same polarity and amplitude of the transmitted (incident) wave. In reality the rope would not be in a frictionless environment (and Z_T would not be infinitely high). In that case the reflected wave would still have the same polarity as the incident wave but the amplitude would be less.

The magnitude and polarity of the reflected wave are described by the reflection coefficient, ρ (Greek letter r), as shown in Eq. (6.14):

$$\rho = \frac{Z_T - Z_{\text{line}}}{Z_T + Z_{\text{line}}}. \qquad (6.14)$$

FIGURE 6-22
Positively reflected wave (Z_T is an open circuit).

The reflection coefficient can have values between -1 and $+1$. If $Z_T > Z_{line}$ (i.e., as Z_T approaches ∞, as in the example above), then

$$\rho \cong \frac{Z_T}{Z_T} = 1,$$

which means that the reflected wave will be exactly the same amplitude and have the same polarity as the incident wave.

Next we consider what happens if Z_T (in Figure 6-18) is a short circuit instead of an open circuit. At first the exact same thing occurs as described above when the switch is shut. That is (assuming the same initial conditions as above, all caps are discharged, etc.), the capacitors and inductors take their turn charging up and building up magnetic fields, V_{line} is applied to the line, current I_{line} flows, and $Z_{Cn} = V_{line}/I_{Cn}$. So the instantaneous line impedance is equivalent to Z_{Cn}. A different result occurs at the end of the transmission line. Since $Z_T = 0\,\Omega$ and inductors L_7 and L_8 again want to maintain their current flow, I_{line} flows straight through the short, Z_T.

Since the current through L_7 and L_8 is maintained (even for just an instant) and since the voltage drop across an inductor with a constant current flow is zero, capacitor C_4 sees the short and begins to discharge through L_7 and L_8 (helping to maintain their current flow) and on through the short, Z_T. A short moment later C_4 is at the same potential as C_5 and Z_T ($0\,V$), while L_7 and L_8 have managed to maintain their current. The capacitors continue to discharge one after the other (C_3 then C_2, etc.) and each inductor pair maintains its current until finally all capacitors are shorted (and all the inductors look like a short if they have the same constant current). In the final analysis, $V_{line} = V_{ZT} = 0\,V$ and therefore $Z_{line} = 0/I_{line} = 0\,\Omega$ and $I_{line} = V_S/R_S$.

FIGURE 6-23
Negatively reflected wave (Z_T is a short circuit).

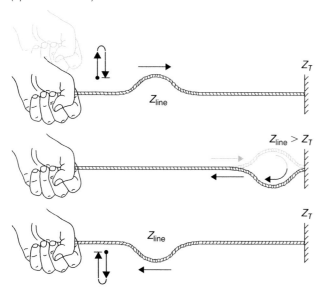

A mechanical analogy of a wave reflected negatively from a "dead short" is shown in Figure 6-23. If a positive wave is launched into a rope that is rigidly fixed at the end, the wave will be negatively reflected. In a perfectly lossless environment, the reflected wave will be of the same magnitude but opposite polarity as the incident wave.

In the electrical example above, the negatively reflected wave has the same magnitude as but opposite polarity to the voltage stored on the capacitors. As the negative wave hits each capacitor, it is forced to give up its charge (as current), which helps maintain the current flow through the nearest inductors and all the way down the line through the short at the end of the line. This negatively reflected wave is again represented mathematically by the reflection coefficient (Eq. [6.14]), but in this case since $Z_T < Z_{\text{line}}$ (i.e., as Z_T approaches 0), $\rho = -1$, as shown in Eq. (6.15):

$$\rho = \frac{-Z_{\text{line}}}{Z_{\text{line}}} = -1. \qquad (6.15)$$

Now let's say for argument that the characteristics of our transmission line are such that when we calculate $Z_{Cn} = V_{\text{line}}/I_{Cn}$ at each capacitor/inductor section, $Z_{Cn} = 50\,\Omega$. Let us also set $R_S =$ to $50\,\Omega$. Now what happens when $Z_T = 50\,\Omega$? As you can suppose by this time, at the moment the switch is shut, the capacitors take their turn getting charged (and the inductors are building their fields). Since each $Z_{Cn} = 50\,\Omega$, then Z_{line} is also $50\,\Omega$. Since R_S and Z_{line} are equal (and act as a voltage divider), $V_{\text{line}} = \frac{1}{2}V_S$. Once the wave front has propagated down the line and reaches Z_T, which is also $50\,\Omega$, I_{line} continues to flow into Z_T as if nothing different has occurred and $V_{ZT} = V_{\text{line}} = \frac{1}{2}V_S$. As long as Z_T is purely resistive, then everything is at steady state and $Z_{\text{line}} = Z_T = 50\,\Omega$. Also no voltage is reflected back toward the source because no change in voltage occurred on the capacitors and no current change occurred in the inductors.

In this case, since $Z_T = Z_{\text{line}}$ the reflection coefficient is 0 ($\rho = 0$) as shown in Eq. (6.16):

$$\rho = \frac{0}{Z_T + Z_{\text{line}}} = 0. \qquad (6.16)$$

The mechanical analogy is shown in Figure 6-24, in which none of the wave energy is reflected but is perfectly absorbed into the load at the end of the line.

From these examples we can conclude that Z_{line} is in effect only during voltage transitions and is the result of the voltage and the current transients that flow to charge the line capacitance to the new voltage and to build the magnetic fields in the inductors. We can also see that, if the impedance Z_T is not the same as Z_0, then a reflection will occur, but if the impedance Z_T is the same as Z_0, then no reflection will occur. Furthermore it takes a finite amount of time for a wave front to propagate from one end of a transmission line to the other, and if a reflection does occur it takes another

FIGURE 6-24
No reflection (Z_T absorbs wave energy).

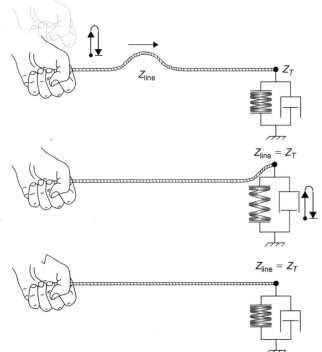

finite amount of time for the reflection to propagate back to the source. What happens at the source is the next topic.

Ringing

When $\rho \neq 0$ between any adjacent impedances, reflections will occur. This is true both from driver to transmission line and from transmission line to load (and back). If there is little or no loss along the transmission line, the reflected waves will bounce back and forth between the driver and the load if they are not matched to the transmission line (or if the transmission line is not matched to them). When viewing a particular point along the path, for example at the output pin of a gate or amplifier, the repeated reflections will be evident as ringing. Ringing is a direct result of reflections, which in turn are due to impedance mismatches.

One of the problems with ringing is that the voltage at any point along the line is effectively out of control, since ringing causes voltage overshoots and undershoots (see Figure 6-26 later). Overshoots can actually damage active devices that have input voltage limitations and will radiate greater EMI than normal signals. Overshoots and undershoots can cause digital circuits to be falsely triggered if the reflected voltage swings across switching thresholds. In analog circuits the interactions between a continuous wave signal and its reflections creates standing and/or traveling waves that can degrade the signal of interest.

The magnitude and frequency of the ringing depend on the speed of the wave through the transmission line, the length of the line, and the reflection coefficient at each impedance discontinuity. We take a detailed look at ringing using the circuit shown in Figure 6-25.

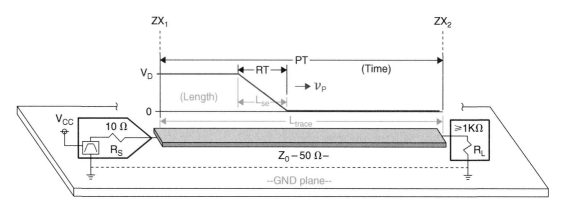

FIGURE 6-25
Representation of signal propagation on a PCB trace.

The circuit consists of a driver that is powered by V_{CC} and has a low output impedance, R_S ($10\,\Omega$); a transmission line with a characteristic impedance, Z_0 ($50\,\Omega$); and a receiver with a high input impedance, R_L (usually $1\,\mathrm{k}\Omega$ or higher).

The dashed lines indicate the interfaces of the mismatched impedances and are labeled as ZX_1, and ZX_2. The dimensions in green represent length, and the dimensions in blue represent time. The circuit in the figure can be used to represent an analog or digital circuit, but we will consider the digital application.

Consider the following:

- RT is the rise time, the time it takes for the output of a driver to transition from a minimum value to a maximum value. RT is specific to individual devices and is given in the data sheets.
- L_{trace} is the length of a trace (transmission line) on the PCB.
- v_P is the propagation velocity of a wave and is determined by Z_0 which is determined by ε_r and the transmission line dimensions (trace width and distance to the ground plane).
- PT is the propagation time, the time it takes for the transition to propagate from one end of the transmission line to the other.
- L_{SE} is the effective length of the rising edge (also called *transition distance* or the *spatial extent of the transition* [Bogatin 2004, p. 215] or *edge length* [Johnson and Graham 1993, p. 7]).
- Length and time are related by the propagation velocity of the wave, v_p (units of distance/time), where $PT = L_{trace}/v_p$ (units of time) and $L_{SE} = v_P \times RT$ (units in distance).
- If length of the trace, L_{trace}, is longer than the spatial extent of the rising edge, L_{SE}, then the rising edge will fit entirely within the length of the trace and the reflection voltage will be an amplitude-scaled copy of the entire rising edge, for which the scaling is determined by the reflection coefficient, ρ. Another way of looking at the same thing is if the RT (rise time) is faster than the PT (propagation time), then the rising edge will have time to be fully reflected.

Electrically Long Traces

The goal is to design PCB traces such that they do not allow conditions to exist under which propagation times are too slow (compared to signal RTs) or a trace's length is too long (compared to a signal's spatial extent). When these conditions cannot be met, the trace is considered to be "electrically long" and must be treated as a transmission line. Proper treatment of a transmission line means controlling the impedance of the line over the entire length of the line and matching the impedance of the line with the source and load impedances so that reflections do not occur.

The obvious question is, when is a trace too long (or when is the RT too fast)? The magnitude of reflections and the ringing frequency are governed by the down and back (round trip) time of the reflections. Much of the literature states that the propagation time, PT, should be less than one half of the rise time (i.e., $PT < \frac{1}{2}RT$) or that the length of the trace should be less than one half of the special extent of a rising edge (i.e., $L_{trace} < \frac{1}{2}L_{SE}$). These relationships define the limits, not the goal. The shorter trace lengths are or the slower the RT is, the better off you will be. The examples below illustrate this in greater detail. After the examples general design recommendations are provided.

Figure 6-26 shows what happens when PT is too long compared to RT. The data in the figure were generated using the transmission line model found in PSpice and the PT was set four times longer than the RT (instead of being <½RT). Refer to Figures 6-25 and 6-26 during the discussion.

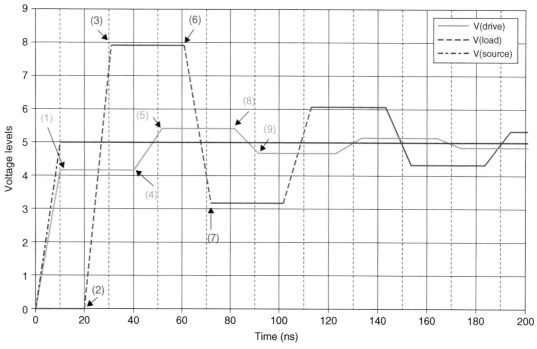

FIGURE 6-26
Ringing on an electrically long transmission line.

0. At time $t = 0$ ns the logic gate output (V_{source}) switches to $V_{CC} = 5$ V_{DC} and begins to increase in voltage at the output (V_{drive}) (start of rising edge). The first capacitor in the transmission line is uncharged and acts like a short to GND.

1. At $t = 10$ ns the gate has finished switching and the voltage at the output of the driver, V_D, is $V_D = V_{CC} \dfrac{Z_0}{Z_0 + R_S} = 5[50/(50 + 10)] = 4.17$ V because of the voltage divider established by R_S and Z_0. At this point the beginning of the rising edge is halfway to the load, and the tail end of the rising edge is just leaving the load side of R_S.

2. At $t = 20$ ns the beginning of the rising edge reaches the load resistor, R_L. Since there is an impedance mismatch between Z_0 (50 Ω) and R_L (1 kΩ), there is a positive reflection that begins immediately to head back to R_S. The reflected voltage is added to the rest of the rising edge as it continues to arrive at R_L. The reflection coefficient looking into the load from the transmission line is $\rho = (1000 - 50)/(1000 + 50) = 0.90$.

3. By $t = 30$ ns the trailing end of the rising edge reaches the load. The voltage at the load (V_{load}) is now the sum of its previous voltage (0 V) plus the value of the incoming voltage (4.17 V) plus the reflected voltage ($4.17 \times 0.90 = 3.75$ V) for a total of $4.17 + 3.75 = 7.92$ V. The reflected voltage (3.75 V) is well on its way back toward the source.

4. At $t = 40$ ns the rising edge that was positively reflected from the load begins to arrive at the source, R_S. The voltage at the source begins to rise (pts 4 to 5). However, since the impedance of the transmission line is greater than that of the source resistor, a negative reflection immediately begins to head back to the load. The reflection coefficient from the transmission line looking into R_S is

$$\rho = \frac{10 - 50}{10 + 50} = -0.67.$$

5. At $t = 50$ ns the 3.75 V reflected off from the load has completely reached R_S. The voltage at the load side of R_S is the sum of its original value (4.17 V) and the incoming reflected voltage (3.75 V) plus the voltage being rereflected back to the load ($-0.667 \times 3.75 \text{V} = -2.50$ V) for a total of $4.17 + 3.75 + -2.5 = 5.42$ V. And the -2.50 V is on its way to the load.

6. At $t = 60$ ns the -2.50 V reaches the load and begins to lower the load voltage from its previous value of 7.92 V. Because of the impedance mismatch between the transmission line and the load, a reflection is immediately launched again. The reflection coefficient is still $+0.90$, so the reflection will have the same polarity as the incident wave. Since the incident wave is the -2.50 V reflected off from R_S, the load will reflect back a negative voltage. As the incoming -2.50 V runs into a positively reflected negative voltage, the overall voltage at the load (pts 6 to 7) drops significantly since ρ is high (0.90).

7. At $t = 70$ ns the -2.50 V reflected off from the source has completely reached R_L where the voltage is the sum of its original value (7.92 V) and the incoming reflected voltage (-2.50 V) plus the voltage being rereflected back to the load ($-2.50 \text{V} \times 0.90 = -2.25$ V) for a total of $7.92 + -2.50 + -2.25 = 3.16$ V). And of course the -2.25 V is on its way to the source.

8. At $t = 80$ ns the negative voltage that was reflected (positively, i.e., leaving the sign intact) from the load begins to arrive at R_S. Since the incoming voltage is negative, the voltage at R_S begins to fall. But since there is still a negative reflection coefficient (-0.667) from the transmission line looking into R_S, the wave that immediately begins to bounce off from R_S is now positive and heads back to the load.

9. At $t = 90$ ns the -2.25 V reflected off from the load has completely reached R_S. Again the voltage at R_S is the sum of its previous value (5.42 V) and the incoming reflected voltage (-2.25 V) plus the voltage being rereflected back to the load (-0.667×-2.25 V $= +1.50$ V) for a total of $5.42 + -2.25 + 1.50 = 4.67$ V). And of course the $+1.50$ V is on its way to the load.

The reflections continue back and forth but decrease in value each trip. The losses occur because the energy that is not reflected at each impedance interface is absorbed into the source and load resistors. Eventually, the reflections become too small to notice and we say that it has reached steady state. The time to reach steady state is called the *settling time* and the shorter the settling time, the better.

If the length of the trace, L_{trace}, is much shorter than the special extent of L_{SE} (as represented in Figure 6-27), then the rising edge will not fit within the length of the trace and will reach the driver before the driver has even completely reached is steady-state value. If the trace is very short, the reflection voltage will be reflected many times and repeatedly fold back onto itself as the driver output climbs to its steady-state value. Since the voltage at an interface is the sum of its existing voltage, the incoming reflection, and the reflected reflection, the effects of the reflections become "smeared" into each other. By the time the driver has fully reached its final value most of the reflections have come and gone and only the last, smaller overshoots and undershoots are evident.

FIGURE 6-27
Representation of an electrically short trace.

Figure 6-28 shows what happens when PT is much shorter than RT on an electrically short trace. The reflections occur as previously described, but as shown in the graph many of the reflections have occurred by the time the driver reaches its full output level. Recall that the voltage at each impedance interface is the sum of its previous voltage plus the incoming voltage plus the reflected voltage, but since the reflections happen fast (relative to the rise time), each reflection never has a chance to reach its full voltage level (only a fraction of the rising edge at that time), so the reflections are much smaller and therefore the peaks (overshoots) and valleys (undershoots) are also much smaller

FIGURE 6-28
Negligible ringing on electrically short traces.

(hardly noticeable), while the driver output is still rising. Also the ring frequency is higher with the shorter trace.

Critical Length

As mentioned above an electrically short trace is one for which the propagation time is less than one half of the rise time (i.e., $PT < \frac{1}{2}RT$) or the length of the trace is less than one half of the special extent of a rising edge (i.e., $L_{\text{trace}} < \frac{1}{2}L_{\text{SE}}$). The length of a trace or transmission line for which these conditions are just barely met is called the *critical length*. Figure 6-29 shows the voltage levels at critical length. Ringing still occurs, but the peaks never level off and the reflections settle sooner. But again the one-half rule is a limit and not a goal. Examples of determining critical length and designing transmission lines are given in the design section below and in Example 4 in the PCB Design Examples.

Transmission Line Terminations

If we cannot make the rise time slower and/or the length of the trace shorter, then we will have noticeable reflections and ringing. The only other way to stop the reflections and ringing is to eliminate the impedance mismatches that are causing them by properly terminating the ends of the transmission line with the proper source and/or load resistors.

FIGURE 6-29
Reflections when PT = ½RT.

We can make $R_S = Z_0 = R_L$ by using a resistor in series with the source and a resistor in parallel with the load.

If the impedances are all matched then there will be no reflections, as shown in Figure 6-30. However, only half the voltage will reach the load because a voltage divider results with R_S equal to R_L so $V_{load} = \frac{1}{2}V_{source}$. This lower voltage at the load may not reach required logic thresholds, preventing affected digital circuits from functioning.

An alternative is to put a resistor in series with the driver such that the impedance that the transmission line sees looking at the driver and series resistor is equal to Z_0. So if Z_0 is 50 Ω and the driver output resistance is 10 Ω, then the transmission line will be matched to the driver by putting a 40-Ω resistor in series with R_S. An example of the result is shown in Figure 6-31. Even when PT is >½RT, only one reflection occurs (the fiat section on V_{drive} between 10 and 20 ns) and it is absorbed into the 40-Ω resistor and R_S, so the reflection dies there. An advantage of this type of termination technique is that the voltage at the load is also much closer to the ideal voltage. The momentary hold on V_{drive} is usually not a problem, but it can be a problem in high-speed clock circuits for which the steady-state on (off) time is about the same duration as the rise time.

To match impedances between the source and the transmission line, place a resistor in series with the driver such that $R_{series} = Z_0 - R_s$.

FIGURE 6-30
No reflections when all impedances are matched.

FIGURE 6-31
Reducing reflections by using a resistor in series with the source.

To match impedances between the transmission line and the load, place a resistor in parallel with the load such that $R_{parallel} = (R_L Z_0)/(R_L - Z_0)$.

PCB ROUTING TOPICS

There are four areas for electrical considerations when routing your PCB: placing parts, PCB layer stack-up, bypass capacitors, and trace width and spacing width.

Parts Placement for Electrical Considerations

Chapter 5 addressed parts placement with manufacturability in mind. Here we consider parts placement with electrical performance in mind. Usually the two goals complement each other, but occasionally they conflict. When conflicts do occur an attempt should be made to resolve the conflict in a way that is mutually beneficial. If that is not possible, electrical considerations usually have priority over mechanical considerations unless doing so will result in a mechanical failure of the board. For example, it may be necessary to manipulate the assembly or soldering processes or place parts on both sides of the board in order to meet manufacturability requirements and meet electrical performance requirements. Whatever solution is chosen the PCB needs to be manufacturable and operational.

Aside from manufacturability goals the first approach to placing parts for electrical considerations is usually determined by the function of the circuit. This is especially true for analog circuits where a signal enters the PCB, flows through the circuitry in more or less a single path (including feedback networks), and then leaves the board. Since analog circuits are susceptible to noise the goal is usually to place the parts to minimize the possibility of degrading the signals. This usually means keeping the parts as close together as possible so that the traces can be as short as possible and keeping the signal path as straight as possible (not zigzagging back and forth or from one side of the board to the other). This approach may increase the size of the board, however, and is not always possible.

With digital circuits it is also desirable to keep related parts close together and lines short, but because digital circuits often contain many parallel paths and branches and may contain wide data busses it may be nearly impossible to do so. Sometimes the best that can be accomplished is to keep parts that are functionally related closer together or place parts together that have the highest speed clocks and rising edges in order to minimize the length of related signal lines.

Mixed signal boards are even more challenging, in that both analog and digital circuits and high-power circuits (such as switching regulators) exist on the same board. In these cases the PCB should be segregated into different areas as shown in Figure 6-32. The topology may vary but the idea is to keep the higher power and other noisy circuitry closer to the connector if possible. This

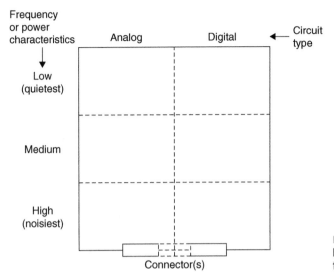

FIGURE 6-32
Board layout recommendations for noisy circuits.

limits the amount of return plane that is utilized by these circuits and therefore minimizes the amount of return plane that is common between them and the rest of the circuitry. Digital and analog circuits should also be kept apart from each other to minimize the effects of switching noise on analog circuitry. When dividing the PCB in this way, it is usually necessary (and beneficial) to set up split and isolated plane layers as shown in Figure 6-15.

PCB Layer Stack-Up

Although PCB Editor's Layout Cross Section dialog box allows you to define a stack-up, the PCB layer stack-up and thicknesses are really assigned when the board is ordered from the manufacturer. But the PCB layer stack-up and thickness need to be defined early in the design stage before working in PCB Editor, since the stack-up will determine how many layers to enable and how many power and ground planes to establish. The strategy for stacking up a PCB depends on a number of things such as the capabilities of your board manufacturer, the circuit density (both routing and parts), the frequency (analog) and rise/fall times (digital) of the signals, and the acceptable cost of the board.

As circuits become more dense, additional routing and plane layers are required. Digital circuits commonly require more layers than analog circuits because digital circuits typically consist of parts with greater numbers of pins per chip and because they have a higher number of parallel interconnections. High-speed circuits (whether analog or digital) may require a greater number of layers even if they are not very dense. This is because multilayer boards can provide better impedance control and shielding since they can have additional ground planes. The more layers a board has, the more it costs, but once a PCB stack-up exceeds four layers, the cost increase per layer usually becomes less for additional layers (to a point). Additionally the benefits of shielding and impedance control can considerably outweigh the increased cost of extra layers.

There are many possible ways to design the stack-up. If the board will be operated at low speeds and in an isolated environment almost anything you can dream up will work. However, as signal speeds increase the layer stack-up and trace routing become increasingly important because a poorly designed stack-up can lead to reflections and excessive EMI radiation (affecting external circuits as well as causing self-inflicted cross talk). A well-designed layer stack-up (and proper routing) not only minimizes the energy it radiates, but also can make your circuit relatively immune from external sources of radiation. In designing a layer stack-up there are only a few guidelines to follow, but they are important. The following few paragraphs and the examples demonstrate the guidelines.

Since PCBs are constructed of double-sided cores bonded together with prepreg, multilayer PCBs usually contain an even number of layers. Odd-layered boards can be made, but in most cases there is no cost benefit to adding only one layer instead of a layer pair (e.g., if you need five layers you may as well go with six). The extra layer can be an extra return/ground plane and the symmetry of an even number of layers helps minimize board warpage.

A signal layer should always be adjacent (and close) to a plane layer (preferably a return/GND plane) to minimize loop inductance, which minimizes electromagnetic radiation and cross talk. Power planes should also be adjacent (and close) to a return plane as this adds interplane capacitance, which helps minimize power supply noise and radiation. If you have to choose between a signal layer being adjacent to the return plane and a power plane being adjacent to the return plane, choose the signal layer. You can add more bypass and bulk capacitors between the power planes to make up for the loss in interplane capacitance.

The following stack-up examples are offered as a reference only. There are many more combinations possible, but only a few are shown here (please see the references for additional examples and details). The final board thickness in the examples is 0.093 in. and the copper is 1 oz (1.35 mils) thick. The availability of certain dimensions will depend on the manufacturer's capabilities and specific processes.

The thicknesses listed in the figures are given to provide perspective. You can use Tables 4-3, 4-4, and 4-5 to get an idea of the different combinations of cores and prepreg types used to make up different layer thicknesses, but remember that the thicknesses from the tables are preassembly thicknesses, and finished thicknesses may vary since traces sink into the soft prepreg, while plane layers do not (see the Advanced Circuits Web site, www.4pcb.com, for examples of layer thicknesses).

A signal's return current will be on the ground or power plane that is closest to the signal line (if possible), and the relative dimensions determine the trace/plane routing pairs. Ideally all return currents will need to end up at the ground/return pin on the PCB's power connector. It may be beneficial to stack up and route your PCB to *try* to make it easier for the return currents of critical high-speed traces to be paired up with certain ground planes (so that they are not forced to go through capacitor leads or vias to get back to the ultimate return point), but there is no guarantee they will follow it. But remember that to an AC signal there is no real difference between the power plane and the ground plane because they are shorted together with bulk and bypass capacitors. So when signal traces transition from one layer to another, the return currents may not have an easy path from one ground plane to another and may choose to return to the source on a power plane until they find a convenient path back to the preferred ground plane through a bypass capacitor. While the return current is not flowing on the preferred plane, the impedance of a transmission line can be significantly different from what is expected. To assist the return current in staying with the signal line when it changes layers, return current bridges can be installed by using free vias connected to the ground planes or capacitors with fan-out vias near vias used to transition signals from one layer to another.

There are several symbols that need to be defined with regard to the stack-up figures. The *H* in the figures implies horizontal routing and the *V* implies vertical routing. In some of the figures, an *R* is used to imply nonspecific routing direction. Concentrating the routing strategies in horizontal or vertical directions makes the routing process more efficient in high-density designs and reduces the number of vias required to complete the connections. The symbol HS is used to imply high-speed signals that are usually buried between plane layers for extra shielding. The GND symbol is used to represent any ground or return plane (for signal or power). The symbols PWR or ±PWR represent any power plane, such as $+5V$ V_{CC} for digital circuits or $+V$ and $-V$ for analog circuits.

There are three basic types of transmission lines: microstrip, stripline, and coplanar. You may not always need a transmission line, but no matter how you route your board one of these types will be represented (even if it is not a close resemblance). Other types of transmission line configurations can be realized by other types of stack-ups, which are not shown here as they will depend on the capabilities and processes of your board manufacturer.

Figure 6-33 shows three different four-layer stack-ups. Indicated thickness is in mils.

Figure 6-33(a) is one of the most common four-layer stack-ups for simple digital or analog PCBs. The land patterns and traces are placed on the outside (top and bottom) of the board and the power and return planes are inside. Placing the traces on the outside enables postfabrication inspection and troubleshooting.

Figures 6-33(b) and 6-33(c) place the power and return planes on the outside of the board and the traces on the inside. This arrangement helps shield the traces from outside EMI and helps contain self-generated radiation between the planes and consequently minimizes radiated EMI. The configuration in Fig. 6-33(c) can be used to route low-density analog circuits that require $+/-V$ for

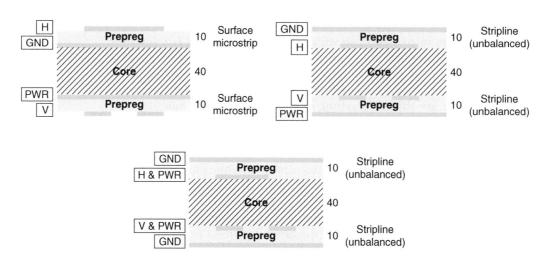

FIGURE 6-33
Typical four-layer stack-ups.

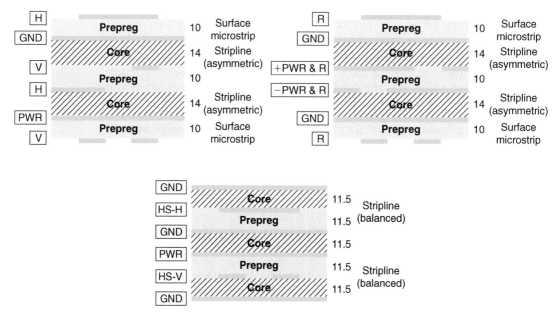

FIGURE 6-34
Six-layer stack-up examples.

dual-supply op amps. Power is routed to the amplifiers with wide traces or copper pours and shares the layers used for routing signals.

Figure 6-34 shows examples of six-layer PCB stack-ups. Figure 6-34(a) is a typical stack-up for digital circuits or analog circuits that do not require dual-power supplies. This arrangement provides four routing layers and two plane layers. The inner routing layers can be used for the higher speed signals since they are shielded by the ground and power planes. For analog circuits that require dual power planes, the stack-up in Figure 6-34(b) provides a highly functional stack-up with two full routing layers, two partial routing layers (which share layers with the +PWR and −PWR layers), and ground planes that are adjacent to all signal and power layers and provide shielding for the inner signal traces. Alternatively the shared power and routing layers can be dedicated to power only, but this limits the board to two routing layers. The stack-up in Figure 6-34(c) provides two well-shielded, balanced stripline routing layers for high-speed digital circuits and adjacent ground planes for each of the signal and power layers. A limitation of this stack-up is that there are only two routing layers.

Figures 6-35 and 6-36 show examples of 8- and 10-layer stack-ups, respectively. Only a few examples are shown, but by using the different stack-up strategies of the previous examples and capitalizing on the various finished board thicknesses available from most board manufacturers, it can be seen that the higher layer stack-ups can provide a multitude of routing possibilities.

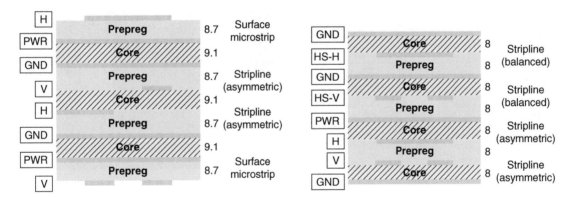

FIGURE 6-35
Examples of eight-layer PCB stack-ups.

Bypass Capacitors and Fan-Out

Bypass Capacitors serve two main functions, namely to short high-frequency noise to ground and to act as current reservoirs. Consequently there are three basic methods to fan out power pins. The first method usually occurs when the autorouter is used to create the fan-outs automatically. The autorouter creates individual fan-out vias for the IC's power and ground pins, and it creates individual fan-out vias for each pin on the bypass capacitor (resulting in no direct trace connection between the IC and its capacitor). In general this is acceptable. The second method is to deliberately route the power pin to the bypass capacitor before the fan-out via and to the power plane as Figure 6-37(a) shows, and

FIGURE 6-36
A 10-layer PCB stack-up.

the third method is to route the power pin to the power plane first by placing the fan-out via between the power pin and the bypass capacitor as shown in Figure 6-37(b).

At first glance there may appear to be no difference electrically speaking, but the differences can be significant at high frequencies and fast rise times. Additionally, other issues such as the method of assembly (wave vs. reflow soldering) and the available board real estate often influence orientation and routing of bypass capacitors. Sources in the literature do not all agree with which method is best, but more often Figure 6-37(a) is recommended for analog circuits and Figure 6-37(b) for digital circuits.

(a)

(b)

FIGURE 6-37
Power pin fan-out methodologies: (a) power pin to bypass capacitor to via, (b) power pin to via to bypass capacitor.

Trace Width for Current-Carrying Capability

When current flows through a conductor, it will heat up due to I^2R losses. Wider traces exhibit less resistance and therefore less heating. *To determine the minimum trace width* required to minimize heating, determine the maximum current a trace will carry and the thickness of the copper you will use on your board. Use Eq. (6.17) to calculate the minimum trace width,

$$ w = \left(\frac{1}{1.4 \times h} \right) \times \left(\frac{I}{k \times \Delta T^{0.421}} \right)^{1.379} \tag{6.17} $$

where w is the minimum trace width (in mils), h is the thickness of the copper cladding (in oz/ft²), I is the current load of the trace (in A), $k = 0.024$ is used for inner layers and $k = 0.048$ is used for outer (top or bottom) layers, and ΔT is the maximum permissible rise in temp (°C) of the conductor above ambient temperature. You can also use the graph shown in Figure 6-38, which was derived from curves in IPC-2221A.

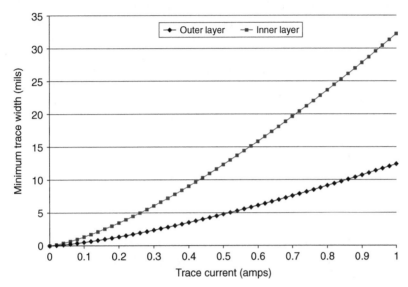

FIGURE 6-38
Minimum trace widths for 1 oz copper for $\Delta T = 10°C$.

Note: The glass transition temperature for FR4 is -125 to $135°C$, so as ambient temperature increases, the allowed ΔT decreases and minimum trace width increases. Even if the board will be at room temperature, giving you a $\Delta T = 108°C$, it does not hurt to restrict the temperature rise intentionally just to be safe (e.g., specify $\Delta T = 20°C$ even if it could be much higher).

Note: Equation (6.17) was derived from the IPC-2221A standard. Please refer to the standard for the actual equation and additional details.

You can usually use any of the standard technology files for most small signal applications. Figure 6-38 shows the minimum trace widths for l-oz inner and outer layer copper with $\Delta T = 10°C$. With 6-mil traces you can run up to about 300 mA on inner traces and about 600 mA on the outer traces, but for manufacturability and reliability considerations, they should be as wide as practical. Usually the applications of concern are power supply lines for large circuits and power supply boards in general.

Trace Width for Controlled Impedance

Controlled impedance PCB can be significantly more expensive than standard process PCBs. Before going through the effort and expense of designing controlled impedance transmission lines on the PCB, it is a good idea to determine if they are needed by determining if traces are electrically long (and whether controlled impedance is required).

For digital systems, at a minimum, we want the propagation time, PT, to be less than the rise time of the driver (i.e., PT < ½RT or RT > 2PT). From the discussion under Reflections another way to look at it is that the length of the trace, L_{trace}, should be less than one half of the special extent (edge distance) of the rising edge (i.e., $L_{\text{trace}} < \frac{1}{2}L_{\text{SE}}$ or $L_{\text{SE}} > 2L_{\text{trace}}$). To determine if this condition is met we need to determine RT and PT. RT (or fall time—FT) can be obtained from the data sheet for the device that is driving the trace and will have units of time (e.g., nanoseconds). Note that, since both the rise and the fall times must fall within limits, the smaller value should be used in the calculations. The RT and FT of several logic families are listed in Appendix C.

To calculate PT we need to know the intrinsic propagation delay (t_{PD}), which has units of time/distance. The intrinsic propagation delay for various transmission line configurations is given in Tables 6-6 and 6-7. Bracketed numbers [1] to [12] refer to the numbered references at the end of the chapter.

TOPOLOGIES, Z_0, t_{PD}, TRACE WIDTH, AND TRACE SEPARATION DESIGN EQUATIONS FOR VARIOUS TRANSMISSION LINES

Units for h, w, and t can be mils, centimeters, inches, etc., as long as they are consistent.

Note that in general the values of intrinsic capacitance (C_0) and intrinsic inductance (L_0) are given by [1]

$$C_0 = t_{\text{PD}}/Z_0$$

and

$$L_0 = Z_0^2 C_0/1000,$$

where C_0 is in picofarads per inch, t_{PD} is in picoseconds per inch, Z_0 is in ohms, and L_0 is in nanohenrys per inch.

For values of C_0 and L_0 that are specific to particular transmission line topologies, please see the appropriate references as listed.

There are two families of transmission lines described here, microstrip and stripline, including their subfamilies, which are as follows:

Microstrip:

1. Surface microstrip,
2. Surface differential microstrip,
3. Embedded microstrip, and
4. Embedded differential microstrip.

Stripline:

1. Balanced (symmetric) stripline,
2. Unbalanced (asymmetric) stripline,
3. Broadside coupled differential stripline (symmetric), and
4. Edge coupled differential stripline (symmetric and asymmetric).

Next calculate the propagation time, PT, using Eq. (6.18),

$$PT = L_{\text{trace}} \times t_{\text{PD}} \tag{6.18}$$

where PT is a trace's one-way propagation time, L_{trace} is the length of the trace as measured on the PCB, and t_{PD} is the intrinsic propagation delay from Tables 6-6 and 6-7.

If it is determined that the trace is electrically too long, then either it needs to be shortened or the impedances need to be matched. To determine the maximum allowable trace length use Eq. (6.19).

Since we want $PT < \frac{1}{2}RT$ and $PT = L_{\text{trace}} \times t_{\text{PD}}$, then

$$L_{\text{trace}} < \frac{RT}{2t_{\text{PD}}}. \tag{6.19}$$

For analog systems we determine the critical length of a trace with respect to the wavelength rather than the rising edge. The wavelength, λ, is determined using Eq. (6.20),

$$\lambda = \frac{v_P}{f}, \tag{6.20}$$

where v_P is the intrinsic propagation velocity ($v_P = 1/t_{\text{PD}}$) and f is the frequency of the signal on the trace.

Various trace length limits are stated in the literature: anywhere from $L_{\text{trace}} < 1/6\lambda$ to $L_{\text{trace}} < 1/20\lambda$. IPC-2251 recommends $L_{\text{trace}} < 1/15\lambda$, where L_{trace} is the length of the trace as measured on the PCB and λ is the wavelength of the highest frequency component of the signal (the shortest wavelength).

Tables 6-6 and 6-7 can be used to determine t_{PD} (ps/in.) for critical length calculations for both analog and digital circuits. If it is determined (using Eq. [6.19] or [6.20]) that a trace is electrically long, then source and load resistance and the transmission line impedance need to be controlled.

To design a controlled impedance transmission line on a PCB use Tables 6-6 and 6-7 to determine the trace width, w (in.). The trace thickness, t (oz/ft²), dielectric thickness, h (mils), and dielectric constant, ε_r (unitless), are determined by your board manufacturer.

Table 6-6   Microstrip Transmission Line Configurations

Microstrip transmission lines

	Topology	Characteristics	Intrinsic propagation delay
Surface		**Characteristic impedance** $$Z_0 = \frac{k}{\sqrt{\varepsilon_r + 1.41}} \ln\left(\frac{5.98h}{0.8w + t}\right) \text{ ohms}$$ $k = 87$ for $15 < w < 25$ mils [1, 2, 3, 4] $k = 79$ for $5 < w < 15$ mils [3, 4] **Restrictions** $0.1 < w/h < 3.0$ [1] $1 < \varepsilon_r < 15$ [1] **Design equations** Trace routing width to use in PCB Editor $$w = 7.475h \times e^{(-Z_0\sqrt{\varepsilon_r + 1.41})/k} - 1.25t$$ (use $k = 87$, then check against width rules, use $k = 79$ if necessary)	$$t_{PD} = 84.75\sqrt{0.475\varepsilon_r + 0.67} \text{ (ps/in.)}$$
Surface differential		**Characteristic impedance** $$Z_0 = \frac{87}{\sqrt{\varepsilon_r + 1.41}} \ln\left(\frac{5.98h}{0.8w + t}\right) \text{ ohms}$$ (same as surface microstrip) [5, 6, 7] **Differential impedance** $$Z_{diff} = 2Z_0\left[1 - 0.48e^{\left(-0.96\frac{d}{h}\right)}\right] \text{ ohm}$$	$$t_{PD} = 84.75\sqrt{0.475\varepsilon_r + 0.67} \text{ (ps/in.)}$$

Restrictions

None specifically noted except as applies to the surface microstrip.

Design equations

Trace routing width to use in PCB Editor

$$w = 7.475h \times e^{(-Z_0\sqrt{\varepsilon_r + 1.41})/87} - 1.25t$$

Trace separation:

$$d = \ln\left[2.08\frac{-1.04Z_{diff}}{Z_0}\right]\left(\frac{-h}{0.96}\right)$$

Characteristic impedance

$$Z_0 = \frac{87}{\sqrt{\varepsilon_r' + 1.41}}\ln\left(\frac{5.98h}{0.8w + t}\right) \text{ (ohms) [2,8,9]}$$

where $\varepsilon_r' = \varepsilon_r\left[1 + e^{\left[\frac{-1.55H}{h_2}\right]}\right]$

OR

$$Z_0 = \frac{87}{\sqrt{\varepsilon_r + 1.41}}\ln\left(\frac{5.98h}{0.8w + t}\right)\left(1 - \frac{h_1}{0.1}\right) \text{ (ohms) [1]}$$

Restrictions [8, 10]

$0.1 < w/h_2 < 3.0$

$1 < \varepsilon_r < 15$

$$t_{PD} = 84.75\sqrt{\varepsilon_r'} \text{ (ps/in.)}$$

OR

$$t_{PD} = 84.75\sqrt{0.475\varepsilon_r + 0.67} \text{ (ps/in.) [1]}$$

Embedded **Topology**

Continued

Table 6-6 Microstrip Transmission Line Configurations (Continued)

Microstrip transmission lines	Characteristics	Intrinsic propagation delay
	Line widths: 0.127(5 mils) to 0.381 mm (15 mils) Dielectric thickness: 0.127 (5 mils) to 0.381 mm (15 mils) $40 < Z_0 < 90\,\Omega$ **Design equations** Trace routing width to use in PCB Editor $w = 7.475h_2 \times e^{-x} - 1.25t$ where $$x = \frac{-Z_0\sqrt{\varepsilon_r' + 1.41}}{87} \text{ for } Z_0 \text{ from [2, 8, 9]}$$ OR $$Z_0 = \frac{-Z_0\sqrt{\varepsilon_r + 1.41}}{87\left(1 - \dfrac{h_1}{0.1}\right)} \text{ for } Z_0 \text{ from [1]}$$	

Embedded differential	Topology	Characteristic impedance
		$Z_0 = \dfrac{87}{\sqrt{\varepsilon_r + 1.41}} \ln\left(\dfrac{5.98h}{0.8w + t}\right)\left\|\left(1 - \dfrac{h_1}{0.1}\right)\right.$ (ohms) [5] $t_{PD} = 84.75\sqrt{0.475\varepsilon_r + 0.67}$ (ps/in.) [5]

Differential impedance

$$Z_{diff} = 2Z_0 \left[1 - 0.48e^{\left(-0.96 \frac{d}{h_1 + h_2 + t} \right)} \right] \text{ (ohms)}$$

Restrictions [5]

Same as embedded microstrip

Design equations

Trace routing width to use in PCB Editor

$$w = 7.475 h_2 \times e^{-x} - 1.25t$$

where

$$x = \frac{-Z_0 \sqrt{\varepsilon'_r + 1.41}}{87 \left(1 - \frac{h_1}{0.1} \right)} \quad [1]$$

Trace separation of pair:

$$d = \ln \left[2.08 - \frac{1.04 Z_{diff}}{Z_0} \right] \left[\frac{-(h_1 + h_2 + t)}{0.96} \right]$$

Table 6-7 Stripline Transmission Line Configurations

Stripline transmission lines

	Topology	Characteristics	Intrinsic propagation delay
Balanced (symmetric)	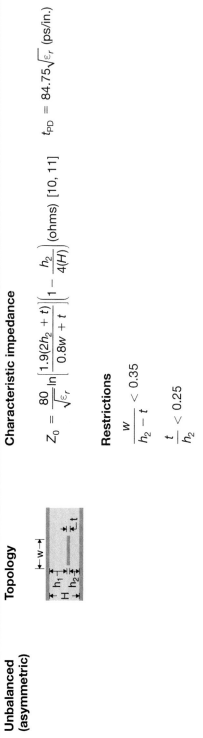	**Characteristic impedance** $Z_0 = \dfrac{60}{\sqrt{\varepsilon_r}} \ln\left(\dfrac{1.9H}{0.8w + t}\right)$ (ohms) [10, 11] **Restrictions** Line widths: 0.127 (5 mils) to 0.381 mm (15 mils) Dielectric thickness: 0.127 (5 mils) to 0.381 mm (15 mils) $40 < Z_0 < 90\,\Omega$ **Design equation** Trace routing width in PCB Editor: $w = 1.25\left[1.9H \times e^{\left(\frac{-Z_0\sqrt{\varepsilon_r}}{60}\right)} - t\right]$	$t_{PD} = 84.75\sqrt{\varepsilon_r}$ (ps/in.)
Unbalanced (asymmetric)	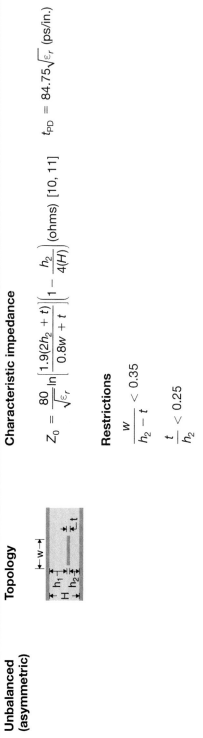	**Characteristic impedance** $Z_0 = \dfrac{80}{\sqrt{\varepsilon_r}} \ln\left[\dfrac{1.9(2h_2 + t)}{0.8w + t}\right]\left[\left(1 - \dfrac{h_2}{4(H)}\right)\right]$ (ohms) [10, 11] **Restrictions** $\dfrac{w}{h_2 - t} < 0.35$ $\dfrac{t}{h_2} < 0.25$	$t_{PD} = 84.75\sqrt{\varepsilon_r}$ (ps/in.)

Design equation

Trace routing width in PCB Editor

$w = 2.375(2h_2 + t)e^{-x} - 1.25t$

where

$$x = \dfrac{Z_0\sqrt{\varepsilon_r}}{80\left(1 - \dfrac{h_2}{4H}\right)}$$

Stripline transmission lines

	Topology		Intrinsic propagation delay

Broadside coupled differential stripline (symmetric) [12]

Required trace width (in) for desired Z_0 and/or Z_{diff}.

Characteristic (between conductors) impedance

$$Z_0 = \frac{82.2}{\sqrt{\varepsilon_r}}\ln\left(\frac{5.98d}{0.8w + t}\right)(1 - e^{(-0.6h)})\ \text{ohms}$$

Restrictions

None given in the references

Design equations

Trace routing width to use in PCB Editor

$w = 7.475d \times e^{-x} - 1.25t$

where

$$x = \frac{Z_0\sqrt{\varepsilon_r}}{82.4(1 - e^{-0.6h})}$$

$t_{PD} = 84.75\sqrt{\varepsilon_r}$ (ps/in.)

Edge coupled differential stripline (symmetric) [12]

For symmetric ($h_1 = h_2$) or asymmetric ($h_1 \neq h_2$)

Topology

Characteristic impedance

For symmetric ($h_1 = h_2$)

$$Z_0 = \frac{60}{\sqrt{\varepsilon_r}}\ln\left(\frac{1.9H}{0.8w + t}\right)\ \text{(ohms)}$$

$t_{PD} = 84.75\sqrt{\varepsilon_r}$ (ps/in.)

Continued

Table 6-7	Stripline Transmission Line Configurations (Continued)	
Stripline transmission lines	**Characteristic impedance**	**Intrinsic propagation delay**
	For asymmetric ($h_1 \neq h_2$) $Z_0 = \dfrac{80}{\sqrt{\varepsilon_r}}\ln\left[\dfrac{1.9(2h_2 + t)}{0.8w + t}\right]\left[1 - \dfrac{h_2}{4(H)}\right]$ (ohms) [10, 11] **Differential impedance (both)** $Z_{diff} = 2Z_0\left[1 - 0.374e^{\left(-2.9\frac{d}{H}\right)}\right]$ **Design equations** Trace routing width to use in PCB Editor For symmetric ($h_1 = h_2$) $w = 1.25\left[1.9H \times e^{\left(\frac{-Z_0\sqrt{\varepsilon_r}}{60}\right)} - t\right]$ For asymmetric ($h_1 \neq h_2$) $w = 2.375(2h_2 + t) \times e^{-x} - 1.25t$ where $x = \dfrac{Z_0\sqrt{\varepsilon_r}}{80\left(1 - \dfrac{h_2}{4H}\right)}$ Trace separation in layer stack-up (for symmetric or asymmetric) $d = -0.347\ln\left[2.67\left(1 - \dfrac{Z_{diff}}{2Z_0}\right)\right]$	

There is another type of transmission that is not in Tables 6-6 and 6-7: The coplanar transmission line is shown in Figure 6-39. Typically $d < h$, and w is relatively wide compared to the other configurations. The copper along each side of the trace is a topside return plane and is typically $>5w$ in both directions. The bottomside return plane is not present in some applications.

The equations are not well documented in the literature for this type of coplanar configuration except in the *Transmission Line Design Handbook*, by Brian C. Wadell (1991). The equations are not presented here because they are rather unwieldy, and this type of transmission line is not widely used on FR4. If you need to use this type, please see the reference for a full description of its design, use, and limitations. An approximation of this type of configuration is sometimes attempted using guard traces and copper pours, but because of the relative dimensions (i.e., $h < d$ or $h \approx d$), an accidentally designed surface microstrip is usually the result.

FIGURE 6-39
A coplanar transmission line.

Trace Spacing for Voltage Withstanding

There are two reasons for controlling the spacing between traces: (1) to ensure adequate voltage-withstanding capability (insulation resistance) between high voltage lines and (2) to minimize cross talk between signal lines. Table 6-8, which is an abridged version of a similar table given in IPC-2221A (Table 6-1), shows the required trace spacing for various voltage ranges on internal and external layers. The spacing on external traces depends on both the voltage and the external coating of the board.

Table 6-8	Minimum Conductor Spacing (Mils)			
Voltage between conductors (V_{DC} or V_{p-p})	Internal traces	External traces		
		Bare	Soldermask only	Conformal coating
0–15	2	4	2	5
16–30	2	4	2	5
31–50	4	24	5	5
51–100	4	24	5	5

After IPC-2221A

Trace Spacing to Minimize Cross Talk (3w Rule)

The default trace spacing used in PCB Editor's Constraint Manager is shown in Figure 6-40(a), in which the edge-to-edge spacing between traces is typically one conductor width. If a trace is susceptible to cross talk from adjacent traces then it should be kept a minimum of two trace widths (edge to edge) from the

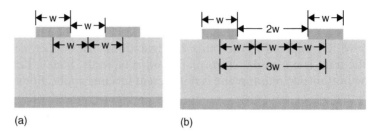

(a)   (b)

FIGURE 6-40
Trace spacing methods: (a) typical trace spacing, (b) 3w spacing to minimize cross talk.

other traces, as shown in Figure 6-40(b). This is referred to in the references as the $3w$ rule because the center-to-center spacing is $3w$, as indicated in Figure 6-40(b). At $3w$, the traces are out of reach of about 70% of each other's magnetic field if the traces are controlled impedance transmission lines (or reasonable approximations). Keeping the traces $10w$ apart at the centers will keep the traces out of about 98% of each other's field (Montrose 1999, p. 210).

To adjust the trace-to-trace spacing in PCB Editor select the Constraint Manager button on the tool bar to display the Constraint Manager as shown in Figure 6-41. Select the Spacing tab and the Line icon as shown. Enter the edge-to-edge spacing in the Default row or click the '+' box and enter edge-to-edge values for specific nets.

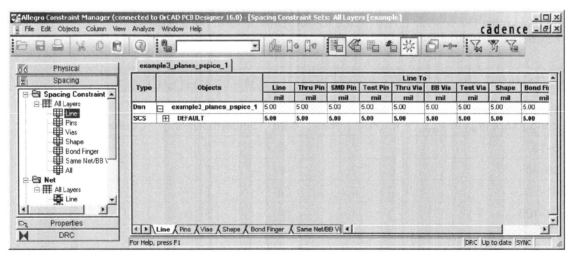

FIGURE 6-41
Setting the trace spacing in PCB Editor's Constraint Manager.

Traces with Acute and 90° Angles

Routing high-frequency analog or high-speed digital traces with acute or 90° angles has long been discouraged, but not everyone agrees anymore as to how much of a problem it really is (see Jonson and Graham 1993, p. 174; Montrose 2000, p. 220; Bogatin 2004, p. 317; and Brooks 2003, p. 383). The argument is that the trace width increases by a factor of 1.414 at the corner of the trace (as shown in Figure 6-42) and causes a change in the characteristic impedance (due to an increase in capacitance) of the trace. As discussed above impedance mismatches cause reflections. The reflections in turn cause ringing in digital circuits with fast rise times and standing or traveling waves in high-frequency analog circuits. In theory then, 90° corners should be avoided—at least when routing controlled impedance transmission lines.

A look through the references shows that, when using high-end network analyzers and time domain reflectometers to measure the reflection from a 90° corner, the effects are evident (Bogatin 2004, p. 318). However, the frequencies at which the reflections occur are high (into the upper gigahertz for traces greater than 50 mils wide and tetrahertz for traces less than 10 mils), and compared to other sources of impedance discontinuities (such as vias), the effects are insignificant.

FIGURE 6-42
Trace geometry of a sharp 90° corner.

Another thing to consider is that (with PCB Editor anyway) the corners are not as sharp as shown in Figure 6-42. Recall from Chapter 1 that pads and traces are drawn as flashes or draws, respectively, by photoplotters with apertures of a given diameter. While square apertures can be used (and are for square pads), PCB Editor uses round apertures for traces in the Gerber codes. A 90° corner produced by PCB Editor is shown in Figure 6-43(a). Although the corner may look fairly sharp, a close-up of the corner shown in Figure 6-43(b) shows that the outer edge is rounded and only the inner corner is square. This results in a smaller increase in width than with a square corner.

The difference between a 12-mil trace and a 17-mil trace (at the sharp corner in Figure 6-42) is about 11.5 Ω for a dielectric constant of 4.2 and a core thickness of 10 mils, and the difference between the 12-mil trace and a 14.5-mil trace (at the rounded corner in Figure 6-43) is about 6.2 Ω. However, the effect

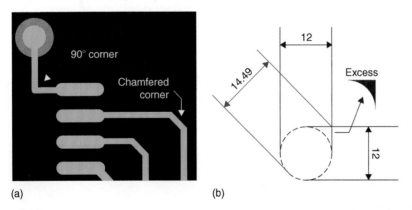

(a)   (b)

FIGURE 6-43
Geometry of a 90° corner in Layout: (a) 90° and chamfered corners, (b) excess copper at a 90° corner.

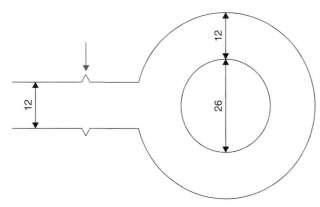

FIGURE 6-44
Excess area compared
to a typical via (Brooks
2003, p. 383; Johnson
and Graham 1993,
p. 174; Montrose 1999,
p. 220; Bogatin 2004,
p. 315).

of extra width (and change in impedance) is very small. The excess area shown in Figure 6-43(b) is 7.73 mils[2] for a 12-mil trace. If the equivalent area is divided by 2 and each piece is placed on either side of a straight segment of a 12-mil trace (as indicated by the arrow in Figure 6-44) it is clear that the excess area created by the 90° corner is insignificant compared to other factors such as vias and land patterns. The literature (Brooks 2003, p. 385) suggests that even acute angles of 135° can be used up to about 1 GHz.

USING PSPICE TO SIMULATE TRANSMISSION LINES

PSpice is used in Chapter 7 to develop PSpice subcircuit models for creating new Capture parts with simulation capabilities. In this section PSpice is used to simulate transmission lines. When used in conjunction with the design equations from this chapter and the simulations in the PCB Design, Examples, PSpice can be of help in designing PCB-based transmission lines and in understanding transmission lines in general.

The circuit shown in Figure 6-45 is used here to simulate a high-speed digital circuit (it was also used previously to generate the plots used to explain ringing). The circuit consists of a driver gate with an output impedance of 10 Ω, a

FIGURE 6-45
A basic transmission line simulation circuit.

50-Ω transmission line, and a receiving gate with in input resistance of $1\,k\Omega$ and input capacitance of $15\,pF$. The PSpice parts used in the circuit are

V2 is VPULSE (from the Source library), the driver.
R1 is R (from the Analog library), the source resistance (output impedance) of the driver.
T1 is T (from the Analog library), the transmission line.
R2 is R (from the Analog library), the load resistance.
C1 is C (from the Analog library), the load capacitance.

Figure 6-45 also shows the various reflection coefficients at the impedance mismatches, where the reflection coefficient looking into T1 from R1 is

$$\rho = \frac{Z_0 - R1}{Z_0 + R1} = \frac{50 - 10}{50 + 10} = +0.667,$$

the reflection coefficient looking into R1 from T1 is

$$\rho = \frac{R1 - Z_0}{R1 + Z_0} = \frac{10 - 50}{10 + 50} = -0.667,$$

and the reflection coefficient looking into R2 from T1 is

$$\rho = \frac{R2 - Z_0}{R2 + Z_0} = \frac{1000 - 50}{1000 + 50} = +0.90.$$

To simulate a transmission line (T1) the characteristic impedance, Z_0, must be specified and either the transmission delay (TD) must be specified or the frequency (f) and number of wavelengths (NL) must be specified. Use TD for digital signals, and use f and NL for analog signals.

Simulating Digital Transmission Lines

The value TD is the same as PT (propagation time) from the earlier discussion and is calculated by

$$TD = t_{PD} \times \text{Length}_{trace}$$

where t_{PD} is found from Tables 6-6 and 6-7. For the surface microstrip t_{PD} is

$$t_{PD} = 85\sqrt{0.475\varepsilon_r + 0.67}.$$

In laying out a digital circuit (for instance the digital design example in the PCD Design Examples), it is important to know the critical trace length so that traces can be kept short enough and parts can be placed accordingly. The critical trace length is calculated by

$$\text{Length}_{trace} < \frac{RT}{k \times t_{PD}},$$

where k is a safety factor and is essentially the ratio PT/RT. The maximum trace length recommended is $k = 2$; larger values of k (shorter traces) are better.

For ALS family logic RT is approximately 2 ns. Using $\varepsilon_r = 4.2$ the critical trace length is then 7.3 in. Table 6-9 shows the effective k value and TD for various transmission line lengths.

Table 6-9	TD Calculations Considering Various Trace Lengths		
	Long	**Critical**	**Safe**
Length$_{trace}$ (in.)	30	7.3	3.5
k (approximate)	½	2	4
TD (ns)	4.1	1	0.24

Knowing RT (2 ns) and TD (from Table 6-9), we can use PSpice to simulate the various transmission line lengths. Set up the PSpice simulation as shown in Figure 6-46. To display the Simulation Settings dialog box select Edit Simulation Profile from the PSpice menu. Note: If the step time is too large, the simulation may become unstable and you will not get good results. A maximum step size of $^1/_{1000}$ of the total run time usually produces satisfactory results. Click OK.

FIGURE 6-46
PSpice simulation settings.

To start the simulation, press the blue triangle on the menu bar, or press F11 on the keyboard, or select Run from the PSpice menu. The results are shown in Figures 6-47 to 6-49.

FIGURE 6-47
Simulation results for a long line ($k = \frac{1}{2}$).

FIGURE 6-48
Simulation results for a critical length line ($k = 2$).

FIGURE 6-49
Simulation results for a short line ($k = 4$).

Simulating Analog Signals

For analog signals f is the frequency on the trace and NL is the length of the trace in relative wavelengths (e.g., NL would be 0.25 for a quarter wavelength trace at frequency f). To determine NL you need to know the wavelength, λ. You can calculate the wavelength using Eq. (6.21):

$$\lambda = \frac{v_{PD}}{f} = \frac{1}{f \times t_{PD}}$$ (6.21)

where

v_{PD} is the propagation velocity (distance/time) of a wave through a dielectric.
f is the frequency of the wave.
t_{PD} is the propagation time (time/distance) as described above.

So for a 66-MHz signal traveling through the same surface microstrip from the above example ($\varepsilon_r = 4.2$), $\lambda = 110.8$ in. The critical length for traces carrying analog signals varies depending on which book you read but is often cited as being $\lambda/6$, $\lambda/15$, or $\lambda/20$ (or somewhere in between). As with the digital signals, the shorter the trace, the better.

REFERENCES

Bogatin, E. 2004. *Signal Integrity—Simplified*. Upper Saddle River, NJ: Pearson Educational.
Brooks, D. 2003. *Signal Integrity Issues and Printed Circuit Board Design*. Upper Saddle River, NJ: Pearson Educational.

IPC-2141. 1996. *Controlled Impedance Circuit Boards and High Speed Logic Design.* Northbrook, IL: IPC-Associated Connecting Electronics Industries.

IPC-2221A. May 2003. *Generic Standard on Printed Board Design.* Northbrook, IL: IPC-Associated Connecting Electronics Industries.

IPC-D-330.

Johnson, H., and Graham, N. 1993. *High-Speed Digital Design: A Handbook of Black Magic.* Upper Saddle River, NJ: Prentice Hall.

Montrose, M. I. 1999. *EMC and the Printed Circuit Board: Design, Theory, and Layout Made Simple,* 2nd edn. New York: IEEE Press.

Montrose, M. I. 2000. *Printed Circuit Board Design Techniques for EMC Compliance: A Handbook for Designers,* 2nd edn. New York: IEEE Press.

Serway, R. A. 1992. *Physics for Scientists and Engineers with Modern Physics,* 3rd edn. Orlando, FL: Harcourt Brace Jovanovich.

Wadell, B. C. 1991. *Transmission Line Design Handbook.* Norwood, MA: Artech House.

NUMBERED REFERENCES

[1] IPC-2251. November 2003. *Design Guide for the Packaging of High Speed Electronic Circuits,* p. 32. Northbrook, IL: IPC-Associated Connecting Electronics Industries.

[2] IPC-2141. 1996. *Controlled Impedance Circuit Boards and High Speed Logic Design,* p. 12. Northbrook, IL: IPC-Associated Connecting Electronics Industries.

[3] Montrose, M. I. 2000. *Printed Circuit Board Design Techniques for EMC Compliance: A Handbook for Designers,* 2nd edn., p. 101. New York: IEEE Press.

[4] Montrose, M. I. 1999. *EMC and the Printed Circuit Board: Design, Theory, and Layout Made Simple,* 2nd edn., p. 172. New York: IEEE Press.

[5] IPC-2251. November 2003. *Design Guide for the Packaging of High Speed Electronic Circuits,* p. 36. Northbrook, IL: IPC-Associated Connecting Electronics Industries.

[6] Montrose, M. I. 2000. *Printed Circuit Board Design Techniques for EMC Compliance: A Handbook for Designers,* 2nd edn., p. 107. New York: IEEE Press.

[7] Montrose, M. I. 1999. *EMC and the Printed Circuit Board: Design, Theory, and Layout Made Simple,* 2nd edn., p. 177. New York: IEEE Press.

[8] Montrose, M. I. 2000. *Printed Circuit Board Design Techniques for EMC Compliance: A Handbook for Designers,* 2nd edn., p. 103. New York: IEEE Press.

[9] Montrose, M. I. 1999. *EMC and the Printed Circuit Board: Design, Theory, and Layout Made Simple,* 2nd edn., p. 177. New York: IEEE Press.

[10] IPC-2251. November 2003. *Design Guide for the Packaging of High Speed Electronic Circuits,* p. 33. Northbrook, IL: IPC-Associated Connecting Electronics Industries.

[11] Montrose, M. I. 2000. *Printed Circuit Board Design Techniques for EMC Compliance: A Handbook for Designers,* 2nd edn., p. 105. New York: IEEE Press.

[12] IPC-2251. November 2003. *Design Guide for the Packaging of High Speed Electronic Circuits,* p. 35. Northbrook, IL: IPC-Associated Connecting Electronics Industries.

Making and Editing Capture Parts

Capture provides many libraries of parts that you can use to build your schematic, perform simulations, and generate PCB layouts. However, you will need to make your own parts at some point. There are a couple of ways of working with the part libraries. You can build custom parts and save them to existing or new libraries, modify existing parts and save them to a library, or modify and save parts to just a specific project. Whether you decide to build custom parts or modify existing ones, you need to know about the libraries, the different ways of making parts, how different parts are packaged, what types of pins to use with different parts, and how to connect them properly for board layout and simulation purposes.

THE CAPTURE PART LIBRARIES

The full version of Capture contains 141 parts libraries in the Capture libraries, with 108 PSpice libraries and 34 libraries in the main Capture library; and each library contains many parts. As mentioned in Chapter 2, two parts libraries are supplied with the software, both of which are located in the Capture folder. One type of parts and a PSpice folder are located inside the Capture Library folder. The parts located immediately inside the Capture Library folder are schematic parts that have no footprints or PSpice models associated with them, but the parts located in Capture's PSpice folder have PSpice models and footprints associated with them. For the purpose of board layout, it does not matter from which library you select parts, because it is a simple matter of assigning (or changing) a footprint regardless of the library from which it came. Assigning PSpice simulation capabilities to a part is another matter altogether and is discussed later in this chapter.

The libraries are fairly well labeled and, in most cases, particular parts are relatively easy to find. However, there is some overlap between some of the libraries (especially the PSpice libraries), which can make searching for parts in general less than straightforward. For example MOSFET transistors can be found in the discrete.olb and the fairchild.olb libraries, the infineon_x.olb series libraries, and the jpwrmos.olb library, to name a few. Some of the libraries contain mixtures of parts (anlg_dev.olb contains amplifiers, multipliers, and multiplexers), while others contain only certain types of parts (fairchild.olb contains only FETs). So at times you may have to search manually through the libraries to find parts. The first place to begin searching is to open the PSpice library list (lib_list.pdf) located in the lib_list folder in the OrCAD/Doc path. The lib_list lists all of the OrCAD parts in two sections: The first section lists parts by device type and the second section by part library. Using the PDF search tool you can quickly search the entire library list (all 1123 pages).

If you cannot find the part you are looking for in the lib_list or in the actual parts libraries, you may be able to download parts and models from the Internet. Several semiconductor manufacturers generate their own parts and models and provide them free of charge from their Web sites.

If you cannot locate a ready-made part, you will need to make your own. It is highly recommended that you make your own folders and libraries in which to save your parts, rather than adding parts to the libraries supplied with the software. Instructions on how to create Capture parts and PSpice models are provided later. You can find additional information on the subject in the Capture and PSpice user's guides in the document folders and on the frequently asked questions (FAQ) section of the OrCAD Web site.

TYPES OF PACKAGING

One package contains one or more parts. All the parts can be the same (homogeneous), or the parts can be different (heterogeneous), and either type of part can be passive or active.

Homogeneous Parts

Figure 7-1 shows examples of homogeneous packages (a single bipolar junction transistor and a digital IC with four identical NAND gates). In both cases all the parts within a package are identical (even if there is only one). All the gates in the IC are active and have to share the two power pins (7 and 14). Homogeneous parts can be placed independently in a schematic. Setting up the gates to share the power pins at the schematic level and the board level is addressed later.

FIGURE 7-1
Homogeneous—one or more identical parts in a package.

Heterogeneous Parts

Figure 7-2 shows an example of a heterogeneous package. This package has three parts, which can be placed independently on the schematic: one relay coil (*K*) and two sets of contacts (*A* and *B*). Although power can be applied to the coil and to one or more of the contacts, it is a passive device and does not have power pins (strictly speaking).

Pins

When making parts in Capture that will be used to make a PCB design in PCB Editor or perform simulations in PSpice, one of the most important issues in making new parts is properly handling the pins. All the other lines that make up a part are cosmetic, but how the pins are assigned determines if you get design rule check errors when checking your schematic, if you have a PCB laid out correctly, and if a design simulates correctly.

FIGURE 7-2
Heterogeneous relay—dissimilar parts in a single package.

There are eight types of pins. Each pin has a pin number, a pin name, and an order number. Each of these pins also has specific characteristics that determine how it performs in the package with regard to what it looks like in the schematic, how it is "judged" during design rule checks, how routing is governed in PCB Editor, and the like. Additionally, you can assign pin characteristics that determine what the pins look like once the part is placed into a schematic. Table 7-1 shows a summary of the pin types, shapes, and visibility options.

With regard to PCB layout, the pin types are simply power pins and nonpower pins (any of the other types). Power pins require special handling because of their global nature. For example, it needs to be considered if any of the pins will be shared between parts within one package, if they will be visible on all the parts, and if they will be connected to a global net or a nonglobal net. Examples that explain how to deal with pins are given later.

Table 7-1	Pin Types and Attributes	
Pin types	**Pin shapes**	**Visibility settings**
Power/GND	Clock	A pin's number
Three state	Dot	A pin's name
Bidirectional	Dot–clock	The pin itself*
Open collector	Line	
Open emitter	Short	
Input	Zero length	
Output		
Passive		

*Applies to power pins only. All other pins are always visible.

PART EDITING TOOLS

Before we begin making parts, we take a quick look at the part editing tools.

The Select Tool and Settings

SELECT TOOL

The Select tool is used to select pins, text, or any of the graphical objects (lines, arcs, etc.). You can select objects by clicking on them or dragging a box across or around objects (see the Area Select description).

SNAP TO GRID

The Snap-to-Grid function is actually a coarse or fine setting and never actually fully disables the snap-to functionality. If the Snap-to-Grid function is active (coarse setting), you can place or move objects only in 0.1-unit increments— this is the default. If the grid function is not active (fine setting), you can place or move objects in 0.01-unit increments. Except when working with graphics, the Snap to Grid should always be active.

AREA SELECT

Objects can be selected by fully enclosing the object with a selection box (using the Select tool) or just intersecting the object with the selection box. Using the fully enclose method is handy for selecting an object that is surrounded by many closely located objects.

DRAG CONNECT OBJECT

When you move parts on your schematic, nets stay connected to the part and may cross over other nets or part pins may land on nets. Two modes of operation determine how these events change or allow connectivity. If the drag connect mode is off, ▥, the default, then connectivity changes are prohibited. This helps prevent unwanted connections, but if you are placing new parts on the schematic, you need to specifically add new nets from existing nets to the new part to make connections. If the drag connect mode is active, ▥, then connectivity changes are allowed. New or moved parts will automatically be connected to existing nets if the pins touch them. Red dots are used to indicate where connections will be made.

The Pin Tools

PLACE PIN TOOL

The Place Pin tool is used to place pins on a part one at a time.

PLACE PIN ARRAY TOOL

The Array tool is used to simultaneously place multiple pins on a part, such as a 16-bit output port on a microcontroller. The Array tool enables you to establish a base name common to all the pins then automatically number each pin chronologically.

The Graphics Tools

PLACE IEEE SYMBOL

The Place IEEE Symbol tool is used to place IEEE standard symbols, such as arrows, math symbols, or pin states, on your part. The symbols are graphic only and do not add functionality to the part or its pins.

PLACE LINE TOOL

The Line tool is used to place a single orthogonal or diagonal line segment.

PLACE POLYLINE TOOL

The Polyline tool is used to place multisegment lines. Polylines are orthogonal by default but can be made diagonal by holding down the Shift key on the keyboard while drawing the line.

PLACE RECTANGLE TOOL

The Place Rectangle tool is used to place closed rectangular parallelograms.

PLACE ELLIPSE TOOL

The Place Ellipse tool is used to place circles and ellipses.

PLACE ARC TOOL

The Place Arc tool is used to construct arcs and circles.

PLACE TEXT TOOL

The Place Text tool is used to place text objects, which have font, color, and rotation settings.

The Zoom Tools

ZOOM IN

The Zoom In tool zooms in by set increments. You can use the button, press the I key on your keyboard, or select Zoom In from the View menu.

ZOOM OUT

The Zoom Out tool zooms out by set increments. You can use the button, press the O key on your keyboard, or select Zoom Out from the View menu.

ZOOM TO REGION

Use this tool to zoom to a particular region of your design by dragging a box around the area to which you want to zoom.

ZOOM TO ALL

Use this tool to see the entire design.

METHODS OF CONSTRUCTING CAPTURE PARTS

Four methods are used to construct Capture parts. Three of the methods are completed from the Capture Library Manager and the fourth is initiated from the PSpice Model Editor and finished with the Capture Library Manager. PCB footprints can be assigned to any of the parts regardless of how it was constructed, and the footprints can be assigned when the parts are first constructed or assigned later when the parts are actually used in a design.

Methods from Capture Library Manager:

1. Design menu → New part.
2. Design menu → New part from spreadsheet.
3. Tools menu → Generate part.

Method from PSpice Model Editor:

4. File menu → Export to Capture part Library (then use the Capture Library Manager to modify the part's appearance).

Table 7-2 shows which method to use by how the part will be used. The procedure for each method is described later. Later in the chapter, it will be shown how to make and download PSpice models, which can be converted to Capture parts using Methods 3 and 4.

Table 7-2 Methods of Constructing Parts	
Function/purpose of the part	**Method of construction**
Schematic entry only	Method 1 (easier) or Method 2
Schematic entry and PCB layout	Method 1 (easier) or Method 2
Schematic entry and PSpice simulation using functional OrCAD schematic designs (can also assign footprints for PCB layout)	Method 3 (easier) or Method 4
Schematic entry and PSpice simulation using new or existing PSpice models (can also assign footprints for PCB layout)	Method 4

Method 1. Constructing Parts Using the New Part Option (Design Menu)

Constructing parts using the New Part option in Capture is demonstrated through three design examples. The first two examples are used to design homogeneous parts; the first is a transformer (a single-part, passive device) and the second is an operational amplifier (a multipart, active device) with shared power pins. The third example is a heterogeneous, multipart, passive device.

DESIGN EXAMPLE FOR A PASSIVE, HOMOGENEOUS PART

The first design example is of a transformer with an iron core, a single primary, and a center-tapped secondary. The transformer and its schematic representation are shown in Figure 7-3.

To begin, start Capture and from the session frame navigate to File → New → Library as shown in Figure 7-4.

FIGURE 7-3
Single-part, homogeneous transformer and schematic.

The window that opens is the Capture Library Manager. Select the Library icon (Figure 7-5, left), then select New Part from the Design menu (Figure 7-5, right) or right click and select New Part from the pop-up menu.

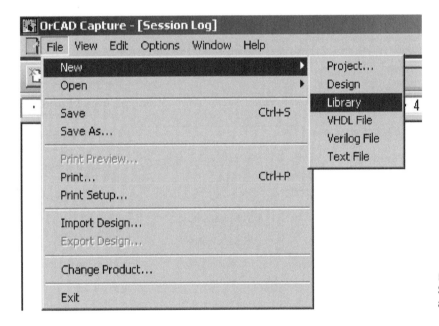

FIGURE 7-4
Starting a new part from a new Capture library.

A New Part Properties dialog box will open as shown in Figure 7-6. Enter a name for the part (for example, Spri_CTsec) and T for the reference prefix. You can also enter a PCB footprint if desired (or known). You can change this later if you are not sure what the footprint is or will be. Leave the Parts per Pkg: as 1 and Package Type as Homogeneous. Select Part Numbering as Numeric and check Pin Number Visible if you want the pin numbers to be visible on the schematic. Click OK.

A part editing window will open with a dotted outline as shown in Figure 7-7. The dotted box defines the boundary of the part. Pins are placed on the

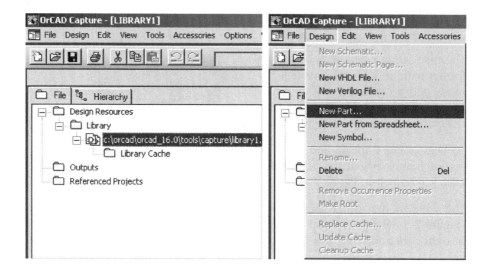

FIGURE 7-5
Beginning a new part from the Library Manager.

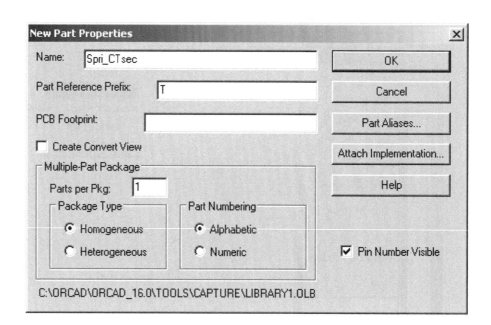

FIGURE 7-6
New Part Properties dialog box.

boundary (as described later) and text and graphics are placed inside the boundaries. You can also add pictures and IEEE symbols if desired.

Snap to Grid:

On   Off

The default grid is 0.1 × 0.1 in., with the upper left corner as coordinate (0, 0). By default pins and graphical objects are placed on the grid. The snap-to-grid setting can be changed by toggling the Snap-to-Grid tool on, ▣, or off, ▣. The grid can be shown as dots or lines. To make changes to the grid settings, navigate to Options → Preferences from the menu bar and select the Grid Display tab as shown in Figure 7-8. Select the desired part and symbol grid properties on the right side of the dialog box.

The first step to make the transformer is to make the part outline 0.4 in. wide by 0.6 in. tall (for comparison, resistors from the Analog library are 0.2×0.2 in., not including the pin lengths). To resize the part outline, left click on the dotted border then click to hold one of the corner squares (handles) and drag the corner to resize the box.

The next step is to add the pins that make the transformer's five leads. To add pins to the part, toggle the pin tool, .

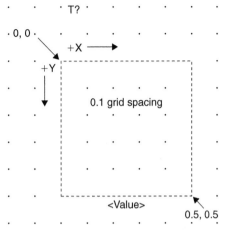

FIGURE 7-7
The new part shown in the part editing window.

Pin tool

The Place Pin dialog box shown in Figure 7-9 will be displayed. Select Short for the pin shape and Passive for the pin type; click OK. Place the first two pins on the left side of the border at positions (0.0, 0.1) and (0.0, 0.5) for the primary winding leads and three pins on the right side at positions (0.4, 0.1), (0.4, 0.3), and (0.4, 0.5) for the secondary winding leads. To quit placing pins press the Esc key on the keyboard or right click and select End Command from the pop-up menu.

The Place Pin tool will automatically increment the name and number of each pin as it is placed. To change a name or number of a pin, double click on a pin

FIGURE 7-8
Grid settings in the Options → Preferences dialog box.

FIGURE 7-9
The Place Pin dialog box.

to bring up the Pin Properties dialog box (which looks exactly like the Place Pin dialog box) and change it as necessary. To simultaneously change the properties of multiple pins, hold down the Ctrl key and left click each pin to select more than one pin. Then right click and select Edit Properties... from the pop-up menu. The Browse Spreadsheet dialog box shown in Figure 7-10 will be displayed. More than one pin must be selected to get the spreadsheet, otherwise the Pin Properties dialog box will be displayed.

FIGURE 7-10
Use the Browse Spreadsheet dialog box to modify pin parameters.

The pin type and pin length cells have dropdown lists that allow you to select the desired pin properties. The other columns require you to enter the values manually. The number and name are the same as the settings you can choose using the Pin Properties dialog box. Naming pins is flexible, but the pin numbers you assign in a Capture part must match the pad name in a PCB Editor footprint. For PCB layout considerations, the pin order is irrelevant. The pin order is extremely important to PSpice, however, and is discussed in Methods 3 and 4.

Grid button   Place Arc tool

The next step is to *place graphics* to build the transformer coils. Turn the Snap to Grid off (toggle the grid button, ▨, so that it turns red). Select the Place Arc tool, ◖. Arcs are defined by three points, as shown in Figure 7-11. The first point you click defines the center of the arc, the second point defines where the arc starts, and the third point defines the end. The arc is drawn counterclockwise from point 2 to point 3, so the location of point 2 depends on the direction you want the arc to face. After you have placed the first couple of arcs, you can copy and paste the rest in place. *To copy graphic objects,* toggle the Select tool, ▸,

and select the objects you want to copy. Left click once to select a single object or use Ctrl + left click to select multiple objects. Copy and Paste the objects as will be shown in Figure 7-13 using typical Windows Copy/Paste techniques. You can also left click once to select an object, then hold down the Ctrl key, left click and hold the part and "drag a duplicate part" out of the first, release the mouse button to place the copy, then release the Ctrl key.

Select tool

FIGURE 7-11
Making coils with the Place Arc tool.

Next, place two parallel lines between the arcs to represent the core. Use the Place Line tool,

Place Line tool   Place Ellipse tool

▸, to draw them. Then make the dots for the dot–coil indicator by using the Place Ellipse tool, ◯. Draw a small circle near one of the coils, then hit the Esc key to exit drawing mode. Change the circle to a filled dot by using the Edit Filled Graphic dialog box shown in Figure 7-12. To display the dialog box, use the Select tool to select the circle then right click and select Edit Properties from the pop-up menu. You can leave the Line Style as is but change the Fill Style to Solid. Click OK.

Figure 7-13 shows the transformer design so far. Note that the pin numbers are on the pin outside the boundary and pin names are inside the boundary with the graphics.

You can *use the* User Properties *dialog box to change the visibility of the pin names and numbers,* shown in Figure 7-14 (left). To display the User Properties dialog box, select Part Properties from the Options menu. Use the User Properties dialog box to make the pin names and numbers invisible. You can also change the visibility of the pin numbers using the check box in the Edit Part Properties dialog box shown in Figure 7-14 (right). *To display the* Edit Part Properties *dialog box,* select Package Properties from the Options menu. The completed part is shown in Figure 7-15.

FIGURE 7-12
Use the Edit Filled Graphic dialog box to fill objects with patterns.

FIGURE 7-13
Completed part with pin names and numbers visible.

FIGURE 7-14
User Properties dialog box from Options → Part Properties menu: (left) User Properties dialog box (Part Properties), (right) Edit Part Properties dialog box (Package Properties).

FIGURE 7-15
Completed homogeneous part (pin names and numbers not visible).

The final task is to save the new part (and the library if it is new). *To save the part*, close the part editing window and click Yes at the Save changes to partname? prompt. *To save the library*, select the Library icon in the Project Manager, right click, and select Save As... from the pop-up menu. Save the library with a new name in your UserLibrary in the Capture/Library path.

DESIGN EXAMPLE FOR AN ACTIVE, MULTIPART, HOMOGENEOUS COMPONENT

The second example shows how to construct the dual op-amp component shown in Figure 7-16. The parts are identical (homogeneous) and share the power pins.

Begin as you did with the transformer, by selecting New Library from the File menu in the Capture session frame. Select the Library icon in the Library Manager, right click, and select New Part from the pop-up menu. In the New Part Properties dialog box (Figure 7-17),

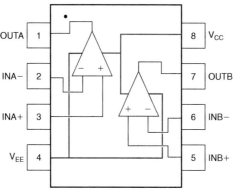

FIGURE 7-16
Data-sheet diagram for a dual op-amp component.

enter a name for the op amp, make U the part reference prefix, and change the parts per package to 2. Leave the package type as homogeneous, assign alphabetic part numbering, and make the pin numbers visible. Click OK when you are finished.

FIGURE 7-17
New Part Properties dialog box for a dual op amp.

Use the Place Pin tool to begin placing pins as shown in Figure 7-18. Label and place pins for part A per the data sheet (Figure 7-16). You can either correctly name and number each pin as you place it or place all the pins then go back and fix the names and numbers after they have been placed. Whichever method you use, set the pin characteristics as shown in Figure 7-19. Remember that, to change a name or number of a pin, double click on a pin to bring up the Pin Properties dialog box and, to change the properties of multiple pins simultaneously, hold down the Ctrl key and left click each pin to select more than one pin. Then right click and select Edit Properties... from the pop-up menu. Finally, add short

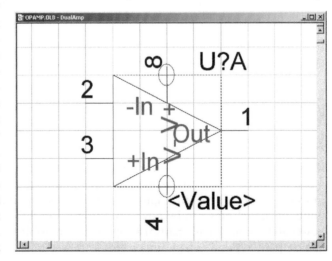

FIGURE 7-18
Dual op amp, part A.

FIGURE 7-19
Pin characteristics for part A of the dual op amp.

Place Line tool

graphic lines between the op-amp body and the power pins using the Place Line tool, . You may need to turn off the Snap to Grid to do this. Remember to turn the Snap to Grid back on when you are finished with the graphics.

In the Part Editing window, make the part border 0.4 × 0.4 in. Then make the triangular body of the op amp using the Place Line tool (see Figure 7-18).

As you can see from Figure 7-18, the op amp looks cluttered because the pin names run into each other and are not optimally spaced with respect to each of the pins. You can fix this by making the pin names *not* visible and adding your own text. First, *turn the pin names off* by setting the Pin Names Visible to False in the User Properties dialog box (from the Options → Part Properties menu). Use the Place Text tool or the Place Line tool to add − and + signs in place of the −IN and +IN pin names. Add text near the power pins to indicate which is the positive supply pin and which is the negative supply pin.

For the time being, that completes part A. There are a couple of ways to get to part B. *To view a specific part of a multipart device*, go to the View menu and select Package to see both parts, as shown in Figure 7-20. Then double click on part U?B. Alternatively you can select Next (or Previous) Part from the View menu, or use Ctrl + N or B on your keyboard to toggle through the parts.

Once you have part B in the editing window, select all of the pins and set the parameters as shown in Figure 7-21 using the procedure described previously.

ASSIGNING POWER PIN VISIBILITY

If you compare Figure 7-19 to Figure 7-21, you see that parts A and B share pins 4 and 8. Shared power pins require special handling to make the parts look and function properly between Capture, PSpice, and PCB Editor. The type and visibility of shared pins can be power and visible, power and nonvisible, or nonpower and visible. Digital parts usually share nonvisible power pins, while active analog parts typically share either visible power pins or nonpower-type pins

FIGURE 7-20
Package view of dual
op-amp part.

Browse Spreadsheet

	Location	Order	Number	Name	Type	Clock	Dot	Pin Length
1	B2	3	4	V-	Power	☐	☐	Zero Length
2	L1	1	6	-In	Input	☐	☐	Short
3	L3	2	5	+In	Input	☐	☐	Short
4	R2	0	7	Out	Output	☐	☐	Short
5	T2	4	8	V+	Power	☐	☐	Zero Length

OK   Cancel   New...   Copy   Remove   Paste   Help

FIGURE 7-21
Pin characteristics for
part B of the dual op
amp.

(such as input or passive pins), which are always visible. The type and visibility are initially established at the part level in the library, but you can change these parameters at the schematic level after you have placed the part into your design.

Several methods can be used for changing the part properties on the schematic, and each method has slightly different effects. If you add up all the combinations of the possible pin settings with the various methods of setting them, you will find over 200 possible combinations! Some of the combinations can cause severe errors that prevent you from creating netlists, others cause various types of warnings, and still others cause no errors or warnings but fail to route correctly on the board. We do not discuss all 200 combinations here, but we look at various scenarios in the PCB Design Examples. For now, to keep things simple and to follow typical convention, we make the shared power pins visible power pins since this is an analog part.

To set a power pin's properties, select pin 8 then right click and select Properties from the pop-up menu (or just double click the pin) to bring up the Pin Properties

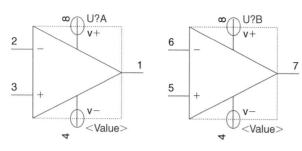

FIGURE 7-22
The finished op amps.

dialog box (see Figure 7-9 for an example). Make sure that the Pin Visible box is checked, an option that is available only for power pins (since all other types of pins are always visible). If you use the User Properties dialog box (Options → Part properties), you can set the visibility of only the pin's names or numbers, not the visibility of the pin itself. The completed op amps are shown in Figure 7-22. Save the parts as described previously.

When you use a part with shared pins in a schematic then create a PCB Editor netlist, you may get the warning, WARNING [MNL0016] Duplicate pin number '4' on 'LM324'. If the part was constructed properly, you can ignore the warning.

DESIGN EXAMPLE FOR A PASSIVE, HETEROGENEOUS PART

In this design example we make a passive, heterogeneous part, namely, a double-pole, double-throw relay, as shown in Figure 7-23. The relay consists of three parts: two sets of identical double-throw contacts and an operating coil with its clamping diode.

Begin by navigating to File → New → Library from the Capture session frame as described in the preceding examples. Select the Library icon and select New Part from either the Design menu or the pop-up menu after right clicking on the icon. Fill out the New Part Properties dialog box as shown in Figure 7-24. Make sure you select 3 parts per package and select the heterogeneous package type. Click OK. From the View menu, go to the Package view to make sure that you have three parts.

FIGURE 7-23
A heterogeneous, multipart relay: (left) relay package, (right) relay schematic.

Begin by constructing the relay coil and clamping diode as part K?-1. Bring up K?-1 in an editing window; double click on the K?-1 window if you are in the Package view or use Ctrl + N on your keyboard to toggle through the parts if you are in the Part view. Resize the border so that it is 0.3 in. wide by 0.6 in. tall. Using Figure 7-25 as a guide, use the Place Pin tool to place two pins on the left side of the part border. Define the top pin as number 2 with name C1 and the bottom pin as number 6 with name C2. Make both pins short, passive lines, and make the names nonvisible. Use the Place Line and Place Arc tools to add the coil, diode, and "1" graphics to the part as described in the previous examples.

The next step is to add the first set of contacts. Toggle to part K?-2. Resize the border to 0.4 in. wide by 0.6 in. tall. Using Figure 7-26 (left) as a guide, add two pins

New Part Properties   ×

Name: K3   [OK   ]

Part Reference Prefix: K   [Cancel]

PCB Footprint: [Part Aliases...]

☐ Create Convert View   [Attach Implementation...]

┌─ Multiple-Part Package ──────────────┐
│ Parts per Pkg: 3   [Help   ]
│ ┌─ Package Type ──┐ ┌─ Part Numbering ──┐
│ │ ○ Homogeneous │ │ ○ Alphabetic   │
│ │ ◉ Heterogeneous│ │ ◉ Numeric   │   ☑ Pin Number Visible
│ └─────────────────┘ └───────────────────┘
└───────────────────────────────────────┘

C:\ORCAD\ORCAD_16.0\TOOLS\PROJECTS\CHAP7_PARTS\RELAY.OLB

FIGURE 7-24
New Part Properties
dialog box for the
heterogeneous relay.

to the right side of the border and one on the left. Set the pin parameters as shown in Figure 7-26 (right) and make the pin names nonvisible. Use the Place Line tool to add the pole and the lines leading from the pin to the throws (the contacts).

Use the Place Polyline tool to draw the throws so that you can make them solid filled. Use the Shift key to make diagonal polylines. After you have the triangle drawn, double click on it to bring up the Edit Filled Graphic dialog box and set the fill to Solid.

Once you have the pole and throws for contact assembly K?-2 finished, copy and paste the graphics into the K?-3 part. To copy all the pieces at one time, hold down the Ctrl key on your keyboard and select each line and polyline one at a

FIGURE 7-25
Relay coil and diode.

(a)   (b)

FIGURE 7-26
Schematic (left) and pin settings (right) for the first set of relay contacts.

	Location	Order	Number	Name	Type	Clock	Dot	Pin Length
1	L2	2	7	P2	Passive	☐	☐	Short
2	R1	0	9	NC2	Passive	☐	☐	Short
3	R3	1	8	NO2	Passive	☐	☐	Short

FIGURE 7-27
Pin parameters for the second set of contacts.

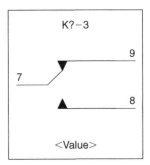

FIGURE 7-28
Package view of the completed heterogeneous relay.

time. Once you have them all selected, right click and select Copy from the pop-up menu (or select Copy from the Edit menu or use Ctrl + C on your keyboard). Go to the next part, K?-3, by typing Ctrl + N on your keyboard then paste the copied graphics into the empty part outline (right click and select Paste from the pop-up menu, select Paste from the Edit menu, or use Ctrl + V on your keyboard). Add three pins using the Place Pin tool as you did for the previous parts. Set the pin parameters as shown in Figure 7-27. You will notice that the order starts at 0 for each part. The order of the pins is of concern only when the part is to be used with a PSpice model for performing simulations. We address that issue in the next sections. The completed, multipart relay is shown in Figure 7-28.

When you place the relay in a Capture schematic, use the Part: dropdown list as shown in Figure 7-29 to place the desired part of the multipart package.

Method 2. Constructing Parts with Capture Using the Design Spreadsheet

Use this method to automatically generate a single or multipart package. It is similar to the first method, except that you define all the pins up front, using

FIGURE 7-29
Selecting a part from a
multipart package in a
Capture schematic.

FIGURE 7-30
The New Part Creation Spreadsheet.

the spreadsheet shown in Figure 7-30. New Part from Spreadsheet generates heterogeneous packages by default, but you can use it to make homogeneous
parts. The parts will be defined as heterogeneous but will be homogeneous in
effect if there are no differences between the parts.

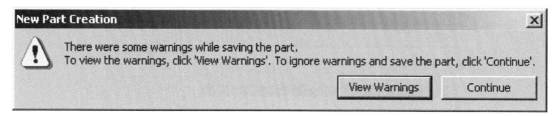

New Part Creation

There were some warnings while saving the part.
To view the warnings, click 'View Warnings'. To ignore warnings and save the part, click 'Continue'.

View Warnings   Continue

FIGURE 7-31
New Part Creation warnings.

To construct a part using the design spreadsheet, open an existing Capture library to which you want to add a part, or start a new library as described previously. In the Library Manager, select the Library icon and select the New Part from Spreadsheet... option, which you can do either from the Design menu or from the pop-up menu by right clicking on the Library icon.

The spreadsheet will initially be blank. Add the pin numbers and parameter settings as needed per the part requirements. The pin settings shown in Figure 7-30 would generate a part identical to the dual op amp designed in a previous example (see Figure 7-16).

Notice that pins 4 and 8 were added twice, once for each part, to define the shared power pins.

After the pin parameters have been entered into the spreadsheet, click the Save button. The information box shown in Figure 7-31 will be displayed, letting you know that there were warnings (because of the duplicate pins). You can view the warnings or click Continue to view the parts.

Figure 7-32 shows the package view of the parts generated by the spreadsheet. By default, the "parts" are initially represented by simple boxes. All you have left to do at this point is to modify the graphics to make the op amps cosmetically correct. In the homogeneous op amp from the previous example, all parts were automatically made identical when you made a graphic in one part. Since this is a heterogeneous part by default (which you cannot change), you have to modify the graphics on every part individually. Delete the boxes that lie on top

FIGURE 7-32
Package view of new parts generated using the spreadsheet.

of the dotted outline and add the graphics as described in the previous op-amp example.

Method 3. Constructing Parts Using Generate Part from the Tools Menu

Use this method to create a new Capture part (or parts) from a PSpice model library or a Capture project. The part(s) can then be used for schematic entry, PSpice simulations, and PCB layout (if you assign a footprint to it). To use this method, a PSpice model library or a functional schematic design needs to preexist. In the section that follows, several methods for developing or obtaining the PSpice libraries and functional schematic designs are described. For the time being we will go ahead under the assumption that a PSpice library is available so that the part creation process is kept separate from the library/schematic creation process.

FIGURE 7-33
Generate Part dialog box from the Tools menu.

Begin by starting a new Capture library or open an existing library from the Capture session frame's File menu. Select the Library icon and then select Generate Part... from the Tools menu. The Generate Part dialog box will pop up, as shown in Figure 7-33.

Select PSpice Model Library (or Capture Schematic/ Design if appropriate) from the Netlist/source file type: list, then select the library (*.lib etc.) or design (*.dsn etc.) from the Netlist/source file: list. Make sure that the Destination part library: path and name are what you want and that the library has a .OLB extension. Make sure that the Create new part option is checked and click OK.

After Capture has completed importing the library/design file you selected, you should see the list of parts in the Library Manager. If the Capture library (.olb) you are working with was originally empty, the new parts will be located directly under the Library icon in the Design Resources folder. But if the library already had some parts in it, the new parts and copies of the previously existing parts will be added to both the Library icon and the Library folder under the Design Resources folder (see Figure 7-34, in which, as an example, the breakout library was added to the analog_p library). The Library folder will also contain a duplicate of the Library icon. This allows you to copy, add, or delete parts to or from the new library. Once you are satisfied with which parts are contained under the Library icon under the Design Resources folder, you can delete the duplicate Library icon and parts in the Library folder, keeping just the library in the Design Resources folder.

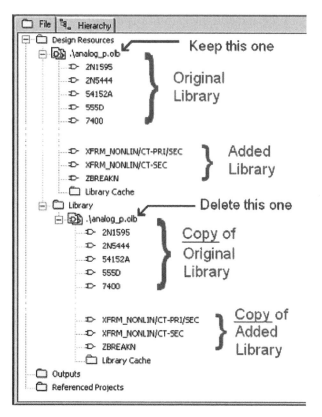

FIGURE 7-34
Structure of Library Manager after copying a library into an existing library.

Once you finish adding parts to the library, you can edit the added part(s), pins, and links to footprints using the Capture Part Editor, and the Part and Package properties dialog boxes from the Design menu as described in previous examples.

Finally, save and close the library. You can now use the library and its parts in new designs for schematic entry, simulations, and PCB layout.

Method 4. Generating Parts with the PSpice Model Editor

You can use the PSpice Model Editor to make Capture parts that can be used in schematic designs, circuit simulations, and PCB layouts. The difference between using the PSpice Model Editor and the Generate Part method (Method 3) is that you can work directly on and with the PSpice models with the Model Editor.

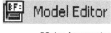

PSpice Accessories

Start the PSpice Model Editor. Go to Start → All Programs → OrCAD 16.0 → PSpice Accessories → Model Editor. You will begin with a blank Model Editor session window. *To open an existing PSpice library* (*name*.LIB), navigate to OrCAD/tools/pspice/library and select Open from the File menu. Select one of the libraries (e.g., bipolar.lib) and click Open; then select one of the models from the model list. Figure 7-35 shows the Q2N696 BJT transistor model from the bipolar library. The Models List window

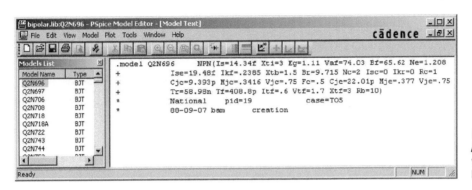

FIGURE 7-35
A PSpice library as viewed from the Model Editor.

pane shows all of the simulation models contained in the library, and it tells you whether it is a primitive model or a subcircuit model. The text window to the right of the Models List displays the "code" that describes the model.

You can construct new models by using existing model listings as examples and/or the model descriptions found in the PSpice reference guide (pspcref.pdf) located in the OrCAD/doc folder. The following sections describe how to download primitive models (models beginning with .model) from the Internet and how to create your own subcircuit models (models beginning with .subcircuit). The models can then be added to a PSpice library from which you can generate Capture part libraries as described. Creating custom primitive models is not described here.

GENERATING A CAPTURE PART LIBRARY FROM A PSPICE MODEL LIBRARY

Beginning with a PSpice library (.lib) open in the Model Editor (similar to Figure 7-35), select Export to Capture Part Library... from the Model Editor's File menu. You will be presented with the Create Parts for Library dialog box shown

in Figure 7-36. In the Enter Input Model Library: text box, use the Browse... button to find the PSpice library (*.LIB) for which you want to make parts. Use the second Browse... button to specify the location for the new Capture library (*.OLB). Remember that the PSpice models and the libraries that contain them are usually stored in the OrCAD/tools/pspice/library path, while the Capture parts that use the models are stored in the ...OrCAD/tools/capture/libraries/pspice path. Click OK once you have the input and output libraries and paths specified.

FIGURE 7-36
The PSpice Create Parts for Library dialog box for Capture parts.

Note: A Capture part library (bipolar.olb) already exists in the Tools/Capture/Library/PSpice folder for the bipolar PSpice model library (bipolar.lib). The bipolar library is used here for demonstration purposes. If you perform the following

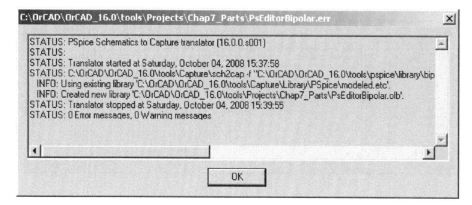

FIGURE 7-37
New part creation
information box.

procedure on an existing library, save the new library to a temporary or user folder so that you do not overwrite the existing library.

When the Model Editor has finished, it will display an information box similar to the one in Figure 7-37. If the Model Editor is successful at generating the Capture part library, the last line will say 0 Error messages, 0 Warning messages. Click OK to close the information box.

The PSpice Model Editor generates Capture parts with correctly named and numbered pins, but the parts are generic boxes because the PSpice Model Editor describes only how the parts function, not what they look like. You need to use the Capture Part Editor to modify the graphical appearance of the parts, as described in the earlier examples. To view the new part library, start Capture and select Open → Library… from the File menu in the Capture session frame.

MAKING AND/OR OBTAINING PSPICE LIBRARIES FOR MAKING NEW CAPTURE PARTS

Before you send a final board design to be manufactured, at some point in the design process you will want to simulate your design. A detailed explanation of PSpice model development and simulation process is outside of the scope of this text and many references are available on the subject. But in the interest of completeness, a brief explanation of how to add PSpice models to your Capture parts is discussed here.

PSpice contains many libraries, but manufacturers continually design new parts, which may not be included with your version of PSpice, so eventually you will want to be able to develop your own Capture parts that have simulation capabilities (and ultimately will be used in a board design).

A PSpice library contains primitive models (such as resistors and diodes) and subcircuit models (such as logic gates and op amps). Primitive models are basically limited to behavioral descriptions of single devices. Subcircuit models can describe the behavior of a single device, multiple devices, or complete circuit designs. Unless specifically noted, *model* in this text refers to both primitive and subcircuit models in general.

With regard to PSpice model libraries, the models that the libraries contain begin as simple text files that have a .mod extension. A model is added to a library by importing it into the PSpice (.lib) library. Once the model is imported into the library, the original .mod file is no longer needed and can be deleted or archived in a separate folder.

DOWNLOADING LIBRARIES AND/OR MODELS FROM THE INTERNET

The easiest way to make a new PSpice model library is to download it from the manufacturer's Web site (when available). You can often download complete libraries, but sometimes OEMs provide only individual models that you can add to your own libraries.

Here is an example. Suppose you were to use an RS2A fast recovery diode (Diodes Incorporated) in your design and need its PSpice model. By going to the Diodes Incorporated Web site (www.diodes.com/products/spicemodels/index.php), you can obtain the model for the RS2A diode as shown in Figure 7-38. Copy the text as is from the Internet Explorer window and paste it into any simple text editor such as Notepad. Save the text file with a .mod file extension (e.g., RS2A.mod) to a convenient folder that is set up for your collection of model files.

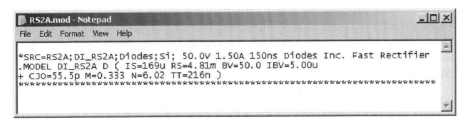

FIGURE 7-38
Spice diode model downloaded from the Web.

Next start a new PSpice library for Diodes Incorporated parts. From the PSpice Model Editor File menu choose New. You will be presented with a blank Models List window pane and an unsaved library (Untitled1.lib). From the Model menu, choose Import. From the Open File dialog box, navigate to and select the RS2A.mod file. Click Open. The Models List window should now contain the new model as shown in Figure 7-39. Save the new library in a user folder with a name such

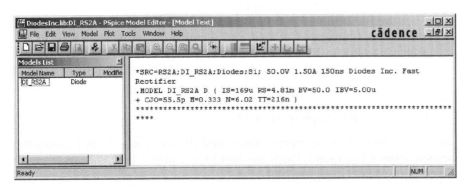

FIGURE 7-39
New RS2A PSpice model in the Model Editor.

as DiodesInc.lib (select Save As... from the File menu). You now have a PSpice library (with only one part in this case) for which you can make a new Capture part. To make a Capture part for the new library, use the Export to Capture Part Library... command from within the PSpice Model Editor as described under Method 4 or use the Generate Part command from within Capture as described under Method 3.

MAKING A PSPICE MODEL FROM A CAPTURE PROJECT

In the event that a PSpice model (or even a generic spice model) cannot be located, you can make your own *.mod file. There are two ways to do this, depending on the type of model you are trying to make. If you need to make a primitive model for a device (e.g., a diode or P-channel MOSFET transistor), you need to compose a .mod file using a text editor then import the model into a library as just described. To make an accurate model, you need to be familiar with model parameters for the part and the "code" that PSpice understands. This is not described here, but you can read about the details in the PSpice Reference document (pspcref.pdf) located in the OrCAD/doc folder. As a starting point you can also look at examples from the PSpice Breakout library (breakout.lib).

If you need to make a model for a nonprimitive part (an IC or a transformer, for example), you can compose a subcircuit model without having extensive knowledge of model parameters or PSpice code. You simply "draw" the circuit using Capture, have Capture write the subcircuit model for you, then save it as a PSpice library. Once you have the .LIB file, you can use Method 3 or 4 to make a Capture part with the model attached to it and attach a PCB Editor footprint if so desired.

An example of how to make a PSpice model and subsequent Capture part is given here for a transformer with a single primary winding and a center-tap secondary, similar to the one in Figure 7-3 (except that the transformer in this example has a PSpice model—a PSpice template—associated with it). Using this procedure you can specify the inductance and DC resistance of the windings and the coupling between the windings for a specific part as described in a data sheet.

The basic process is as follows:

1. Use Capture to draw a circuit that you can simulate. The circuit will consist of inductors, resistors, and coupling coefficients.
2. After the transformer "circuit" is simulated to verify it behaves correctly, simulation sources are deleted and hierarchical ports are added to the schematic, which will become the leads (pins) of the transformer.
3. Use Capture to create a PSpice library netlist file (.lib) for the circuit. Method 3 or Method 4 is used to generate a Capture part from the .lib file. In this example we use Method 4 so that we can look at and modify the model prior to making a Capture part for it.

To make a .subcircuit model, open Capture and choose New → Project... from the session frame's File menu. From the New Project dialog box, choose Analog or Mixed A/D as shown in Figure 7-40.

FIGURE 7-40
New Project dialog box
for a PSpice project.

Select the location of the new project using the Browse... button at the bottom of the dialog box. If you plan on making more models in the future, it is a good idea to create a new folder just for model development. Once you have your models fully developed and tested, you can copy the finished libraries into the normal Capture and PSpice library folders.

After you click OK, the Create PSpice Project dialog box (Figure 7-41) will be displayed. Check the Create based upon an existing project radio button and select either the empty.opj or the simple.opj project template. Different templates may be displayed depending on which version you have. For what we are going to do in this example, it really does not matter which template you start with. Click OK.

In the Project Manager window, expand the Design (.dsn) icon if it is not already expanded, and double click the SCHEMATIC1 folder (see Figure 7-42). The name of the design (DesignName.dsn) will become the default name of the PSpice library (DesignName.lib), and the name of the root folder (FolderName) will become the

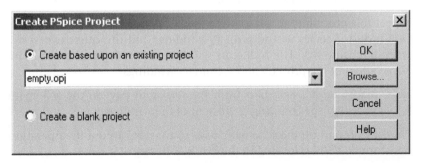

FIGURE 7-41
The Create PSpice
Project dialog box.

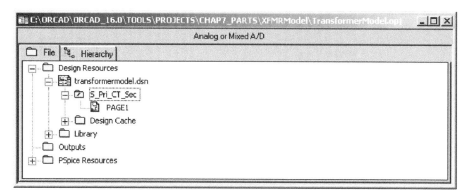

FIGURE 7-42
Project Manager view of
design.

name of the PSpice part model (.subcircuit *FolderName...*), so you want to change
the name of the folder from SCHEMATIC1 to the name you wish for your part.

To rename the schematic folder, select the folder by left clicking once, then right
click, and select Rename from the pop-up menu. Change the name of the schematic folder to the name that you want the part to have (e.g., S_Pri_CT_Sec for
single-winding primary, center-tap secondary).

Double click on the PAGE1 icon to display the schematic page (see Figure 7-42).
Delete any parts or text provided by the template by dragging a box around (or
across) the parts to highlight them, then hit the Delete key on your keyboard.

Place Part button

Place four resistors from the Analog library (which have PSpice models associated with them) on the schematic page (see Figure 7-45 later). The resistors are
used to simulate the DC resistance of the windings and a dummy load resistor.
You can get the resistors from the Place Part dropdown list located on the toolbar at the top of the window frame or by selecting the Place Part button, ▣, on
the toolbar at the right of the schematic page. If you use the Place Part button,
you will be presented with the Place Part dialog box shown in Figure 7-43. Select
ANALOG from the Libraries: list then scroll down the Part List: and select part R.
Notice that the parts in the Analog library have PSpice models and footprints
associated with them, as indicated by the PSpice and Layout (not PCB Editor)
icons located under the part preview box in the lower right corner of the dialog
box. Click OK. Click on the schematic page in four places to place four resistors
(see Figure 7-45 later for reference).

Repeat this procedure to place three inductors (part L) on the schematic page.
One inductor serves as the primary winding, and the other two serve as the
secondary windings. The inductors will be used to define the inductance (the
turns ratios) of the primary and secondary windings.

Next, place one K_Linear part from the Analog library on the schematic page.
K_Linear defines the coupling coefficient of the windings.

Place a VSIN part on the schematic page so that we can test the operation of the
transformer. VSIN is located in the SOURCE library.

FIGURE 7-43
Place Part dialog box.

Finally, place three zero (0) ground references on the schematic page. Click the Place Ground tool, ⬇, then select 0/SOURCE from the Place Ground dialog box, as shown in Figure 7-44. For schematic entry and PCB Editor, you can use any ground symbol, but for PSpice simulations you have to have at least one 0 reference ground.

Place Ground tool

Position and rename the components, wire the circuit, and change the values of the components as shown in Figure 7-45. To change the reference designators (e.g., R1 or Rp) or the component values (e.g., 10 Ω), double click the property you want to change. In the Display Properties dialog box, enter the appropriate value and click OK.

FIGURE 7-44
Place Ground dialog box.

You can also modify a part's properties by double clicking the part (or click once to select it then right click and select Edit Properties... from the pop-up menu). A Property Editor spreadsheet will pop up, as shown in Figure 7-46. If the properties are listed across in rows instead of vertically in columns, you can change the view by selecting the upper left-hand (corner) cell to highlight the entire spreadsheet, then right click and select Pivot from the pop-up menu. You can also specify how many items are listed by using the

FIGURE 7-45
Schematic of the linear step-up transformer.

FIGURE 7-46
Part Property Editor spreadsheet (vertical view).

Filter by: dropdown list just above the spreadsheet cells. To display all pertinent properties, select the <Current properties> option located at the top of the list.

To modify and display the K_Linear coupling properties, double click the coupling part to bring up the spreadsheet. Make sure that either the Current properties or the OrCAD-PSpice filter option is selected. In the L1 cell, type Lp1 (or whatever you named your primary coil) then click the Display button, which is located just above the spreadsheet cells. Enter Ls1 in the L2 cell and Ls2 in the L3 cell and click the Display button for both of these parameters too. The L1, L2, and so on cells establish which coils are coupled as part of the transformer. You can have additional or separate coupling coefficients for different sets of coils, but for this example we have all three equally coupled together using the single linear coupler. Once you finish, close the spreadsheet by clicking the X button in the upper right-hand corner of the spreadsheet.

Now we need to simulate the part with PSpice. You can perform several types of simulations with PSpice. In this example we perform a time domain analysis so that we can see the AC waveforms at the input of the transformer and at the load resistor.

FIGURE 7-47
Simulation Settings
dialog box.

To test the circuit we need to set up a simulation profile. *To set up a simula-tion profile*, choose Edit (or New) Simulation Profile from the PSpice menu. In the Simulation Settings dialog box (see Figure 7-47), select the Analysis tab. In the Analysis type: dropdown list, select Time Domain (Transient). Set the rest of the parameters as shown in Figure 7-47, then click OK.

Place voltage markers on the circuit to specify which voltages to display in the PSpice probe window. Click the Voltage Probe button, , on the toolbar and place a probe on the wire coming from VSIN (green marker) and one (red marker) on the wire going from the secondary winding to the load resistor (RL).

Voltage Probe button

Start the simulation by clicking the Run PSpice button, ▶. PSpice runs the sim-ulation and displays the results in a probe window, as shown in Figure 7-48. The voltage curves show that the transformer functions as a 1:2 step-up trans-former, since the output (red marker curve) is twice as high as the input (green marker curve). Additional tests (e.g., frequency response) could be performed to validate the circuit model further, but these are not discussed here.

Run PSpice button

Since the circuit model has been validated, we now prepare to make a PSpice model of the circuit. Begin by deleting the VSIN source, the load resistor RL, and all the ground references.

Add ports to the schematic, which will serve as the leads of the transformer. Click the Place Port tool, 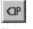. Select the PORTBOTH-L port from the Place Hierarchical Port dialog box, as shown in Figure 7-49. All the port symbols behave identi-cally in Capture, PSpice, and PCB Editor. The only difference is the appearance

Place Port tool

FIGURE 7-48
PSpice probe window for the center-tap transformer.

FIGURE 7-49
Place Hierarchical Port dialog box.

FIGURE 7-50
Hierarchical ports in the circuit design.

on the schematic. Since the transformer is a passive device, we use the symbol that indicates that an applied signal can go in either direction. Click OK and place five of the ports on the schematic page (two for the primary and three for the secondary).

FIGURE 7-51
Create Netlist dialog box.

Reposition, connect, and label the ports as shown in Figure 7-50.

Display the Project Manager window by minimizing the schematic page or *ProjectName*.opj from the Window menu. Select the Design icon then select Create Netlist from the Tools menu.

In the Create Netlist dialog box (Figure 7-51), select the PSpice tab. Check the Create SubCircuit Format Netlist box and the Descend radio button. A default name will be given to the netlist. Modify the path and name as desired. Make sure the name ends with the .LIB file extension. Click OK.

You now have a PSpice library file with one model (the transformer) in it. You can use Method 3 or 4 to generate a Capture part library from this model or add it to an existing part

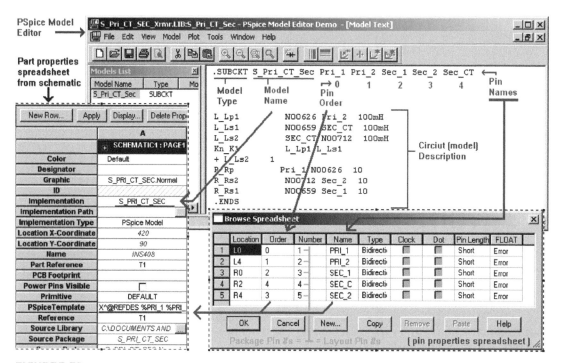

FIGURE 7-52
PSpice transformer model generated from a Capture project.

library. To complete the example we use Method 4 so that we can take a look at the PSpice model generated by Capture.

Start the PSpice Model Editor again. From the File menu, select Open and navigate to the transformer library that you made previously. Click on the S_Pri_CT_Sec model to display the model in the editing window, as shown in Figure 7-52. Notice that the model type is .SUBCKT (a subcircuit). The name of the library is whatever you specified as the netlist file in the Create Netlist dialog box, and the name of the part is the schematic folder name in the Capture design (see Figure 7-42). At this point you could modify the part to add specific requirements that will carry forward into the Capture part. Once the model editing (if any) is completed, generate the Capture part as described previously, using Method 3 or Method 4.

Note that, as shown in Figure 7-52, the pin names and order in the Capture part (as indicated in the Pin Properties spreadsheet) and the PSpice Template (as indicated in the Part Properties spreadsheet) must match the pin names and order in the PSpice model exactly or simulations will fail. The Implementation name in the part properties spreadsheet must also match the model name in the PSpice model file. And finally, the part's pin numbers must match the pin (padstack) numbers in PCB Editor, which is governed by the part's data sheet. If the part's pin number is a letter instead of a number (e.g., A, for the anode of a diode), then the pad in PCB Editor must be named accordingly or the ECO (engineering change order) will fail.

ADDING PSPICE TEMPLATES (MODELS) TO PREEXISTING CAPTURE PARTS

Rather than using Method 3 or 4 to make a new part from the transformer PSpice library, you might wonder why we did not just add the PSpice model to the transformer already created in the first example (for which Method 1 was used). In older versions of OrCAD, this was somewhat of a challenging task (e.g., you have to know what "X^@REFDES %A %B %Y %VCC %GND @ MODEL PARAMS:\n1 IO_LEVEL5 @IO_LEVEL MNTYMXDLY5 @MNTYMXDLY" means). Fortunately, it is a simpler matter with the newer versions.

In this example we see how to add an existing PSpice model to an existing Capture part. We add a basic capacitor model to one of Capture's capacitor parts that has no model associated with it.

A basic capacitor (part C or CAP) from Capture's discrete.olb library has no PCB Editor footprint or PSpice model associated with it. We add a PSpice model to the part now. The location of the basic PSpice capacitor model that we use is in the PSpice Breakout library (breakout.lib).

To add a PSpice model to an existing Capture part, start Capture and select Open → Library and select the library with the part to which you want to add a PSpice model (use discrete.olb for this example). Find the capacitor (C or CAP, for example) in the Capture Library Part Manager and click the part's icon to select it. Right click and select Associate PSpice Model... from the pop-up menu.

At the Import Model Wizard dialog box (Figure 7-53), use the Browse... button in the upper right-hand corner to find the PSpice model you want to associate

FIGURE 7-53
Matching parts with the Model Import Wizard.

FIGURE 7-54
Mapping pins with the
Model Import Wizard.

FIGURE 7-55
PSpice model success-
fully added to a Capture
part.

with the Capture part. The wizard automatically searches through the PSpice library and lists all models in the Matching Models window that have the same number of pins as the Capture part you selected in the Part Library Manager. Select the CBREAK model and click Next.

The wizard then displays the Pin Matching tool shown in Figure 7-54. This is where you connect the PSpice model pin to the Capture part pin. Click Finish when the pins are matched. The information box shown in Figure 7-55 should be displayed, indicating that the capacitor now has a PSpice model attached to it.

CONSTRUCTING CAPTURE SYMBOLS

Four types of symbols are used in Capture: (1) power/ground symbols, (2) off-page connectors, (3) hierarchical ports, and (4) title blocks.

To make a new power symbol, open the capsym.olb library. Select the Library icon under the Library folder. From the Design menu, choose New Symbol. In the New Symbol Properties dialog box (Figure 7-56), enter the name and select the

FIGURE 7-56
Making a new power
symbol.

appropriate radio button from the Symbol Type group box. Click OK. A part editing window and drawing toolbar will be displayed. Use the graphics tools to make the symbol. Close the editing window and save the symbol. Save and close the library.

The construction methods for power symbols, off-page connectors, and hierarchical ports are similar (although each has its own functionality). The title block is different from the other three. The two types of title blocks are default and optional. Every new project has a default title block, which you can specify for each project by going to Options → Design Template... on the schematic page menu. You can also add optional title blocks by going to Place → Title Block..., also on the schematic page menu. The title blocks are located in the capsym library. You can construct your own title blocks and save them with the others or make your own library of power symbols, title blocks, and the like.

FIGURE 7-57
Example of a new title
block used as a parts
list.

To make a new title block, open or make a new Capture library, select the Library icon, and from the Design menu, select New Symbol.... At the New Symbol Properties dialog box (Figure 7-56), select Title Block and click OK. Just as with the power symbols, you will be presented with a part editing window. Use the graphics tools to construct the box (Figure 7-57 shows an example of a parts list) and the Text tool to add titles and headers. To enter text that can be modified on schematic pages double click in a free area within the window to display the User Properties dialog box. Click the New... button to create new text objects. The name field is required but you can either leave the value field empty or enter a default value. Click OK. To make an object field visible select Part Properties from the Options menu and click the Display button in the User Properties dialog box for the field of interest. After you have saved the block, you can add it to any schematic page.

That completes the chapter on making Capture parts and symbols.

Making and Editing Footprints

Footprints provide a means of physically attaching components to your PCB, and they provide electrical connectivity as defined by the netlist generated in Capture. Footprints in PCB Editor are known by several names, including *symbols*, *components*, and *packages*. The PCB Editor library has over 400 symbols, but there will be occasions when you will need to make your own. This purpose of this chapter is to

1. Introduce PCB Editor's symbol libraries.
2. Discuss the composition of a footprint.
3. Provide a detailed explanation of padstacks.
4. Demonstrate how to create a padstack using the Padstack Designer.
5. Provide footprint design examples.
6. Provide design examples for mechanical symbols and flash symbols.

INTRODUCTION TO PCB EDITOR'S SYMBOLS LIBRARY

The footprints are located in the OrCAD_16.0\share\pcb\pcb_lib folder. In that folder are two more folders: a devices folder and a symbols folder. The devices folder contains text files that define logic information for certain component types, which are not discussed here. The symbols folder contains eight types of files. Table 8-1 lists the files and explanations of their functions.

Some files have the same file name but different file extensions, for example, the cap300.dra file and cap300.psm file. In general you work with the .dra files with PCB Editor, then PCB Editor creates the other files that it actually uses in a board design.

For each type of file, PCB Editor operates in a particular mode. When you open an existing drawing (.dra) file, PCB knows which mode to be in as it reads the file and knows the corresponding symbol file (.bsm, .psm, etc.). When you start a new drawing (.dra) file, you select the mode by choosing a drawing type from a list. When you finish working with the .dra file and save it, PCB Editor automatically creates the correct symbol file. This is demonstrated in the examples that follow.

Table 8-1	Library Files and Their Extensions and Functions
File extension	**File type**
name.bsm	A *mechanical symbol* file (derived from the .dra file with the same name)
name.dra	A PCB Editor drawing file
name.fsm	A *flash symbol* file (derived from the .dra file with the same name)
name.log	A text file containing the date and time the symbol or padstack name was last modified or saved
name.osm	A *drawing format* file (derived from the .dra file with the same name)
name.pad	A *padstack definition* file generated/used by Padstack Designer
name.psm	A *package symbol* file (derived from the .dra file with the same name)
name.ssm	A *custom pad shape file* (derived from the .dra file with the same name)

Symbol Types

Package symbols (*name*.dra and *name*.psm) are the component footprints used on your board design that are assigned in Capture. After finishing the footprint drawing, the .psm file is generated when you save the drawing.

Mechanical symbols (*name*.dra and *name*.bsm) are objects, such as mechanical pins, which are not connected to a net, mounting holes on boards, or predefined board outlines.

Flash symbols (*name*.dra and *name*.fsm) are used to define thermal reliefs, which are used to connect padstacks to power and ground planes when the planes are defined as negative plane layers. Positive plane layers do not need flash symbols, because PCB Editor automatically creates thermal reliefs based on trace width and spacing settings as defined in the Constraint Manager (more on that in PCB Design Examples).

Drawing formats (*name*.dra and *name*.osm) are predefined drawing templates that follow standard paper sizes. The six drawing formats range from drawing size A (8.5 × 11 in.) to size E (44 × 34 in.). You can add these to your board designs to give them a professional look and follow industry drawing standards.

Custom pad shapes (*name*.dra and *name*.ssm) are specially designed shapes used to create padstacks that have pad shapes other than round or rectangular/square pads.

Padstack files (*name*.pad) define surface mount and through-hole padstacks and include conductor information, such as inner and outer pad shape and size, and thermal relief sizes and physical information, such as drill diameter and soldermask and pastemask shapes and sizes. Padstack definitions are handled using an application called Padstack Designer, which can be launched from within PCB Editor or launched separately from the Windows Start menu. Padstack Designer is demonstrated in the design examples that follow.

COMPOSITION OF A FOOTPRINT

A footprint symbol is made up of pins (padstacks), graphics (for outlines etc.), and text. As an example Figure 8-1 shows the through-hole resistor symbol res400.dra found in the symbols library. To view it open PCB Editor and navigate to the symbols folder, select the .dra file type and select the res400.dra. Table 8-2 describes what the objects are and to what class and subclass each object belongs. As described earlier the pin (padstack) is actually defined by the Padstack Designer and simply inserted into the symbol drawing along with the graphics and text objects.

FIGURE 8-1
Anatomy of a footprint symbol.

Table 8-2   Types of Symbol Objects and Their Classes

	Object type	Class	Subclass
Pin (padstack)			
Pad	Conductor	Pin	Top
			Bottom
Soldermask_	Nonconductor	Pin	Soldermask_Top
			Soldermask_Bottom
Drill symbol	Drill symbol figure	Manufacturing	Ncdrill_Figure
Pin number	Text	Package geometry	Pin_Number
Graphics			
Place outline	Filled rectangle (frectangle)	Package geometry	Place_Bound_Top

Continued

	Object type	Class	Subclass
Silk-screen outline	Rectangle	Package geometry	Silkscreen_Top
Assembly outline	Line segment	Package geometry	Assembly_Top
Text			
Reference designator	Text	Ref Des	Silkscreen_Top
			Assembly_Top
Value	Text	Component value	Silkscreen_Top
			Assembly_Top
Tolerance	Text	Tolerance	Silkscreen_Top
			Assembly_Top
Part number	Text	User part number	Silkscreen_Top
			Assembly_Top

FIGURE 8-2
Elements of a pad-
stack from the PCB
perspective.

Padstacks

Figure 8-2 shows an example of a multilayer padstack from a PCB's perspective. Padstacks define every aspect of how a component's pins will be fastened to the PCB and how traces will be connected to them. Padstack definitions specify areas for copper pads on outer and inner routing layers, thermal reliefs and clearance areas in plane layers, openings in soldermasks, and solder paste (optional). From the PCB designer's perspective, the drill hole may also be considered part of the padstack definition, but the copper used to plate through holes and vias is not. This is because the plating thickness is controlled by the board manufacturer and is typically insignificant relative to the drill and lead diameters. More about the lead-to-hole diameter is discussed in a later example and Chapter 5.

- Top solder paste
- Top soldermask
- Top copper pad
- Plane layer connected to padstack plating with a 'thermal'
- Inner copper pad
- Plane layer isolated from padstack plating
- Plated through hole
- Bottom copper pad
- Bottom soldermask
- Bottom solder paste

Figure 8-3 shows how a padstack is actually displayed in PCB Editor. When a particular layer is made active (using the Options pane), the pad on that layer is brought to the front of the view, while pads on inactive layers are pushed behind. Padstacks have clearance areas defined that remove copper from plane layers and are visible when the plane layers are visible. The pin number of the padstack is displayed if the Pin_Number subclass box is checked in the Package Geometry class in the Color Dialog panel. Drill holes are also displayed if the Display Plated Holes option is enabled in the Design Parameter Editor as shown in Figure 8-4.

FIGURE 8-3
A padstack as seen in PCB Editor.

Graphical Objects

Graphical objects (lines, arcs, etc.) are placed on footprints and board layouts to show information not defined by padstacks (such

FIGURE 8-4
Controlling drill hole visibility.

as the outline of a part). The objects may be visible on the final board or visible only in the design files. Objects that are visible on the board include silk-screen markings (component outlines, for example), while objects visible only in the design files include things like board outlines, assembly outlines, and boundaries (e.g., keep-out areas where traces and components are not allowed).

Several types of objects can be used in PCB Editor, but usually only three or four types are used when making footprints (i.e., detail objects, place boundary outlines, and occasionally copper areas). Detail objects are used to indicate silk-screen markings that will be visible on the board and on assembly layers to provide information during assembly. Place boundary outlines are used by the DRC utility to maintain required distance between parts while laying out the board. On routing layers, copper areas can be used as heat spreaders or mini-ground planes for components that require them. Examples of how to use these objects are given in the PCB Design Examples.

Text

Like graphics, text objects can be visible on a PCB or visible only within the design files. Text objects on PCBs are often part of the silk screen and may be used for such purposes as component reference designators (part of the footprint silk screen) and displaying board serial numbers or design revisions (part of the board silk screen). Text that is visible only in the design files might be placed on one of the assembly or documentation layers to show board dimensions or special manufacturing instructions.

Minimum Footprint Requirements

As shown in Figure 8-1 many types of text and graphics objects can be used when you make a new footprint. Not all the items shown in the figure are required, but four minimum objects are required on a footprint design. Packages must have the following elements: (1) at least one pin, (2) at least one reference designator, (3) a component outline, and (4) a place-bound rectangle. As shown in Table 8-2, reference designators and component outlines are placed on the silk-screen or assembly layers (usually both), and the place bound rectangle is on the Place_Bound_Top (or Bottom) layer.

Optional Footprint Objects

The following is a list of other elements you can add to a package symbol during the symbol building process:

- Device type (text for the component device type).
- Component value (text for the component value).
- Tolerance (text for the component tolerance).
- Component height (text for the physical height of the component).
- User part number (text for the package part number).
- Route keep-out shapes, identifying areas where etch is not allowed.
- Via keep-out shapes, identifying via keep-out areas.
- Etch (etch lines, arcs, rectangles, shapes, or text added to the symbol).
- Vias.

For further reading on design requirements, please see Chapter 4 for a list of industry standards and Appendix B for more information on packages and footprints.

Good sources for package information can be found at the following Web sites:

- IPC Web site (www.ipc.org)—free Land Pattern Viewer based on the IPC-7351 standard.
- www.fairchildsemi.com/products/discrete/packaging/pkg.html.
- www.national.com/packaging/parts/.
- www.intersil.com/design/packages/hermetic.asp.
- www.diodes.com/datasheets/ap02002.pdf.

INTRODUCTION TO THE PADSTACK DESIGNER

The two basic kinds of padstacks are through-hole padstacks (for leaded components and vias) and surface-mount pads. Through holes are made using padstacks that allow connections from any one layer to any other layer (as in Figure 8-2). Surface-mount pads are isolated from all layers except for the top layer and therefore do not use drilled and plated holes. Surface-mount pads may be on the top or bottom layer (or both in the case of edge connector footprints). Connecting a surface-mount pad to another layer is accomplished using a special padstack called a *fan-out via* (also called a *stringer pad*). A fan-out via is usually not part of the footprint; it is added to a PCB when placing parts and routing traces. You will see how to do this in the PCB Design Examples.

Through-hole padstacks are often named by their shape and size. The seven pad geometries are null, circle, square, oblong, rectangle, octagon, and user-defined shape. For circular through-hole padstacks, one naming convention used is *padXcirYd*.pad, where *pad* indicates padstack, *X* is the pad outer diameter, *cir* is the shape (*sq* for a square top pad or *rec* for rectangular), and *Y* is the pad inner diameter—the drill size (see Figure 8-5). For example, if the outer dimension of a round pad is 62 mils (0.062 in.) and the drill hole is 25 mils, the pad name would be pad62cir25d.pad.

FIGURE 8-5
Pad dimensions.

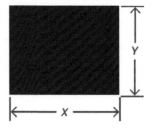

For surface mount pads a typical naming convention is *smdXrecY*.pad (or *smdX_Y*.pad), where *smd* indicates the padstack is for a surface mount device, *X* is the pad width, *rec* (or _) indicates a rectangular pad, and *Y* is the pad height.

Through-hole and surface-mount pads are designed and modified using the Padstack Designer (see Figure 8-6 later). There are two ways to launch the Padstack Designer.

Padstack Designer can be launched from within PCB Editor or separately from the Windows Start menu. In the PCB Design Examples, it will usually be launched from within PCB Editor. In the footprint design examples that follow, Padstack Designer will be launched from the Windows Start menu, where detailed instructions on how to use it are provided in the examples. Pad Designer has a Parameters tab (Figure 8-7[a]) and a Layers tab (Figure 8-7[b]). Each is described briefly next with the figure.

Padstack Designer Parameters Tab

The Parameters tab (Figure 8-6) is where the drill diameter and related parameters are set.

FIGURE 8-6
Padstack Designer
Parameters tab.

The Type options are used to specify whether the padstack is a through-hole, blind or buried, or a single (surface-mount) type.

The Internal Layers options are used to control how unconnected pads on internal layers are processed during photoplotting. Typically the Fixed option is selected for through-hole padstacks unless you wish to create padstacks that have the pads suppressed on layers having no traces to that padstack.

The Units selection box allows you to select either English units (mils and inches) or metric units (millimeters, centimeters, or microns). You can also specify the accuracy of the units in the decimal places box.

The Multiple drill options box allows you to create padstacks with more than one drill hole (e.g., heat spreaders).

The Drill/Slot hole selection area is where the drill details are entered. From these options you specify the type of hole (circular or slotted), whether the hole is plated or nonplated (e.g., for mounting holes), the drill diameter, and tolerance. You can even choose among different nonstandard drill operations, such as laser-drilled or plasma-etched holes.

The Drill/Slot symbol selection area allows you to choose a drill symbol for the padstack, which is then displayed on a drill table on which you can display on your board design.

Padstack Designer Layers Tab

The Layers tab is where the pad shapes and sizes are set. A comparison between a through-hole padstack and a surface-mount padstack is shown in Figure 8-7.

In Figure 8-7(a) the layer stack-up definition consists of a beginning layer (TOP), a default internal layer, and an ending layer (BOTTOM) and the associated soldermasks. In Figure 8-7(b) the layer stack-up definition consists of just a beginning layer (TOP) and the top soldermask. A buried via will have beginning and ending layers, but they will be strictly inner layers.

The bottom of the Layers tab provides places to specify pad shapes on routing layers (regular pads) and plane layers (thermal reliefs and antipads). The pad shapes (and antipads, etc.) are individually assigned for each layer. The dropdown lists and number boxes change depending on which layer is selected. A layer is selected by left clicking on it. The layer currently selected is listed at the bottom.

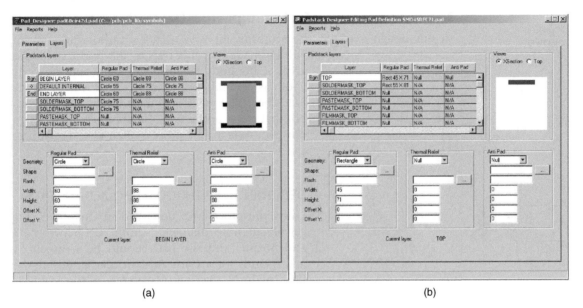

(a)   (b)

FIGURE 8-7
Padstack Designer Layers tab: (a) for through hole, (b) for surface mount.

The default thermal relief for the padstacks provided with the software is circle, and it is usually the same size as the antipad. When you create a plane area on a positive layer PCB Editor automatically creates the thermal relief spokes that connects the pad to the plane. The spoke widths are based on the trace width constraints set in the Constraint Manager for the related net.

Connections to negative planes are different and are accomplished through the use of thermal flashes. Thermal flashes are made using PCB Editor and stored in the symbols library. Then when you need to assign a flash symbol to a pad-stack for a certain layer, you simply click the Browse... button, , and chose the appropriate flash symbol. Creating flash symbols is covered later and using them is demonstrated in the PCB Design Examples.

Browse... button

FOOTPRINT DESIGN EXAMPLES

Three footprint design examples are given. Before starting any footprint design, it is best to have the correct padstacks predefined, as it makes the design process go more smoothly. In each example Padstack Designer is used to locate existing padstacks or design new ones.

The first design example demonstrates how to design a through-hole footprint from scratch. The second design example demonstrates how to modify an existing, surface-mount footprint to create a new one. The last example demonstrates how to use the Symbol Wizard to design a high pin-count pin-grid array footprint.

FIGURE 8-8
A through-hole
component.

Example 1. Design of a Through-Hole Device from Scratch

In this design example, the footprint for a ¼-W, 20%, carbon-film resistor will be made. For this resistor the lead diameter is 25 mils, the body length is 250 mils, and the body width is 100 mils. An example is shown in Figure 8-8. The component is similar to the res400 symbol that exists in the symbols library, but we design our own to learn the process, then we can compare the two.

(Units in mils)

DESIGNING THE THROUGH-HOLE PADSTACK

Using the procedures described in Chapter 5, the calculated hole diameter should be between 33 and 56 mils and the pad should be between 51 and 74 mils for a Class A board design (using the IPC method). Several padstacks in the symbols library meet this requirement, but one that does not exist will be chosen so that the design process can be demonstrated. So, if we decide that, in addition to these requirements, we want an annular ring width of 10 mils, then a padstack with a drill hole diameter of 42 mils and a pad diameter of 62 mils suffices. The typical naming convention in PCB Editor for a pad with these dimensions is pad-62cir42d, which does not exist in the native symbols library. To create this padstack, launch Padstack Designer by selecting Start → All Programs → OrCAD 16.0 → OrCAD PCB Editor Utilities → 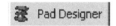. A blank, unnamed padstack will be opened.

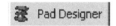
Pad Designer button

In the Parameters tab, select Through as the type and Optional internal layers. Enter the remaining values as shown in Figure 8-9. You can choose any of the available drill symbol figures. The size of the drill figure is not critical but is typically about the size of the drill diameter or a little smaller. Note that the drill symbol will not be visible initially when you ultimately place the component on the board. The drill symbol becomes visible only after you place a drill table in the design.

Before moving on to the next step, it would be a good idea to save your work to this point. When saving padstacks that you plan to use in active designs, you need to save them to the PCB Editor symbols folder or your working folder. Otherwise, if the design is just for practice, you can save it anywhere you like. When you save it for the first time, select Save As… from the File menu. After that you can just select Save from the File menu to save changes.

Next, select the Layers tab. Enter the values shown in Figure 8-10. To enter values for a particular layer select the layer by left clicking the row in one of the three columns to the right. If you left click the row in the name column, the

FIGURE 8-9
Setting up the Parameters tab for a new padstack.

FIGURE 8-10
Setting up the Layers tab for a new padstack.

name will be highlighted so that you can change its name. The active layer will be the Current Layer: as indicated at the bottom.

- **Regular pads**. The values on the internal layers can be identical to the outer layers but often they are slightly smaller (anywhere from 0 to 20 mils difference in diameter depending on the size).
- **Thermal reliefs**. The native padstacks in the symbols library typically use circles that are the same size as the antipad. Circles work for positive planes but not for negative planes. For negative planes you need to use Flash symbols. Flash symbols are described later, so for the time being leave the thermal relief as a circle. The procedure for defining padstacks that have no thermal reliefs (i.e., completely connected to a plane) is described later.
- **Antipads**. The clearance between a pad (or hole) and the surrounding copper (e.g., on a plane layer) should be similar to or larger than the trace spacing constraints you will likely use in your design. A typical clearance diameter is 10 to 20 mils larger than the pad diameter. Note that, if you make an inner pad much smaller than the outer layer pads, the antipad on the inner layer should still be larger than the largest pad in the padstack; otherwise undesirable capacitive coupling can occur between the larger pad and an adjacent plane layer (see the discussion in Chapter 5).
- **Soldermasks**. Soldermask openings are usually a little larger than the outer pads so that registration tolerance errors will not cover any of the pad. The exact amount really depends on the capabilities of your board manufacturer, but in general the soldermask opening is typically 10 to 20 mils larger than the outer pads.

When you complete the entries shown in Figures 8-9 and 8-10 save your design.

DESIGNING A THROUGH-HOLE FOOTPRINT SYMBOL

When a new symbol drawing is first opened, the drawing area is quite large by default. So before opening a new drawing, it is a good idea to have an idea of what the footprint dimensions will be, so that it is known how large the drawing area should be. From the preceding description, the body length is 250 mils and the width is 100 mils. The footprint needs to be larger than that to account for the lead extensions, padstack dimensions and placement, and any other details that you might want to add to the drawing.

Using the procedure described in Chapter 5, the padstacks are placed 400 mils apart. The diameter of each pad is 62 mils, so the drawing area should be another 200 mils or so wider. Silk-screen and assembly details and text are also added, so at the very least the drawing area should be 1000 mils square. The next thing to consider is where the origin of the part will be. Through-hole components often have their origins on pin 1. Surface-mount components typically have their origins at the body center. Since this will be a through-hole component, we put the origin on pin 1. Starting at the lower left corner the (x, y) coordinate will be $(-1000, -1000)$, and we make the working area 3 in. wide by 2 in. tall.

To begin making a new footprint, open PCB Editor from the Windows Start menu. PCB Editor will open the last design you worked on by default. Start a new footprint drawing by selecting File → New... from the menu bar. The New Drawing dialog box will be displayed (see Figure 8-11). Enter a name in the Drawing Name: box for the new footprint and click the Browse... button to choose a location for the new footprint. If you have a project folder setup for your design, you can save it there. Otherwise save it in the symbols library. Select Package symbol from the Drawing Type: list (we will use the package symbol wizard in a later example) and click OK. Note that, at the beginning of the chapter, it was noted that PCB Editor will be in a specific mode depending on what type of drawing you are making; this is where the mode is set.

FIGURE 8-11
New Drawing package symbol dialog box.

When the new design is opened, a PCB Editor window will be displayed that looks just like a board design window.

The first thing that needs to be done is to set up a drawing area that is practical for this design. Select Setup → Design Parameters... from the menu bar. At the Design Parameter Editor dialog box, select the Design tab.

Select Other from the Size: list and enter the Extents values as shown in Figure 8-12. Click Apply (but not OK yet). Select the Display tab. Enable the grid (if it is not already enabled) by checking the Grids on box, and click the Setup grids... button to display the Define Grid dialog box shown in Figure 8-13. For the time being set spacing for all of the layers to 25 mils. Entering values in the All Etch area will automatically change all the Top and Bottom values so that you do not have to set them individually. Click OK to dismiss the grid dialog box then click OK to dismiss the Design Parameters dialog box and update the work area.

FIGURE 8-12
Setting up the drawing environment.

The next step is to place the padstacks. Select Layout → Pins from the menu bar to display the Select a padstack: dialog box as shown in Figure 8-14.

Scroll down until you find the padstack that was just made. If it is not displayed, make sure the Database box is checked. If you did not make the new padstack, you can just pick one that is close in size; pad60cir42d is a good choice. Click OK.

A padstack will be attached to your mouse pointer. Place the first padstack by clicking and releasing your left mouse button at location (0, 0). Another padstack will be attached to your mouse pointer. Place the second padstack at location (400, 0) then right click and select Done from the pop-up menu.

FIGURE 8-13
Setting the grid spacing.

FIGURE 8-14
Selecting a padstack.

Zoom Points button

Line tool

Zoom in so that the two padstacks take up about 75% of your display (use your mouse wheel if you have one or use the Zoom Points button, 🔍, on the toolbar).

Next, draw the body outline on the top silk-screen layer. To draw the silk screen, select the Add Line tool, 🗹, select the Package Geometry class and Silkscreen_Top subclass from the Options pane, and select a Line width of 10 mils or so, as shown in Figure 8-15.

Note: By default, lines (including rectangles) have a width of zero. When PCB Editor processes the artwork lines, those with a width of zero are ignored unless you specify a default width on the artwork control form. Additionally you can change the width of only lines, not rectangles. So if you want wide silk-screen lines, you need to use lines.

FIGURE 8-15
Setting silk-screen options.

Begin the body outline by clicking and releasing the left mouse button at coordinate (75, −50). With the Add Line tool still active change the coordinates display from Absolute to Relative by toggling the YXMode button located at the lower right-hand corner of the screen. Now move the pointer 100 mils in the Y direction (up) from the starting point (watch the mouse coordinates indicator at the bottom, the left arrow in Figure 8-16) and left click and release to place a vertex. Move the mouse 400 mils to the right (X direction) and place another vertex. Move the mouse 100 mils down and place another vertex. Move the cursor back to the starting point, left click to place the last vertex, then right click and select Done from the pop-up menu. The completed silk screen will be shown in white in Figure 8-20.

FIGURE 8-16
Coordinates tools.

Note: *To change the color of objects* in the design, select the Color button, , on the toolbar. Select the desired class from the left window, select the desired color from the color palette, and left click the colored square for the item you want to change in the subclass list at the right. Click the Apply button and then OK.

Color button

Next we add the body outline for the assembly layer by copying the silk-screen outline. *To copy an object*, select the Copy button, , on the toolbar. Left click the silk-screen outline. Move the mouse up and away from the existing outline and left click to place the new outline. Right click and select Done from the pop-up menu to dismiss the Copy tool. The copy is on the silk-screen layer, so we change it to the assembly layer.

Copy button

FIGURE 8-17
Changing outline properties.

To change the layer of an object, select Edit → Change from the menu bar. Move the mouse over to the Options tab to display the pane, if is not already displayed. Select the Assembly_Top sub-class under the Package Geometry class and change the line width to 1 (see Figure 8-17). Left click on the copied body outline to make the changes. The outline will be changed but remain high-lighted. Notice too that PCB Editor will tell you: Changed 1 items out of 1 items found, in the command window. Right click and select Done to complete the change.

Move the new assembly outline to the same position as the silk-screen outline. *To move an object*, select the Move button, , on the toolbar and left click the assembly outline; the outline will be attached to the pointer. Move the assembly outline over the silk-screen outline and left click to place it, right click, and select Done from the pop-up menu. The completed assembly outline will be shown in blue in Figure 8-20.

Move button

The next object to draw is the *place boundary outline*. Change the Non Etch grid spacing to 5 mils (Setup → Design Parameters → Display → Setup Grids). Begin the outline by selecting the Shape Add Rectangle button, , on the toolbar (see the 3-D model section in Chapter 10 for details on the differences between shape types). In the Options pane (Figure 8-18), select the Package Geometry class and the Place_Bound_Top subclass, and select Static solid as the fill type.

Shape Add Rectangle button

FIGURE 8-18
Place outline boundary options.

Draw a box around the footprint so that the outline is as close to the padstack and detail objects as possible, while keeping them inside the place outline. The place outline will be shown in Figure 8-20.

Now we add text objects to the footprint. Table 8-2 and Figure 8-1 list some of the text objects that can be placed in a design, but as stated there, only one is required (i.e., at least one reference designator). See Figure 8-20 while following the steps here. We begin by placing a reference designator on the silk-screen layer.

Label Refdes button

To place a reference designator on the silk-screen layer, select the Label Refdes button, ▦, from the toolbar. From the Options pane, select the Silkscreen_Top subclass (under the Ref Des class) and text block 2 for 31-mil high text (select a larger number for larger text). Left click above the place outline (off to the left a bit) to place the text marker. Type R* and then right click and select Done from the pop-up menu.

CONTROLLING TEXT SIZE

The text block selection applies predefined text formats. To view or modify the formats, select Setup → Design Parameters from the menu bar and select the Text tab. Click the Setup Text Sizes button to display the Test Setup dialog box (shown in Figure 8-19). From the dialog box you can change the existing text formats or add your own custom text blocks.

Add Text button

Next we place a couple of text objects on the assembly layer. Repeat the preceding process except select the Assembly_Top subclass and select text block 1. Left click above the place outline and to the right of the silk-screen reference designator. Again type R* and then right click and select Done from the pop-up menu. To add text object for the component value, select the Add Text button, ▦, from the toolbar, select the Component value class and the Assembly_Top subclass from the Options pane, and again select title block 1 if it is not selected. Left click below the reference designator on the assembly layer, type VAL, then right click and select Done. Repeat this process to enter a part number label. Again use the Add Text tool but select the User Part Number class and the Assembly_Top subclass. Select title block 1 and type ***, right click, and select Done.

The finished footprint is shown in Figure 8-20. Save the footprint by selecting File → Save from the menu bar. PCB Editor will tell you Symbol '*name*.psm' created in the command window.

That completes the first design example. You can compare this to the res400 symbol included in the native PCB Editor symbols library.

Example 2. Design of Surface-Mount Device from an Existing Symbol

The native PCB Editor library contains one surface-mount capacitor footprint symbol and one surface-mount resistor symbol for general use. These capacitor and resistor footprints are roughly the 2010 and 1805 sizes, respectively. Many

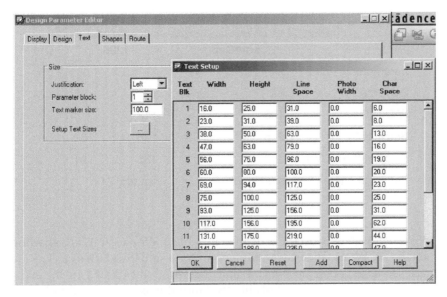

FIGURE 8-19
Text block options.

FIGURE 8-20
The finished through-hole
footprint symbol.

others are on the market. We could start from scratch and build our own; but since they are all similar, we start with an existing one and just modify it. In the next demonstration, we start with the existing smdcap footprint and modify it to make a size 1206 footprint.

MODIFYING AN EXISTING SYMBOL: DETERMINING DESIGN REQUIREMENTS

Start PCB Editor and open the smdcap.dra located in the symbols library. Table 8-3 compares the smdcap symbol to the requirements of a 1206 chip capacitor as defined by IPC using the IPC Land Pattern Viewer. See the end of this chapter for how to use this free tool. The CAP3216X140N model was used as a guide for this example. Note that, since the origin is the body center, the outline coordinates will be ±½ the listed values.

Table 8-3	Surface-Mount Capacitor Properties			
	smdCap		**1206**	
Parameter	X	Y	X	Y
PAD size	87	50	71	45
PAD C/C	195		118	
Body (L/W)	245	90	126	63
Place outline (L/W)	250	90	185	91
Padstack name	smd50_87		smd45_71	

The closest equivalent padstack as defined in Table 8-3 for a size 1206 capacitor is smd44rec72. We use that as a starting point for a smd45rec71 padstack. Start Padstack Designer, open smd44rec72 and Save As... smd45rec71 in the symbols library. Since this is an smd padstack, you can leave the Parameters tab alone, but select the Layers tab and change top and soldermask dimensions as shown in Figure 8-21. Save your changes (File → Save from menu).

REPLACING A PADSTACK DEFINITION

Next we will change existing smd44rec72 padstacks in the footprint drawing to the new smd45rec71 padstack.

Select Tools → Padstack → Replace from the menu bar. Select pin 1. At the Options pane, click the Browse... button to look for the new pad (Figure 8-22).

At the Select a Padstack dialog box, select the new one and click OK.

Show Element button

At the Options pane, select a pin or leave as * for all pins. Click the Replace button. The pads should be visibly different. To verify that the new pads are correct, select the Show Element button, , and left click on a pad. The information window will give you details about the padstack.

Next we need to move the padstacks. Set the grid to 1 mil by 1 mil resolution (Setup → Grids...) and make sure XYMode is set to Absolute. Since pad center-to-center spacing is 118 mils, the left pad should be located at (−59, 0) and the right pad located at (59, 0). Use the Move tool and change the positions of the pads (turn off all other layers if it helps). Set the Find filter to Pins only if pins is unchecked.

Next we need to modify the outline objects. Use Figure 8-23 as a guide (which is based on IPC-7351 guidelines). When working with the outline objects it can be helpful to change the layer colors and turn off all but the outline on which you are working.

We begin with the silk-screen outline. Turn off all layers except Component Geometry/Silkscreen_Top. To resize the outline, select the Shape Select tool, 🔲, left click the object to select it. The line becomes dashed and square handles are displayed. Move your pointer over the outline, and it turns to a double-ended arrow. Use the double arrow to grab a line and slide it to the proper location. Do this on all the sides to produce the proper outline size.

Shape Select tool

As stated already you cannot change the width of the lines that make up rectangles. In some footprints silk-screen outlines are made with lines and others (such as this one) are made with rectangles. Since the rectangles have an actual width of 0, they will be ignored during artwork generation unless you specify a default value for all zero width lines in the Undefined line width: entry in the Film Options area of the Artwork Control Form dialog box (opened from the Manufacture → Artwork menu). In this example thicker Lines were added to the rectangle to emphasize the silk-screen detail on the component sides. To do so select the Add Line, change the line width to 10 mils in the Options pane, and draw lines on the component sides as shown in the figure.

FIGURE 8-22
Replacing a padstack definition.

Next manipulate the assembly outline using the procedure just described. The assembly detail shown in Figure 8-23 is drawn so that it is the actual size of the component and additional lines are drawn to show the end caps and their relationship to the pads.

Finally, modify the place outline. The place outline shown in the figure is defined by the courtyard recommendations in the IPC model. This approach provides more accurate placement spacing parameters, which the DRC can use for spacing rules and to promote manufacturability.

Finally move the text objects as appropriate for the new symbol size. For viewability Figure 8-23 does not show the assembly text objects.

FIGURE 8-23
The finished SMD capacitor.

Save the drawing and make sure that Symbol 'smdcap1206.psm' created is displayed in the command window. That completes the design example.

Example 3. PGA Design Using the Symbol Wizard

Using the preceding procedures allows you to construct the exact footprint symbol you want, but for large or complicated footprints there is an easier approach. The Symbol Wizard performs many of the tasks automatically and can be used to create simple or complex footprint symbols. The resultant symbols contain all the necessary elements for a valid footprint symbol, but it may not meet your specific symbol requirements with regard to all the possible text and detail options. You can then use the procedures described in the second example to modify the automatically generated footprint symbol to meet your specific requirements. In this example the Symbol Wizard is used to construct a pin grid array (PGA) footprint (Figure 8-24).

FIGURE 8-24
Bottom view of a pin grid array (PGA) package.

The Symbol Wizard is used in this example to construct a footprint for the PGA. The first step is to obtain a data sheet for the part. The dimensions for a generic 48-pin, 8 × 8 PGA are shown in Figure 8-25, left. Each pin has a 22-mil diameter and is spaced 100 mils from the others. A padstack with a 32-mil hole and 50-mil-diameter pads suffices for the 22-mil pin (Figure 8-25, right). Before starting the footprint design, you need to have the correct padstack present in the symbols library. A pad50cir32d padstack is included with the symbols library and is used for this design.

FIGURE 8-25
PGA design parameters.

To begin the design process, open PCB Editor and select File → New... from the menu. At the New Drawing dialog box (Figure 8-26), enter a name for the drawing (e.g., PGA_48pin) and select Package symbol (wizard) from the Drawing Type: list. Use the Browse... button to select a directory (the symbols library is default). Click OK to start the wizard.

At the Package Symbol Wizard dialog box (Figure 8-27), select the PGA/BGA radio button and click Next>.

At the Template dialog box (Figure 8-28), leave the Default Cadence supplied template radio button selected and click the Load Template button, then click Next>.

At the General Parameters dialog box, leave the default settings (mils) and click Next>.

FIGURE 8-26
New PGA wizard.

FIGURE 8-27
Select the package type.

FIGURE 8-28
Select a drawing template.

At the Pin Layout dialog box (Figure 8-29), enter 8 for the vertical and horizontal pin counts. Select the Perimeter matrix radio button and enter 2 for Outer rows: and 0 for Core rows: as shown in the figure. The Total number of pins: should indicate 48, as required. Click Next>. At the next Pin Layout dialog box, leave the default settings (Number right and letter down and JEDEC standard). Click Next>.

At the next Array Parameters dialog box (Figure 8-30), leave the Lead pitch at 100 mils, but change the Package width and length to 860 mils, and click Next>.

At the Padstacks dialog box (Figure 8-31), click the Browse... button(s) and select pad50cir32d for the default padstack and pad50sq32d for the pin 1 padstack. Click Next>.

At the Symbol Compilation dialog box, leave the default settings and click Next>. At the Summary dialog box, click the Finish button to complete the design.

The new footprint symbol is shown in Figure 8-32.

The new symbol contains Ref Des text objects on the Package Geometry Assembly_Top and Silkscreen_Top sublasses (Ref Des class) and four outline details on the Package Geometry class: Assembly_Top, Silkscreen_Top, Place_Bound_Top, and Dfa_Bound_Top subclasses (dfa stands for *design for assembly*). The place outlines are constructed of filled rectangles, and the silk-screen and assembly outlines are constructed

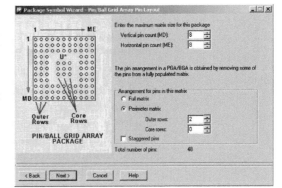

FIGURE 8-29
Setting up the pin layout.

FIGURE 8-30
Setting up the array parameters.

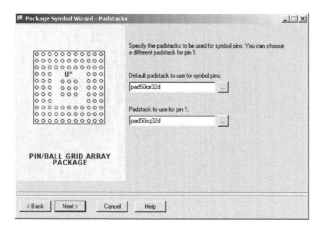

FIGURE 8-31
Select default pin types.

FIGURE 8-32
PGA package generated by the wizard.

of lines. All four outlines are shown as squares with zero width. We modify the outline objects to be more consistent with the IPC recommendations. Begin by changing the grid spacing to 10 mils (Setup → Grids…).

First the Dfa_Bound_Top and the Place_Bound_Top outlines are expanded. Turn off the other outlines (use the Color button or the Options pane). Select the Select Shape tool as described previously. Left click on one of the place outline shapes to select it. Place your pointer over the outline and use the double-ended arrow to grab the outline edge. Drag the edge 40 mils outward from its current position. Do this for all four sides on both of the place outlines. When you are finished resizing the place outlines, turn them off and instead display the assembly top outline.

FIGURE 8-33
Modifying the outline object.

We now add a beveled corner to the assembly outline at the corner at pin 1, as shown in Figure 8-34. Begin by zooming into the upper left corner of the symbol. Select the Add Line tool; at the Options pane, select Package Geometry/Assembly_Top and select Line and 45 as the Line Loc: options. Draw a diagonal line in the upper left corner as shown at (1) in Figure 8-33.

Next we need to trim off the excess part of the outline. To do so vertices need to be added so that they can be used as the trim points. To add a vertex to a line, select Edit → Vertex from the menu bar. Left click at the intersection of the diagonal line you just drew and the existing outline (points at (2) in the figure) and move the vertex out 20 mils or so to create an angle. Do this at each end of the new line.

Next delete the line segments (shown at the Xs near (3) in Figure 8-33). To delete part of a line segment disable any current tool (right click and select Done from the pop-up menu—if it is available; if it is not, no tools are selected). Hover your pointer over one of the line segments so that just the segment is dashed. Left click the segment to permanently select it. Select the Delete tool from the toolbar and only the segment will be deleted. If you select the Delete tool first then select the line, the entire outline will be deleted. Finally move the remaining line segments (the (4)s in the figure) back to the ends of the diagonal line (use the Vertex tool not the Move tool). Perform the same steps on the silk-screen outline and change the silk-screen outline width to 5 mils.

FIGURE 8-34
Final PGA design.

The only text objects added by the wizard were the Ref Des objects on the Assembly_Top and Silkscreen_Top subclasses. You can add your own as described in the preceding examples (e.g., value and part number on the assembly layer).

Select File → Save to save the drawing and update the .psm file. The completed footprint is shown in Figure 8-34.

FLASH SYMBOLS FOR THERMAL RELIEFS

As described in Chapter 2, thermal reliefs are used to provide resistance to heat flow from a plated through hole to the copper plane to which it is connected during solder operations. Figure 8-35 (left) shows how this is accomplished. A pad, belonging to a plated padstack, is connected to the larger plane through spokes. Between the spokes (and between the pad and the plane) are void areas where the copper is removed. The spokes provide electrical conductivity while the voids provide a barrier to heat flow. Flash symbols are used to define the voids on negative plane layers. Figure 8-35 (right) shows the flash symbol in the negative image view and is described further.

FIGURE 8-35
Thermal relief and its flash symbol: (left) copper plane, (right) the negative image flash symbol.

Thermal relief connections between plated through holes and copper areas on positive planes are automatically generated by PCB Editor so flash symbols need not be defined for positive layers. The inner diameter (ID) of the thermal relief is defined by the pad diameter while the outer diameter (OD) is defined by the diameter of the (thermal relief) circle in the padstack definition set in the Padstack Designer. The spokes are generated in the board design by PCB Editor using trace width specifications defined in the Constraint Manager as is shown in the PCD Design Examples.

The realization of thermal reliefs on negative planes is another story. Recall that, with negative plane layers, what you see is what is removed, so thermal reliefs are generated by creating flash symbols (shown in Figure 8-35, right) that define the void areas. Once a thermal flash symbol is attached to a padstack, PCB Editor interprets it and creates the positive view that you see with the "what you see is what you get" (WYSIWYG) view presented during the board design. When manufacturing artwork is created during the last phase of a board design, the negative planes and flash symbols are properly processed as negative Gerber images.

Most of the padstacks in the symbols library have no flashes assigned to them. During the board design process flashes need to be assigned to the padstacks as the need arises. When you construct your own footprints and padstacks, you can follow the same practice or assign flashes right away. Either way you need to know how to construct flash symbols.

A flash symbol is just another type of drawing you construct using PCB Editor. Later, a flash symbol is made for the padstack designed in the first example.

Before we begin with the drawing, we "engineer" the thermal relief. Recall from the first example that the padstack we designed (pad62cir42d) had the characteristics summarized in Table 8-4.

Table 8-4	Pad Dimensions
Parameter	**Mils**
Drill-hole diameter	42
Outer-pad diameter	62
Outer-antipad diameter	82
Outer-pad to antipad clearance	10
Outer annular-ring width	10
Inner-pad diameter	60
Inner-antipad diameter	80
Inner annular-ring width	9
Inner-pad to antipad clearance	10

Typically the flash's inner diameter (ID in Figure 8-35, right) is the same as the regular pad for that layer, and flash's outer diameter (OD in Figure 8-35, right) is the same as the antipad diameter plus or minus a couple of mils. The spoke width is calculated per IPC-2222 (pp. 16–17) and depends on the pad diameter and the number of spokes. The spoke width is calculated as

$$W = 60\%(P/n)$$

where

> W is the spoke width.
> P is the pad diameter.
> n is the number of spokes.

If we assume that only internal layers have negative planes, then only the internal layers need a thermal flash. If it turns out that this is not the case, flashes can be added later.

Then for the pad62cir42d padstack, the flash inner diameter is 60 mils, the outer diameter is 80 mils, and the spoke is 9 mils wide.

FIGURE 8-36
Starting a new flash symbol drawing.

To construct a flash symbol, open PCB Editor. Start a new flash symbol drawing by selecting File → New... from the menu. At the New Drawing dialog box, enter a name for the symbol (e.g., TR_80_60_9 for thermal relief with outer diameter = 80 mils, inner diameter = 60 mils, and spoke width = 9 mils), then select Flash symbol from the Drawing Type: list, as shown in Figure 8-36. Click OK.

Turn on the grids and check to make sure that the origin coordinates (0, 0) are at the center of the drawing area. If not adjust the drawing area by giving the Left X and Lower Y Extents negative values (Setup → Design Parameters from menu, Design tab).

Next, select Add → Flash from the menu bar, as shown in Figure 8-37.

As shown in Figure 8-38, enter the design values for the pad diameters and spoke definitions into the Thermal Pad Symbol Defa... dialog box, and click OK.

The completed symbol should look like Figure 8-35 (right). Save the symbol and make sure you see Symbol 'tr_80_60_9.fsm' created in the command window.

FIGURE 8-37
Add a flash symbol to a drawing.

Now we can add this symbol to the pad62cir42d padstack definition. Use Padstack Designer to open pad62cir42d.pad again. Select the DEFAULT INTERNAL

Browse button

layer and select Flash from the Thermal Relief list. Click the Flash: Browse... button, , and from the Select flash symbol: dialog box (Figure 8-39), select the Tr_80_60_9 flash symbol and click OK. Figure 8-40 shows the flash symbol assigned to the padstack. You can save the new definition with the same name or rename it to something like pad62cir42f to indicate that it has a flash symbol assigned to it.

FIGURE 8-38
Setting up the flash symbol parameters.

FIGURE 8-39
Selecting the flash symbol.

You can now use this padstack definition in the footprint design you did in the first footprint design example. To do so, open the footprint drawing and change the padstack definition to pad62cir42f instead of pad62cir42d. Use the procedure in the second design example to change the padstack (Tools → Padstack → Replace from menu bar). This completes the flash symbol design example.

MECHANICAL SYMBOLS

Mechanical symbols represent physical objects you can place in your board design that are not part of the netlist and therefore have no connectivity. They can be things like mounting holes (mechanical pins), nonconnecting copper etch objects, design outlines, dimensions, and areas (e.g., keep-in and keep-out). Several mechanical symbols are located in the symbols library for you to look at for ideas. The mechanical symbols have a .bsm file extension.

One important mechanical symbol that you may need to have is the mounting hole.

FIGURE 8-40
The new flash symbol assigned to a padstack definition.

Mounting Holes

Mounting holes can be used for attaching the PCB to mounting hardware (such as stand-offs) or for attaching hardware to the PCB (such as heat sinks). The four basic hole types are shown in Table 8-5; in them holes can be made with or without lands (pads), with or without plating, or in any combination of the two. Mounting holes that are plated can be attached to a net or isolated from all nets.

When mounting holes are attached to a net that is assigned to a plane layer, the mounting hole can be connected to the plane through thermal reliefs or with full contact, just like any other plated through hole. When mounting holes are attached to a net, they should be included on the schematic using a Capture part with an assigned footprint, which consists of a connection-type padstack. When mounting holes are not attached to a net, they are not included in the schematic and are added to the board design in PCB Editor. Some designers do not like to have connected mounting holes on the schematic, and it is possible to add them to the board in PCB Editor only. But it is the author's opinion that this is a messy way to do it, since you have to override DRC errors to do it, and

Table 8-5	Basic Hole Types	
	Plated	Nonplated
With pads	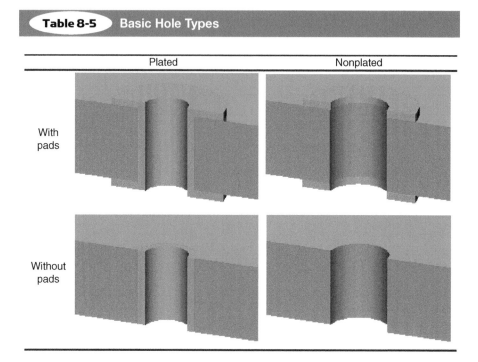	
Without pads		

since it is connected to a net, it really is part of the circuit and should be on the schematic (if only for documentation purposes).

The PCB Editor contains several mounting-hole symbols, which have the name mtgXXX, where XXX indicates the hole diameter in mils. These are nonplated holes with no lands (without pads). The IPC standards call this an *unsupported hole*.

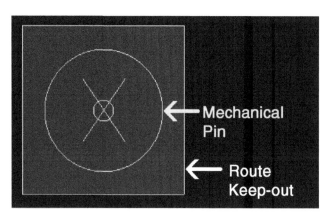

FIGURE 8-41
A mechanical pin.

The mounting hole mtg125.dra is shown in Figure 8-41. The drawing consists of a Pad125 padstack and a route keep-out area. The Pad125 padstack is a mechanical pin rather than a connect pin.

The mechanical pin properties are shown in Padstack Designer in Figure 8-42. To view the pin launch, go to the Padstack Designer from PCB Editor by selecting Tools → Padstack → Modify Design Padstack... from the PCB Editor menu bar. Figure 8-42 (left) shows the drill land plating settings in the Parameters tab and Figure 8-42 (right) shows the layers setting on the Layers tab.

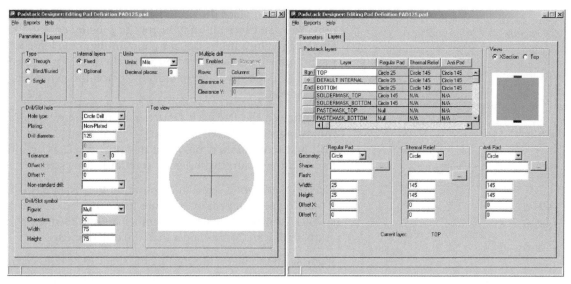

FIGURE 8-42
Padstack definition for a mechanical pin.

The drill diameter is 125 mils and the pads are 25 mils. This means that the pads will be drilled out during the manufacturing process. You will also notice that this is a nonplated hole, which means that this hole will be drilled toward the end of the manufacturing process, separately from the plated holes. A clearance (antipad) is included and is 20 mils larger than the drill diameter, so that any copper planes are always kept away from the hole (this allows for fabrication tolerances).

FIGURE 8-43
Starting a new mechanical symbol drawing.

Creating Mechanical Drawings

To start a new mechanical drawing, open PCB Editor and select File → New... from the menu bar. At the New Drawing dialog box, enter a name for the drawing and select Mechanical symbol from the list as shown in Figure 8-43. Click OK.

You will be presented with a typical PCB Editor drawing window. Add lines, text, and dimensions as needed to complete your drawing.

To add mechanical pins, select Layout → Pins from the menu bar or select the Add Pin button, , on the toolbar. From the Options pane (see Figure 8-44) select the Mechanical pin option instead of the Connect pin (since no net is attached to it). Click the Padstack: Browse... button. At the Select a Padstack: dialog box, you can select any padstack or mounting hole you like. When you select a padstack and click OK in the dialog box then move your cursor to the drawing area, you will have the selected pin (padstack) attached to your pointer. Left click in

Add Pin button

FIGURE 8-44
Options for adding a mechanical pin.

the drawing to place the pin. Right click and select Done to quit. You can place an array or row of pins by entering X: and Y: Qty values in the Options pane. In that case left click in the drawing area where you want the first pin located and PCB Editor will automatically place the remaining pins for you.

You can use the Padstack Designer to make the types of mounting holes shown in Table 8-5. A reference table of drill and screw sizes is provided in Appendix D as an aid to designing mounting holes for standard screw sizes.

For nonplated holes select that option on the Parameters tab. If you do not want pads on the hole, make the pad diameters smaller than the drill size. Remember though to make the antipads at least 20 mils larger than the drill diameter.

If you want to make a mounting hole that can be connected to a plane or net, the mounting hole will simply be a plated padstack. You can use them or design your own as described previously.

When a board design has both plated and nonplated holes, you can combine the drill information for both holes types into one NC drill file or have PCB Editor generate separate drill files. A second set of drill instructions is needed because nonplated holes are drilled after all plating processes are complete, whereas plated through holes are drilled before the plating process; therefore a completely separate set of instructions is required.

The standard, plated drill file has the naming convention *name*.drl file and nonplated drill files have a *name*-np.drl naming convention. You specify combined or separate drill files by selecting Manufacture → NC → NC Drill... from the menu bar. At the NC Drill dialog box (shown in Figure 8-45), uncheck the Separate Files option to generate a combined drill file for all types or select the option to make separate files.

FIGURE 8-45
Specifying combined or separate NC drill files.

Placing Mechanical Symbols on a Board Design

Examples of how to place a mounting hole (a mechanical symbol) on a board are given in Chapter 10 and Example 2 of the PCB Design Examples. Essentially it is the same process described previously for adding a mechanical pin to a mechanical symbol drawing, except that you are placing it in your board design. Open your board design and select Place → Manually from the menu bar to display the Placement dialog box, as shown in Figure 8-46. Select the Add Mechanical Symbols from the placement list (make sure that both the Library and Database options are checked in the Advanced Settings tab). Check the box next to the symbol you want and left click on the board area at the desired location to place the mechanical pin. Repeat the process for each symbol you want placed on your board.

BLIND, BURIED, AND MICROVIAS

Blind and buried vias are not typically designed into footprints but are added to a PCB during the layout and routing process. Examples of PCB design using blind vias is given in the PCB Design Examples (Example 3) but is discussed briefly here since it is related to padstack design.

FIGURE 8-46
Placing a mechanical symbol on a board design.

Figure 8-47 shows how microvias and blind and buried vias can help route high-density ball grid arrays (BGAs). Using these via types, a BGA can be placed on the top of a board and another surface-mounted device can be placed directly opposite the BGA on the bottom of the board. To accomplish this, typical through-hole vias cannot be used. Fan-out and routing of the BGA are accomplished using blind and buried vias. Blind vias are visible on only one side of the PCB and connect traces on the one outer layer to inner

FIGURE 8-47
Via technologies and their applications.

traces only, while buried vias are not visible from either side of the board and connect traces only between inner layers.

Blind and buried vias are realized on built-up PCBs. A built-up PCB often has a standard layer stack-up core (section (a) in Figure 8-47) and additional layers are sequentially added to the board (the two (e) sections in Figure 8-47). Plated through-hole vias in the standard core become buried vias ((b) in Figure 8-47). As the outer layers are built up on top of the base core, buried microvias ((c) in Figure 8-47) and blind microvias ((d) in Figure 8-47) can be embedded into the outer layers. During the build-up process, resistors or capacitors can also be buried in the layers. After all the layers have been built up, additional microvias ((f) and (g) in Figure 8-47) and standard plated through holes ((h) in Figure 8-47, tented on the top end) can be added to the entire assembly.

Figure 8-47 shows three types of microvias: (c) and (g) are laser-drilled, plated vias; (d) is a laser-drilled, paste-filled via; and (f) is a plasma-etched, plated via. To learn more about designing built-up boards and microvias see *Coombs' Printed Circuits Handbook* (Coombs 2001).

Blind and buried vias can be constructed in three ways. One way is to use Padstack Designer to design the via padstack directly. In general you use the Padstack Designer the same as was described for the normal through-hole padstacks. If you know exactly what the layer stack-up definition will be in your board design, you can add those layers to a padstack definition in Padstack Designer and remove other layers. In Padstack Designer select Blind/Buried and Optional on the Parameters tab. On the Layers tab, add and name layers to the stack-up as needed. *To add layers to the stackup,* right click on the layer that will be below the one you are adding and select Insert from the pop-up menu. Give the new layer the name used in your board design and select the appropriate pad sizes. For all other layers select Null for the pad values. An example of a blind via design is shown in Figure 8-48.

The second way is to set up vias interactively from the board design using the BBvia tool. This is the easiest way to make a blind or buried via, because all the layer definitions are already set up in the board design (using the Layout Cross Section dialog box) and will be known to the BBvia tool. The BBvia tool is launched by selecting Setup → B/B Via Definitions from the menu, as shown in Figure 8-49. As indicated in the figure, two options are available.

If you select Define B/B via..., you get the dialog box shown in Figure 8-50. This method creates one padstack that goes from the Start Layer to the End Layer and includes all the layers between them.

If you select Auto Define B/B via..., then you get the dialog box shown in Figure 8-51. You can select which layers will be used in setting up the via. This method actually makes several vias. Essentially it makes as many blind and buried vias as necessary to connect each of the adjacent planes (two layers at a time) and to connect the start and end layers to each other and to each of the internal layers.

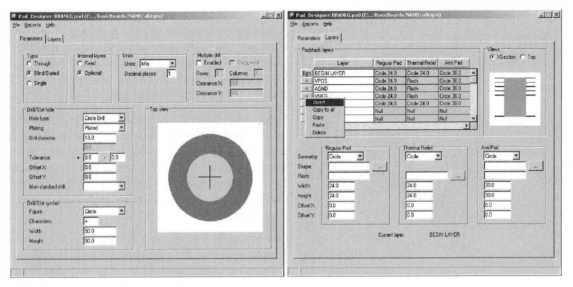

FIGURE 8-48
Blind via definition using Padstack Designer.

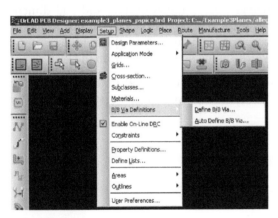

FIGURE 8-49
Starting the BBvia tool.

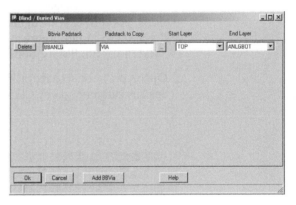

FIGURE 8-50
Define B/B via… option.

Which vias get used in the board design is controlled by the Constraint Manager in PCB Editor as demonstrated in the PCB Design Examples.

USING THE IPC-7351 LAND PATTERN VIEWER

The IPC Land Pattern Viewer is an indispensable tool for designing footprints. The viewer provides drawings and dimensions of hundreds of footprints based on established standards. You can download a freeware version of the software

FIGURE 8-51
Auto Define B/B via...
option.

from the IPC Web site at www.ipc.org. From the home page, look under the Knowledge tab then click on the PCB Tools and Calculators link. On the next page, click on the IPC-7351 Land Pattern Viewer and Tools link. From the next page, click the IPC-7351 Land Pattern Viewer link.

Once you have downloaded and installed the viewer, you can immediately run the application without restarting Windows. When you first open the viewer, you will be presented with an application window, as shown in Figure 8-52. Click the Search button to open a library and begin looking for footprints.

The first step is to select a library from the Search Library: selection list (see Figure 8-53). Five libraries are included by default. You can download additional free and for-fee libraries from the IPC Web site through various links. The SMN7351A .plb library (included with the viewer) contains footprints for most of the discrete and IC packages on the market with nominal outline and pad dimensions. Footprints with tighter tolerances (i.e., at least smaller features) are in the SML7351A.plb library and footprints with larger (most) features are in the SMM7351A.plb library, both of which can be downloaded from the Web site.

Once you have selected a library, five search modes will be listed under the Search Categories: pane. Click a + box to expand a category (see Figure 8-54).

FIGURE 8-52
Initial view of the IPC-7351 Land Pattern Viewer.

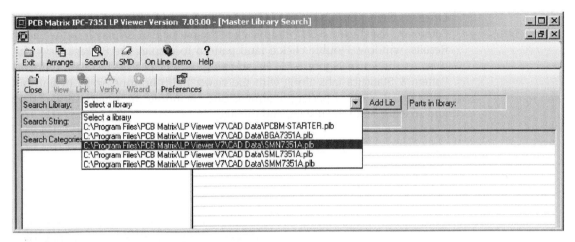

FIGURE 8-53
Selecting a library.

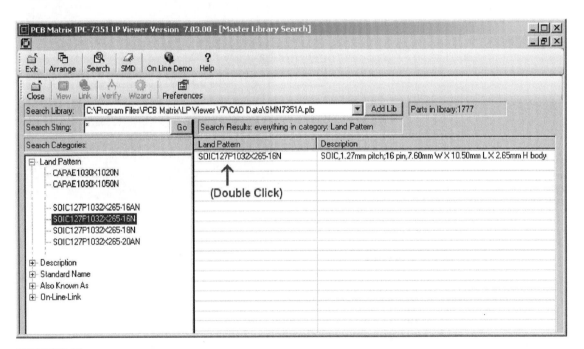

FIGURE 8-54
Selecting a search category.

Regardless of which category you select, the same information is contained in the search results (for the most part), but the footprints are listed differently within each category. The most noticeable difference is the Standard Name category. Rather than listing the footprints by package names, they are listed by industry standard document numbers (e.g., JEDEC MO-187 or MIL-PRF-xxx).

Once you have found the package name, description, or standard in the Search Categories: list (Figure 8-54), double click the package to display it in the Search Results: window. Double click a land pattern to display the Package View window, as shown in Figure 8-55. By selecting the Component, Settings, and Land Pattern & Statistics tabs along with the Viewer radio buttons, you can find out every detail about a package's dimensions, including the suggested pad sizes and spacing, silk-screen sizes, and the like.

FIGURE 8-55
Viewing a footprint land pattern and design parameters.

REFERENCES

Coombs, C. F., Jr. (Ed.) 2001. *Coombs' Printed Circuit Handbook*. 5th ed. New York: McGraw-Hill.

IPC-2222. February, 1998. Sectional Design Standard for Rigid Organic Printed Boards. Northbrook, IL: IPC/Association Connecting Electronic Industries.

IPC-7351. February, 2005. Generic Requirements for Surface Mount Design and Land Pattern Standard. Northbrook, IL: IPC/Association Connecting Electronic Industries.

CHAPTER 9

PCB Design Examples

INTRODUCTION

In Chapter 2 the basic PCB design flow was demonstrated, starting from schematic entry with Capture to board routing with PCB Editor, but no attention was given to board design issues, such as component placement and spacing rules, layer stack-up, or trace routing and spacing rules.

Design examples are used here to illustrate the various PCB design considerations described in Chapters 5 and 6 and how to use the PCB Editor tools to accomplish the design goals. The part count in each example is kept to a minimum (all designs have fewer than 10 parts and 14 pins/part) so as not to be cumbersome and not to distract the reader from understanding how to route and set up layers. In-depth discussion on PCB design for manufacturability is covered in Chapter 5 and signal integrity and routing are covered in Chapter 6.

The following examples use parts that may not be included with your version of OrCAD. You can make the parts yourself using the procedures discussed in the previous chapters, or you can copy the parts and footprints provided in the example libraries located on the Web site for the book.

The *first example* is a simple analog design using a single op amp. The design shows how to set up multiple plane layers for positive and negative power supplies and ground. The design also demonstrates several key concepts in Capture, such as how to connect global nets, how to assign footprints, how to perform design rule checks, how to use the Capture part libraries, how to generate a bill of materials (BOM), and how to use the BOM as an aid in the design process in Capture and PCB Editor. The design also shows how to perform important tasks in PCB Editor, such as finding and selecting specific parts and modifying padstacks. Intertool communication (such as annotation and back annotation) between Capture and PCB Editor is also demonstrated.

The *second example* is a mixed, digital/analog circuit. In addition to the tasks demonstrated in the first example, the design also demonstrates how to set up split planes to isolate analog and digital power supplies and grounds. Other tasks include using copper pours on Routing layers to make partial Ground planes, using copper pours on Plane layers to make nested Power and Ground planes, and defining anticopper areas on Plane and Routing layers.

The *third example* uses the same mixed, digital/analog circuit from the second example but demonstrates how to use multiple-page schematics and off-page connectors to organize and simplify large circuit designs and to incorporate PSpice simulations into a PCB layer design. It also demonstrates how to construct multiple, separated Power and Ground planes and a Shield plane to completely isolate analog from digital circuitry. The use of guard rings and guard traces is also demonstrated.

The *fourth example* is a high-speed digital design that demonstrates how to design transmission lines, stitch multilayer Ground planes, perform pin/gate swapping, place moated ground areas for clock circuitry, and design a heat spreader.

OVERVIEW OF THE DESIGN FLOW

Regardless of which type of board is made, certain steps must be executed in the design flow process. The following is an outline of the process.

1. Initial design concept and preparation:
   A. Generate initial drawings.
   B. Collect data sheets.
   C. Take inventory of packaging and footprint needs.
   D. Search through the Capture libraries to find the parts. For any parts that are unavailable, construct the parts using the Capture Part Editor or the PSpice Model Editor.
2. Set up the design project in Capture:
   A. Draw the schematic (placing and connecting parts).
   B. Perform an annotation to clean up numbering.
   C. Make sure multipart packages are properly utilized.
   D. Make sure global power nets are properly connected.
   E. Assign related components to groups ("rooms") to aid in part placement in PCB Editor.
   F. Perform a Capture design rule check (DRC) to verify that the circuit schematic has no issues. Correct any errors and repeat DRCs as needed.
   G. Generate a bill of materials to identify PCB assigned and missing footprints.
   H. Search through the PCB Editor libraries to find and assign footprints. For any footprints that are unavailable, obtain the data sheets for recommended land patterns and design the footprints using PCB Editor and the Padstack Designer.
   I. Generate a PCB Editor netlist and open the board design.
3. Define the board requirements:
   A. Board dimensions and mounting hole locations.
   B. Part placement considerations (height restrictions, assembly method).
   C. Noise and shielding requirements.
   D. Component mounting technology (SMT, THT).
   E. Trace width and trace spacing requirements.
   F. Required vias and fan-outs (size and tenting, etc.).
   G. Number of Power/Ground planes and Routing layers.

4. Basic board setup:
   A. Physical:
      a. Create the board outline.
      b. Place mounting holes.
      c. Define part and routing restriction areas.
      d. Add dimension documentation (optional).
   B. Preliminary parts placement:
      a. Use Search and Place tools to place selected parts and groups.
      b. Perform a board DRC to check for footprint and placement problems.
   C. Layer setup:
      a. Set up Power and Ground planes.
      b. Set up Routing layers.
      c. Assign ground and power nets to Plane layers.
      d. Define thermal relief parameters.
      e. Set which vias to use for fan-outs, routing vias, jumpers, and the like.
      f. Perform a DRC to check for layer problems.
   D. Final parts placement:
      a. Make sure spacing rules are not violated.
      b. If using split or moated Plane layers, make sure parts are placed accordingly.
      c. Check orientation of polarized components (caps, diodes, etc.).
5. Preroute specific nets using manual and restricted autorouting:
   A. Perform power and ground fan-outs.
   B. Preroute critical nets manually.
   C. Perform DRC to make sure that no errors have occurred. Fix errors.
6. Autorouting:
   A. Set up the autorouter.
   B. Run the autorouter.
   C. Perform DRC to make sure that no errors have occurred. Fix errors.
7. Finalizing the design:
   A. Postrouting inspection:
      a. Sharp (acute) angles.
      b. Long parallel traces (cross-talk issues).
      c. Via locations.
      d. Silk-screen markings.
   B. Board cleanup:
      a. Unroute and then reroute problem traces.
      b. Perform a final DRC.
   C. Synchronization with Capture (back annotation).

A detailed example of how to create the artwork and NC files for the design is included in Chapter 10.

EXAMPLE 1 DUAL POWER SUPPLY, ANALOG DESIGN

The first example demonstrates the basics of schematic entry in Capture, including finding and placing parts, connecting parts and using global power nets, and generating a bill of materials to assist in the design process. The example continues with PCB design in PCB Editor and demonstrates how to specify board requirements, such as trace width and spacing, and setting up the overall board design; how to define layers; and how to perform manual and automatic routing.

To follow the example exactly as it is presented here, you need to copy the footprints for this example from the Web site for this book. Copy the contents of the symbols file folder located in the Projects/Analog_Ex folder (Figure 9-1).

FIGURE 9-1
Location of the PCB
Editor footprint library.

Although all the schematic and design files for this example are also included on the Web site, you need not copy them, because all the Capture parts in the example are already included with the OrCAD software. However, if you choose to copy the design files simply copy the entire ANALOG_EX folder into your project folder. If you have not set up a project folder, now would be a good time to set one up. You can set it up anywhere your computer has access to it, and OrCAD should have no problem with it.

Initial Design Concept and Preparation

Before you start a PCB design process, you will likely have some sort of preliminary design concept jotted down. Perhaps PSpice simulations of sections of the design have even been performed. The design concept for this example is shown in Figure 9-2.

FIGURE 9-2
Analog circuit for design Example 1.

The circuit is very simple, but it contains enough parts that it encompasses the same steps required for larger, more complicated designs. The circuit is a basic amplifier that consists of an active component (the op amp) and several passive components (resistors and capacitors). The circuit also contains an off-board connector that supplies dual rail power to the board and connections for input and output signals. Through-hole components include the connector and power supply filter capacitors, surface-mounted devices include the op amp, its bypass caps, and the signal conditioning and gain resistors.

Once you have the basic design down, you need to make a list of all the parts needed to build the circuit, including the off-board connector. Search through your favorite parts catalog to find the parts and download the data sheets from the manufacturers. It is helpful to make a spreadsheet that details the parts and footprints to keep track of the design details throughout the design flow, so that once you receive the board from the manufacturer, the parts fit properly. An example spreadsheet is given in Table 9-1. During the entire design process, you can continually add information to the spreadsheet to document and organize all aspects of the design process. If a problem does occur, the documentation can help identify where the fault occurred.

Table 9-1	Basic Parts List and Mounting Requirements	
Reference	**Value**	**Mounting package**
J1	6 pin	Through-hole, 0.200" pin spacing
C1, C2	10 μF	Through-hole, axial lead
C3, C4	0.1 μF	SMD, size 2010
U1	LM741	SOIC-8
R1–R3	1 k, 2 k	SMD, size 1805

Setting Up the Project in Capture

To begin a new design project, start Capture and, from the session window's File menu, choose New → Project. At the New Project dialog box (Figure 9-3), enter a name for the project. Select the PC Board Wizard radio button. You can also make PCB designs if you select the Analog or Mixed A/D button, which sets up PSpice simulation templates for the project. Since we are not performing PSpice simulations in this example, PC Board Wizard is used. Select the desired location for your project, then click OK.

FIGURE 9-3
Setting up a new project with the PCB Project Wizard.

At the next Project Wizard dialog box, do not enable project simulation, click Next. At the next dialog box, add the Connector, Discrete, and OPAmp part libraries as shown in Figure 9-4. To add a library, select it (by clicking on it) in the left box and press the Add>> button. After you have added the libraries you want, click the Finish button.

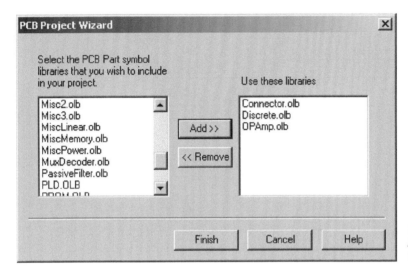

FIGURE 9-4
Adding parts libraries to the project.

If you are using the demo edition, you will be shown an information box that says, The Demo Edition does not support saving to a library with more than 15 parts.... Click OK each time it asks (you will need to click once for each library that you added to the project).

DRAWING THE SCHEMATIC WITH CAPTURE

Figure 9-5 shows the design goal. You can use it as a reference through the design example or modify it to your liking. If you are using the demo version, remember that you cannot save a design in PCB Editor that has more than 10 parts or parts with more than 14 pins. This design example meets these requirements so that you can save your work.

If a schematic page does not automatically open, expand the Design icon and the schematic folder in the Project Manager window, then double click the Page1 icon.

PLACING PARTS

Before placing any parts onto the schematic page, make sure that the place grid is enabled. If parts are placed off grid, you will not be able to connect wires to the part's pins. Use the Search tools just described to locate and place parts on the schematic page. Table 9-2 lists the parts used in this project and the libraries in which they are located.

FIGURE 9-5
Analog circuit schematic design in Capture.

Table 9-2	Capture Library Parts List	
Reference	**Capture part**	**Capture library**
J1	CON6	C:\...\CAPTURE\LIBRARY\CONNECTOR.OLB
C1, C2	CAP POL	C:\...\CAPTURE\LIBRARY\DISCRETE.OLB
C3, C4	CAP NP	C:\...\CAPTURE\LIBRARY\DISCRETE.OLB
U1	LM741	C:\...\CAPTURE\LIBRARY\OPAMP.OLB
R1–R3	R	C:\...\CAPTURE\LIBRARY\DISCRETE.OLB

Place Part
tool button

To place parts click the Place Part tool button, ▱; select Part from the Place menu; or hit the P key on your keyboard. In the Place Part dialog box (see Figure 9-6), select the OPAMP library or the EVAL library (if using the demo version), and scroll down until you find the LM741 or the uA741 op amp, respectively. Either one will work; it is just a matter of which one you prefer.

You can place one of each, compare them, then delete the one you do not like. If a library is not shown in the Libraries: box, you can add it to the list, as explained next. Once you find and select the part, click OK.

If a library is not displayed in the Libraries: window, you need to add it to the list. *To add a library* to the Libraries: list in the Place Part dialog box, click the Add Library... button on the Place Part dialog box (Figure 9-6). Use the Browse File dialog box to locate the desired OLB library. For building schematics and PCB layouts, you can use parts from either the main Capture library folder or the PSpice subfolder. If you are performing PSpice simulations, select parts only from the PSpice folder.

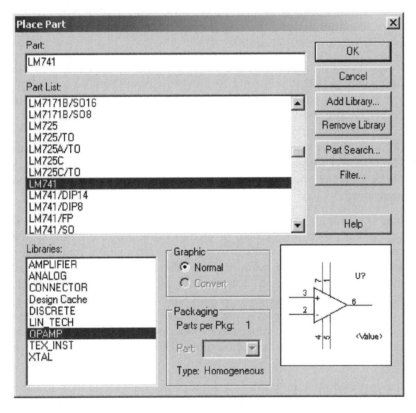

FIGURE 9-6
Choosing parts from the Place Part dialog box.

From the DISCRETE library, place two polarized capacitors (CAP POL), two nonpolarized capacitors (CAP NP), and three resistors (R). From the CONNECTOR library, place a six-pin connector (CON6), and from the OPAMP library, place LM741.

CONNECT PARTS WITH WIRES (SIGNAL NETS)

Place Wire tool

To wire the circuit, use the Place Wire tool, ; select Wire from the Place menu; or hit W on your keyboard to activate the Wire tool. To attach wires between pins, click once to start a wire, move the pointer to the next pin and click once to attach the wire, and continue routing or double click to end the wire. Connect the circuit as shown in Figure 9-5.

MAKING POWER AND GROUND CONNECTIONS

There are three ways of adding power connections to active parts, depending on the part's type of power supply pins. A part's power supply pin can be a power-type pin and nonvisible, a power-type pin and visible, or a nonpower-type pin (such as a passive or input pin), which is always visible. The term *visible* specifically refers to whether the pin is visible to the Wire tool. However, in the general case, a nonvisible pin is also invisible from the user's perspective. Digital parts typically have nonvisible power pins, while analog parts, particularly op amps, commonly use either visible power pins or one of the nonpower-type pins (which are always visible).

If a part's power supply pin is a power pin and is not visible, you cannot connect a wire (a net) to it directly. A nonvisible power pin is a net and it is global. You connect a part's power pin to a power symbol by giving the pin and the power symbol the same name. To make the connection, you need to place a power symbol, which is also global, somewhere on the schematic. Power symbols are always visible and are wired to either an off-board connector or a PSpice power supply. To make the names the same, you can change the name of the power symbol, the power pin, or both. An example of how to do this is given next.

If a power supply pin is a power pin and it is visible, you can either take advantage of the power pin's global properties using power symbols or make direct connections to it with wires. If you use the pin's global nature, the pin name and the power symbol name must be the same, as described already. If you make a direct connection to the power pin with a wire, you need not consider the naming convention. If you have a multipart package (e.g., a quad op amp with shared power pins), all the parts within the package that are placed on a schematic must have their power supply pins connected in the same way. So either all must be global or all must have wires connected to them.

If a part's power supply pin is a not a power pin (see LT1028 in the LIN_TECH library, for example), you must use a wire to connect the pin to some other object, such as a power symbol or an off-board connector. If you place more than one part from a multipart package that has nonpower-type power pins, connections need to be made to only one of the part's power supply pins (although you can make connections to all of them if you want). See Chapter 7 for more information on pin types.

The LM741 op amp used in this example has visible power supply pins, so we could use their global properties by using power supply symbols to make connections to the off-board connector, but we will add the power supply symbols to the amplifier to make it evident from the schematic what power the amplifier uses. Whether you used the LM741 from the OPAMP library or the uA741 from the EVAL library, both power supply pins are visible power pins. The difference in their appearances is that the uA741 has zero-length power pins and the LM741 has line-length power pins. A zero-length pin is still "visible" to the Wire tool but not to the user.

To place power symbols, click the Place Power tool button and select one of the power supply symbols (VCC in this example) from the Place Power dialog box, as shown in Figure 9-7. Click OK and place the symbol onto the schematic.

FIGURE 9-7
Placing global power symbols.

The names of the power pins on the op amp and the names of the power symbols must be the same. You can change the name of the symbol or the pin or both. Some parts have no visible labels for their power pins. *To check a pin's name and type,* select the pin (if it is visible), right click, and select properties at the pop-up menu to get a Property Editor dialog box for the pin (see Figure 9-8).

To change the name or type of a nonvisible power pin, select the part on the schematic, right click, and select Edit Part from the pop-up menu. You will be given a Capture Part Editor window (see Figure 9-9). Double click the pin whose name you want to change to bring up the Pin Properties dialog box as shown in Figure 9-9. Enter the new name in the Name: text box and click OK. Repeat the process for the other power supply pin.

Close the Part Editing window. Click Update Current when Capture asks Would you like to update only the part instance being currently edited, or all part instances in the design? In this case, since there is only one LM741, it does not matter if you click Update Current or Update All. But if you had several LM741s and you did not want all of them to have their power pin names changed, you would click Update Current.

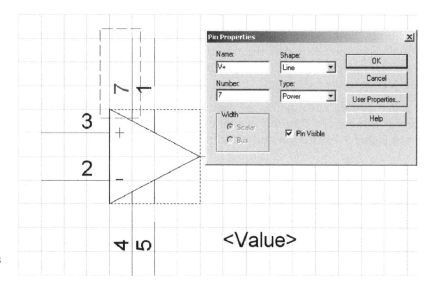

FIGURE 9-8
Using the part Property Editor to change a pin's name or type.

FIGURE 9-9
Determining a power pin's type and name.

Note: When you change a part on the schematic, the link between the design cache and the part library is broken. See the section at the end of the chapter for details on managing the design cache.

Instead of changing the name of the power pin on the part, you can change the name of the power supply symbol. *To change the name of the power supply symbol*, double click the symbol's name on the schematic. The Display Properties dialog box will pop up, as shown in Figure 9-10. The op amp's positive power supply pin name is V+ , so enter V+ in the Value: text box then click OK. Place another power supply symbol and change its name to V− using the same procedure. Copy, place, and connect the V+ and V− power symbols as shown in Figure 9-5.

Next, add a ground symbol. *To place a ground symbol*, click the Place Ground tool button, and select one of the ground symbols from the Place Ground dialog box. Click OK and place the symbol onto the schematic. Place and connect ground symbols as shown in Figure 9-5.

Note: For designs that will be simulated with PSpice, you must use a GND symbol named 0 for the circuit to simulate correctly. You can use any of the symbols as long as they have the name 0. For PCB designs that will not be simulated, you can use any of the GND symbols and name them whatever you want. A separate net will be instantiated for each distinct GND name. See Examples 2 and 3 for techniques on how to establish multiple GND systems.

FIGURE 9-10
Use the Display Properties dialog box to change a power symbol's name.

Preparing the Design for PCB Editor

Once all of the connections have been made, the next step is to prepare the design for making a netlist.

Things to do:

- Make sure all of the footprints are assigned.
- Assign parts to groups (called rooms).
- Perform an annotation.
- Clean up the design cache.
- Perform a DRC in Capture.

There are several ways to find out which (if any) parts have footprints assigned to them and if they are the right ones. You can check each part one at a time by double clicking a part to bring up the properties spreadsheet and look at the PCB Footprint cell, as shown in Figure 9-11. Or you can check all the parts at the same time by selecting all the parts in the schematic (drag a box across the schematic using the left mouse button), then right click and select Edit Properties... from the pop-up menu. The spreadsheet will show the properties of all the parts that were selected. To toggle the spreadsheet view between the

FIGURE 9-11
Part properties spreadsheet.

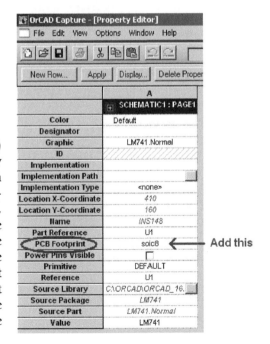

column and row organization, right click the upper left corner or the spreadsheet and select Pivot from the pop-up menu.

The spreadsheet method is fast and easy for small circuit designs, but it may be impractical for large circuits. An alternative is to generate a customized BOM that lists all the footprints in an Excel spreadsheet. The spreadsheet can be printed and used as a reference while you are searching through the footprint symbol library for the right footprints.

To generate the custom BOM, go to the Project Manager, select the Design icon, and then select Bill of Materials... from the Tools menu (see Figure 9-12). At the Bill of Materials dialog box, add the text \tFootprint in the text box labeled Header: and add the text \t{PCB Footprint} in the text box labeled Combined property string:. Put a check mark in the Open in Excel box. You can specify the location and name of the BOM using the Browse... button at the bottom. Click OK when you have finished the setup.

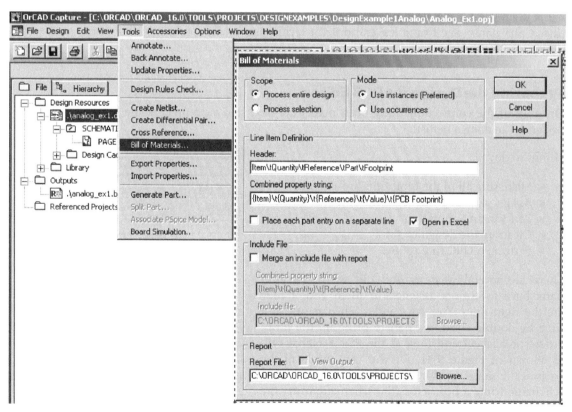

FIGURE 9-12
Generating a bill of materials to include PCB footprints.

Table 9-3	Bill of Materials with Available PCB Editor Footprints	
Reference	**Capture part**	**PCB Editor footprint**
C1, C2	DISCRETE/CAP POL	ck17-10pf
C3, C4	DISCRETE/CAP NP	smdcap
J1	CONNECTOR/CON6	conn6xx
R1, R2, R3	DISCRETE/R	smdres
U1	OPAPM/LM741	soic8

The next step is to assign footprints to the components. To find footprints in PCB Editor's footprint libraries, navigate to the C:\OrCAD\OrCAD_16.0\share\pcb\pcb_lib\ symbols folder (see Figure 9-1). Footprints have the *.dra and *.psm file extensions. When you find the footprint you are looking for, either memorize it or copy the name (without the extension) to the Windows clipboard.

Go back to Capture. To assign the footprint to a part, display the properties spreadsheet for the part to which you want to assign the footprint (double click the part to display its properties spreadsheet). Type or paste the footprint name (without the file extension) into the PCB Footprint cell in the properties spreadsheet (see Figure 9-11).

If several parts have the same footprint (the capacitors and resistors, for example), you can change all of them at the same time. *To assign a footprint to multiple parts simultaneously,* hold down the Ctrl key on your keyboard while you select each component, right click, and select Edit Properties... from the pop-up menu. Select the PCB Footprint column (or row if the spreadsheet is pivoted) to select all of the PCB footprint cells then right click and select Edit... from the pop-up menu to display the Edit Property Values mini spreadsheet shown in Figure 9-13. Paste the footprint name that you copied from PCB Editor into the PCB Footprint cell then click OK. Once you have all of the footprints assigned, you can generate another bill of materials, so that you can see all of the footprints at the same time to make sure you have not missed any. Figure 9-14 shows the final BOM listing with all the footprints assigned.

You can also use the Find tool to automatically select a part or parts of a certain type. To use the Find tool select Find from the Edit menu; the Find dialog box will be displayed as shown in Figure 9-15. If you enter R1 in the Find What: text box, R1 will be highlighted on the schematic and centered on the screen. If you enter R*, then all resistors will be selected. You can use the Find tool to look for other things, such as text and nets, as indicated in Figure 9-15.

FIGURE 9-13
Assigning footprints to multiple parts with the Property Editor.

	A	B	C	D	E
1	ANALOG_EX1.DSN	Revised: Tuesday, August 05, 2008			
2	Revision: 1				
3					
4	Bill Of Mat	Page1			
5					
6	Item	Quantity	Reference	Part	Footprint
7					
8					
9	1	2	C1,C2	10uF	ck17-10pf
10	2	2	C3,C4	0.1uF	smdcap
11	3	1	J1	CON6	conn6xx
12	4	1	R1	1k	smdres
13	5	2	R2,R3	2k	smdres
14	6	1	U1	LM741	soic8
15					

FIGURE 9-14
Final BOM listing with all footprints assigned.

FIGURE 9-15
Using the Find tool to
locate parts.

GROUPING RELATED COMPONENTS (ROOMS)

Grouping related components in the schematic can make placing the parts easier
in PCB Editor. Parts are assigned a room number in Capture, and the grouping
information is exported to PCB Editor during netlist generation (and ECO processes). *To add parts to a group*, select the related parts (e.g., J1 and the two filter caps) on the schematic. Right click and select Edit Properties from the pop-up
menu to display the Property Editor spreadsheet (see Figure 9-16). In the Filter by:
dropdown list, select Cadence-Allegro. Select the ROOM cell to select that property
for all of the components then right click and select Edit from the pop-up menu.
In the Edit Property Values dialog box, assign an integer number or any alpha-
numeric characters for the group (this will be the room name PCB Editor uses).
Click OK to close the dialog box. The alpha-numeric value will be assigned to all
the parts you selected. Close the spreadsheet.

You can also add the room property to the bill of materials listing by adding
\tRoom to the Header: list and \t{ROOM} to the Combined Property String list.

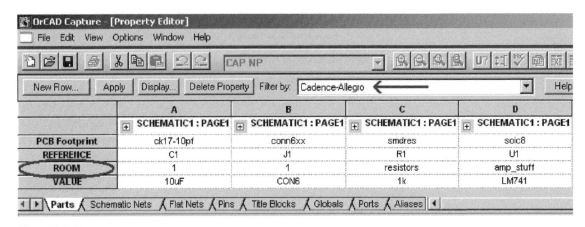

FIGURE 9-16
Using the Property Editor to assign components to groups (Rooms).

Before making the netlist, it is a good idea to tidy up the design by performing an annotation, cleaning up the design cache, and performing a DRC as described next.

ANNOTATION

Performing an annotation chronologically renumbers the part references in your schematic design from top to bottom. *To perform an annotation*, go to the Project Manager and select Annotate from the Tools menu. The Annotate dialog box shown in Figure 9-17 is displayed. Select the items you want updated then click OK. Be sure to check multipart components (hex inverters, for example) to make sure that they are annotated correctly.

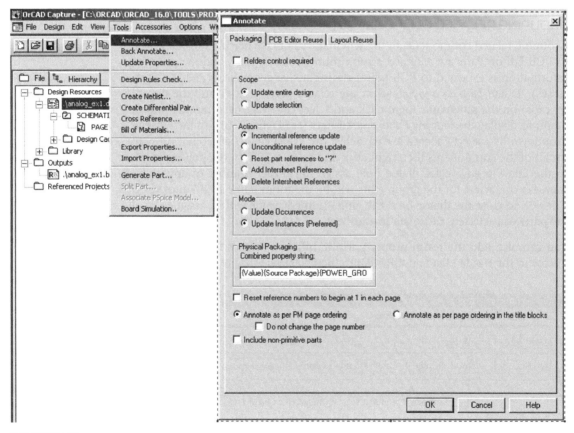

FIGURE 9-17
Performing a design annotation.

CLEANUP CACHE

Cleanup Cache removes unused parts from the design cache. Unused parts pile up in the design cache when you place parts in the design then later delete them. To clean up the design cache, select the Design Cache folder in the Project Manager, right click, and select Cleanup Cache from the pop-up menu or select Cleanup Cache from the Design menu.

PERFORMING A SCHEMATIC DRC IN CAPTURE

It is a good idea to run a DRC before generating a netlist. The DRC checks your design for design rule violations and places error markers on the schematic page. *To perform a DRC* in a Capture project, make the Project Manager window active. From the Tools menu, select Design Rules Check…. The Design Rules Check dialog box is displayed as shown in Figure 9-18. Select the items you want to check and click OK. If you have errors, you can search for the markers by using the Browse DRC Markers command from the Project Manager's Edit menu or from the Find option on the schematic's Edit menu.

FIGURE 9-18
The Capture Design Rules Check dialog box.

You can refine what the DRC looks for and how it reports its findings (as errors, E, or warnings, W) using the ERC Matrix tab in the Design Rules Check dialog box. As shown in Figure 9-19, the matrix lists pin types on the left side and along the slanted side. The yellow W and red E boxes indicate the result of having pins of certain types connected together. For example, an output pin

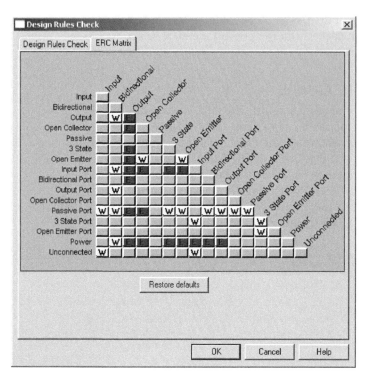

FIGURE 9-19
The DRC error type selection matrix.

connected to another output pin will produce an error during a DRC, but a passive pin connected to an output pin will not cause an error or a warning. You can change the boxes by clicking on them with the left mouse button. By clicking on a button repeatedly, you can toggle through the three options: warning, error, or neither.

GENERATING THE NETLIST FOR PCB EDITOR

Once you have all of your connections finished and footprints and rooms assigned, you can create the netlist for PCB Editor. *To create a netlist for PCB Editor*, select the Design icon in the Project Manager and select Create Netlist... from the Tools menu. In the Create Netlist dialog box (see Figure 9-20), select the PCB Editor tab and specify a netlist name or accept the default name. Select the Create PCB Editor Netlist and the Create or Update PCB Editor Board (Netrev) boxes. You can specify an Input Board File if you have one (such as a board template or preexisting board design), or you can leave it blank. A default Output Board File will be set up for you, or you can specify your own. Select the Open Board in PCB Editor radio button and then click OK. PCB Editor will automatically start.

PROBLEMS DURING NETLIST CREATION

During the netlist creation, you may experience success, success with warnings, or failure. If there are problems, OrCAD will tell you to look at the netrev.lst file to find out what the problems were. The netrev.lst file is located in the folder where the board resides. When you open the file, you will be presented with the File Conversion dialog box shown in Figure 9-21. You can scroll down and look for the error in the preview pane or open the entire document.

In this example an error occurred because the Capture part has one fewer pins than the PCB Editor footprint symbol. Since this is not a fatal error, the netlist process was completed and PCB Editor will be launched. If a fatal error occurs, PCB Editor will still be launched, but the netlist process will not be completed properly and no information will be transferred to PCB Editor. In this case, though, the error will prevent placing the U1, so we need to go back and modify

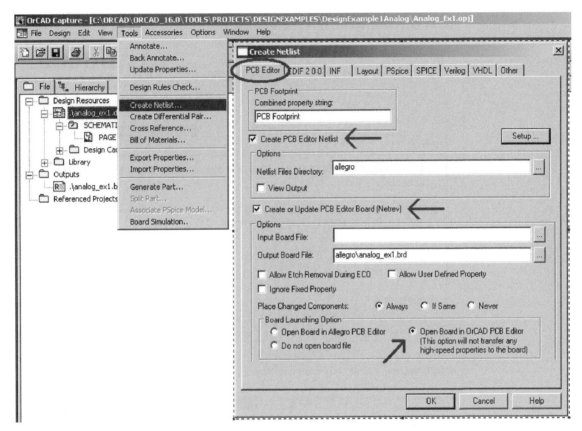

FIGURE 9-20
Creating a circuit netlist for PCB Editor from Capture.

the schematic. We use this error as an opportunity to show how to perform an ECO later in the example.

Setting Up the Board

The first view you have of your new board will be PCB Editor with a black background in the work area. The next steps are to set up the board outline and define some of the basic design parameters.

SETTING UP THE EXTENTS (WORK AREA BOUNDARY)

The default size of the work area is 34 in. × 22 in. (size D). If the drawing area is much larger than it needs to be, that can make navigating around and working on your project harder than if the drawing area size is just right. *To change the size of the work area*, go to Setup → Design Parameters, and change drawing size or change the X and Y Extents on the Design tab of the Design Parameters dialog box, as shown in Figure 9-22. Make sure you click Apply to make the changes take effect, then click OK.

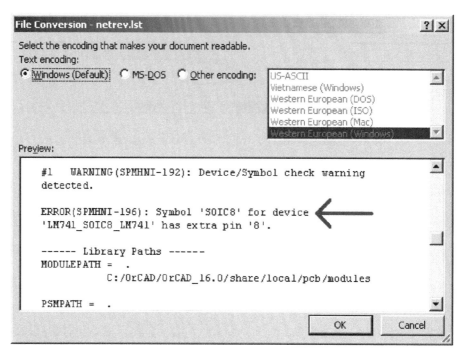

FIGURE 9-21
Using the netrev.lst file to find netlisting errors.

FIGURE 9-22
Setting up the board extents.

CONTROLLING THE GRID

The next step is to turn on the grid. *To turn the grid on,* click the Grid button, .
To change the grid spacing, select Grids... from the Setup menu; the Define Grid dialog
box will be displayed (see Figure 9-23). There are two grid systems: one for copper elements (Etch) and one for everything
else (Non-Etch). You can change the etch
grids on each layer individually or change
them all simultaneously by using the All
Etch entries. Click OK to dismiss the dialog
box when you are finished.

MAKING A BOARD OUTLINE

The next step in laying out the board is to
make the board outline and add mounting holes so that the boundaries of the
board are known.

First we draw the board outline. Zoom in
an appropriate amount so that your final
board size will take up about 75% of
your display's viewing area (large enough
for good resolution but small enough
that you can see the whole thing.). To
change zoom, select one of the zoom
options from the View menu, or select
one of the Zoom buttons on the toolbar.

To place a board outline, select Setup →
Outlines → Board Outline... from the
menu bar. The Board Outline dialog box
will be displayed, as shown in Figure
9-24. In the basic example in Chapter 2,
we just drew a box. This time we draw
an irregular shape using the polygon
tool. Click the Draw Polygon radio button
in the Board Outline dialog box then left
click and release once in the work area at
the starting vertex of the board. Left click
at point 0, 0 to place the first vertex. Left
click to place each vertex of the board
outline. Make the board outline about
5.0 × 3.0 in. (5000 × 3000 mils). Left
click and release at each vertex. To finish
the board, click on the starting point again to close the polygon. The board
outline will be a dashed line with squares (handles). You can use the handles to
resize or change the shape of the board outline. If the board outline is correct,

FIGURE 9-23
Setting the design grids.

FIGURE 9-24
The Create Board
Outline dialog box.

FIGURE 9-25
The finished board out-line and keep-in areas.

then click the OK button to dismiss the Board Outline dialog box. The example board outline is shown in Figure 9-25.

ADDING DIMENSION MEASUREMENTS

You can add dimension lines and measurements to your board for documentation purposes. You can also add temporary construction lines as guides to mark locations for mounting holes and connectors and the like. Dimensions are placed on the BoardGeometry/Dimension layer (class/subclass), so once you are satisfied with the board layout, you can turn off that layer to hide the dimension objects or delete temporary construction lines.

For this example we add dimension lines to mark a mounting hole location. First set the grid spacing for nonetch to 10 mils (see Figure 9-23). Next you may want to change the coordinate indication (XYMode) to absolute or relative depending on the dimension you are using. To change the XYMode, click the R or A button at the bottom of the design window (see Figure 9-26).

FIGURE 9-26
Toggle between absolute and relative XY modes.

To place a dimension line, select Dimension/Draft from the Manufacture menu and from the menu choose the desired dimension type. Figure 9-27 shows the available dimensioning tools. A linear dimension is demonstrated here.

FIGURE 9-27
Dimensioning tools.

Left click to place the first dimension point (point 1 in Figure 9-28), left click to place the end mark (point 2 in Figure 9-28), and finally left click to place the location of the text (point 3 in the figure). Use the *XY* coordinates to precisely place the dimensions. You can also just pick a line object, and the dimension tool will automatically measure it and place the end points for you.

You can also use the Add Line, ![], and Add Rectangle, ![], tools to add construction lines, etc. to your design on the BoardGeometry/Dimension layer.

Add Line   Add Rectangle

You can change dimension parameters such as text size and arrow terminations. *To change the appearance of the dimension lines*, select Manufacture → Parameters from the menu bar to display the Drafting dialog box, from which

FIGURE 9-28
Linear dimension line and pick points.

you can change various aspects of the dimension elements through the dialog boxes shown in Figure 9-29.

Once you place a dimension and end the command, the dimension object is nondynamic. This means that you can move it, but it will not be updated with new dimension information. If you need to change or move a dimension line to measure a new parameter, you have to place a new one.

ADDING MOUNTING HOLES

Next we place mounting holes on the board. This should be done before moving any parts into the board outline. *To place mounting holes*, select Place → Manually from the menu bar and select Mechanical symbols from the Placement list as shown in Figure 9-30. If no symbols are listed, check to make sure you have the Library box checked in the Advanced Settings tab.

Select the desired mounting hole symbol (it must be in the library ahead of time). Left click at the desired location in the work space to place the mounting hole. Only one will be placed per click. To add additional holes, recheck the desired symbol and left click again.

To add multiple mounting holes, you can copy and paste instead of using the placement method. To copy a mounting hole, click the Copy button (or type copy in the command line and hit the Enter key). The command window will say Select elements to copy:. Use the left mouse button to drag a box across an existing mounting hole (to select all elements of it). The command window then will say Pick user-pick origin. Click on the object (or near it) to get a handle on it. Left click to place one or more copies of the mounting hole. When you are finished placing the mounting holes, right click and select Done from the pop-up menu.

To delete a mounting hole click the Delete button (or type delete in the command line and hit the Enter key). The command window will say nothing. Use the left mouse button to drag a box across an existing mounting hole (to select all elements of it). Right click and select Done from the pop-up menu. Alternatively you can use the left mouse button to drag a box across an existing mounting hole (to select all elements of it). Right click and select Symbol → Unplace component.

The mounting holes supplied with PCB Editor are nonplated. Nonplated holes are drilled separately as one of the very last steps of the board fabrication and no plating processes occur after the final drilling. See Chapter 8 on designing mounting holes and Appendix D for standard machine screw and drill hole sizes. To make mounting holes plated, select Tools → Padstack → Modify Design

Drafting

Standard conformance

- ● ANSI ○ ISO ○ BSI
- ○ JIS ○ AFNOR ○ DIN

Parameter editing

Units: ● Inches
○ Millimeters

...	Dimension text
...	Dimension lines
...	Extension lines
...	Balloons

OK   Help

Dimension Text

General parameters

Text block: 9
Place holder: ● Period ○ Comma

□ Use leading
□ Use trailing
□ Align text w
□ Use diamet

Diameter s

Radius sym

Special ha

Arrows

Head
Head
Head

Scale fact

Dimensional tol

□ Use toleran
Type:
Location:

Primary dimensions

Units: Inches

s: 2

Dimension Line

Terminations

Leader: Arrow
Linear Top/Left: Arrow
Linear Bottom/Right: Arrow
☑ Fill

Extension Line

Line Suppression:
Offset Distance from Element:
Distance Beyond Dimension Line:

OK

Bullet Diameter: 0.150
Slash Length: 0.150
Odd Angle Axis: Horiz/Vert

OK   Help

Balloon Parameters

General

Type: Circle
Size: 0.50
Text block: 13
Next value: 1
☑ Auto-increment value

Auto-size

□ Auto-size
Max characters: 2

OK   Help

OK   Help

)0.00 5040.00
nts to place using tree view.
CAP POL_CK17-10PF_10UF / CK17-10PF .
)0.00 4340.00

FIGURE 9-29
Controlling dimension-ing parameters.

Placement

Placement List | Advanced Settings

Mechanical symbols

☐ ☐ 📁 Mechanical symbols
- ☐ ◆ EUROD
- ☐ ◆ EUROS
- ☐ ◆ IBM
- ☑ ◆ MTG125
- ☐ ◆ MTG125B
- ☐ ◆ MTG156
- ☐ ◆ MTG250
- ☐ ◆ MTG4P
- ☐ ◆ MTG4P2
- ☐ ◆ MULTIBUS

Selection filters

Match:

Quickview

● Graphics ○ Text

OK   Hide   Cancel   Help

FIGURE 9-30
Placing mounting holes on the board.

Padstack from the menu bar, left click the mounting hole to select it, right click, and select Edit from the pop-up menu; the Padstack Designer dialog box will be displayed as shown in Figure 9-31. In the Drill/Slot hole section, you can select whether the hole is plated or not.

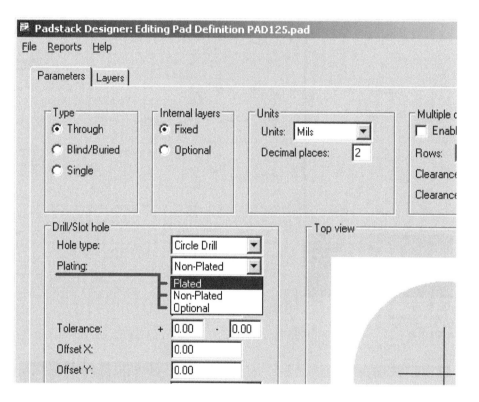

FIGURE 9-31
Using Padstack
Designer to determine
hole plating.

You can leave a mounting hole isolated from the copper on all layers, or if it is a plated mounting hole, you can connect it to a net (e.g., the Ground plane). To connect a plated mounting hole to a net, you need to add a single pin part to the design in Capture then connect the part's pin to the desired net. The mounting hole will become part of the netlist. Then in PCB Editor the mounting hole can be placed in the same manner as all the other parts, as described later, and it will be connected to the proper net. This is demonstrated in the manufacturing design example in Chapter 10. Along with making a part in Capture, you also need to make a footprint with a connect pin (rather than a mechanical symbol with a mechanical pin) for the mounting hole in the PCB Editor footprint library (see Chapter 8 for details).

PLACING PARTS

Placing parts is part art and part science. How you ultimately place the parts on the board depends on both mechanical and electrical factors. Mechanical factors include designing for manufacturability (assembly and soldering processes) and physical board constraints (size, shape, etc.). Electrical factors include functional signal flow, thermal management, signal integrity, and electromagnetic compliance. Usually all these considerations are important, and in some cases, they can conflict with one another. These issues are discussed in greater detail in Chapters 4, 5, and 6 and in the IPC standards. In this example, the parts are simply placed so that the board layout is similar to the flow of the schematic.

Before you begin placing parts, you may want to adjust the placement grid resolution. Go to Setup → Grid and change the Non Etch Grid to 50 mils (or 25 mils for greater resolution, as suggested by IPC).

The three basic ways to place parts are manually, quick place, and through intertool communication (ITC). The easiest method for small projects is to use the manual method, as was shown in Chapter 2 (i.e., select Place → Manually from the menu bar, select the part or parts you want to place, and left click to place them). In this example we see how to use the quick-place method and the Intertool Communication tool to select parts for placement.

Note: Regardless of how you place the parts, you will likely have a problem placing U1 in this example if you used the default op-amp part in Capture and a default eight-pin footprint in PCB Editor. If this occurs we will see later how to fix the problem using the ECO function.

To use the quick-place method, select Place → Quickplace… from the menu bar; the Quickplace dialog box will be displayed (see Figure 9-32). There are many ways of selecting and filtering parts for placement. When using the Placement Filter options, a property has to be assigned to a part or parts for that filter to work (e.g., a room has to be assigned to use the Place by room filter). Note that there are similarities between the Quickplace dialog box and the Placement dialog box. When using the Place by REFDES filter, the Placement dialog box works better (it has more details options and allows you to see what parts match the filters), but the Quickplace dialog box gives more control with the Placement Position section options.

In this example we use the Quickplace dialog box to place parts by rooms. Recall that rooms were assigned to parts in Capture; this is where we use that property. Before we do, though, we need to define areas on the board where the parts with common room numbers will be placed. *To define a room on the board,* select Setup → Outlines → Room Outline… from the menu bar. The Place Outline dialog box will be displayed (see Figure 9-33). Also, if you look at the Options tab, you will notice that the default class/subclass is Board Geometry/ Both_Rooms (also indicated in the Side of Board section, the Both radio button is checked). In the Place Outline dialog box, you can specify the room (group) that was assigned in Capture (i.e., the Room Name). You can also select the room type. The options are HARD, SOFT, INCLUSIVE, HARD_STRADDLE, and INCLUSIVE_ STRADDLE, which determine how the DRCs are issued (see *Using the ROOM and ROOM_TYPE Properties* in alegroplace.pdf, p. 24). With the Place Outline

FIGURE 9-32
Using the quick-place method to place parts.

FIGURE 9-33
Creating a Room outline.

dialog box displayed, draw the outline. When finished, click the OK button to dismiss the Place Outline dialog box.

Once the room outlines are drawn, you can then use the Place by room option to place parts. In the Quickplace dialog box (or the Placement dialog box), select the Place by room option and select the desired room name, then click the Place button. The parts that were assigned to that room name will be placed in the room outline (see Figure 9-34). You can then use the Move tool, , to move the parts around within the room.

Move tool

FIGURE 9-34
Parts placed in a room using quick place.

Note: Remember that U1 will probably not place properly because of the problem mentioned previously. This will be addressed later.

If you select a part and do a Show Element, you will see the ROOM property included as shown in Figure 9-35.

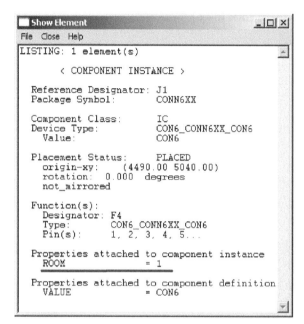

```
Show Element                                    _|□|×|
File  Close  Help
LISTING: 1 element(s)                             ▲

        < COMPONENT INSTANCE >

Reference Designator: J1
Package Symbol:        CONN6XX

Component Class:       IC
Device Type:           CON6_CONN6XX_CON6
  Value:               CON6

Placement Status:      PLACED
  origin-xy:    (4490.00 5040.00)
  rotation:  0.000  degrees
  not_mirrored

Function(s):
  Designator: F4
  Type:       CON6_CONN6XX_CON6
  Pin(s):      1, 2, 3, 4, 5...

Properties attached to component instance
  ROOM              = 1

Properties attached to component definition
  VALUE             = CON6
                                                  ▼
```

FIGURE 9-35

Using Show Element tool to display a component's room assignment.

INTERTOOL COMMUNICATION

Intertool communication is a communication link between Capture and PCB Editor. This link allows you to perform two functions: cross-probing (also called *cross selection*) and cross-highlighting. Cross-probing is used as an aid in placing components in PCB Editor (communication is from Capture to PCB Editor only). Cross-highlighting is a tool that allows you to visually highlight components, pins, or nets in one application and simultaneously highlight it in the other application (communication is bidirectional). To enable inter-tool communications, see Figure 9-47.

USING CROSS-PROBING TO PLACE PARTS

Depending on the complexity of your design it may be necessary to lay out specific parts of the board in a particular order or location on the board. In this case it might be easier to pick the specific parts from the schematic and have PCB Editor place those parts, and this is exactly what cross-probing does. For cross-probing to be enabled, you have to have your schematic open in Capture, your board layout open in PCB Editor, and be in manual placement mode in PCB Editor.

Prepare for cross-probing in PCB Editor by selecting Place → Manually... from the toolbar to display the Placement dialog box. From the dropdown menu, select Components by refdes and, in the Selection filters section, make sure the Match: radio button is selected (see Figure 9-37 later).

To begin cross-probing from Capture to PCB Editor, go to your schematic in Capture. Left click the part you want to place to highlight it (C1 in this example) then right click and select PCB Editor select from the pop-up menu, as shown in Figure 9-36.

Now go back to PCB Editor. That part will be checked in the dialog box and the part will be attached to your cursor, as shown in Figure 9-37. Left click to place the part. You can either click OK (or right click and click Done) to quit or go back to the schematic and select another part as just described.

ROTATING PARTS

To rotate a placed part, select the Move button, left click to select the part, then right click; select Rotate from the pop-up menu. The part will pivot around its origin, which may be pin 1 or the body center. Move the mouse around the pivot point until it is in the correct orientation. Use the Options pane to select a

FIGURE 9-36
Using cross-probing to
place parts.

FIGURE 9-37
The part selected in
Capture is also selected
in PCB Editor.

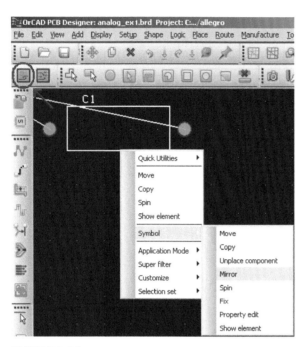

FIGURE 9-38
Mirroring a placed component.

different rotation criteria. Left click to place it, then right click and select Done from the pop-up menu.

MIRRORING PARTS

If you want to move a part to the back side of your board, you mirror it. *To mirror a part,* select the Generaledit button (vice the Etchedit button), left click to select the part, right click and select Symbol → Mirror from the pop-up menu, as shown in Figure 9-38.

As mentioned earlier U1 may not place properly because of the netlist warning that occurred during the netlist creation. The problem is that the part in Capture has only seven pins and the footprint in PCB Editor has eight pins. Every pin in a PCB Editor footprint must have a corresponding pin in the Capture part that is using it. So we either need to add a pin to the Capture part in the schematic or delete a pad in the PCB Editor footprint on the board. While it is possible to do the latter by deleting the connection type pin and installing a mechanical type pin, this would make that footprint useful for only that part or parts like it. You would need a different footprint for other op amps. This is true for SOT23 packages for diodes as well. So we change the part in Capture then perform an engineering change order (ECO) to update the board design.

CHANGING A PART IN CAPTURE

Go back to the schematic in Capture. Left click the op amp to select it. Right click and select Edit Part from the pop-up menu.

Place pin button

In the part editor (see Figure 9-39), click the Place pin button, . In the Pin Properties dialog box, enter NC (no connect) for pin name and the number 8 under pin number. Select the Short and Passive options. Click OK and place the pin somewhere on the part boundary, as shown. You can also add a line to make the pin touch the body if you like or leave it floating, since it is a no-connect (NC) pin. And, in that case, it might be a good idea to add text indicating that, as shown in the figure.

Note: If a part has more than one NC pin, you have to give them different names (e.g., NC1, NC2, etc.) because a unique name is required for each pin; otherwise you will get a DRC error.

Once you finish modifying the part, close it by clicking the X in the upper right corner of the part editor window. Click Update Current (or Update All if there are

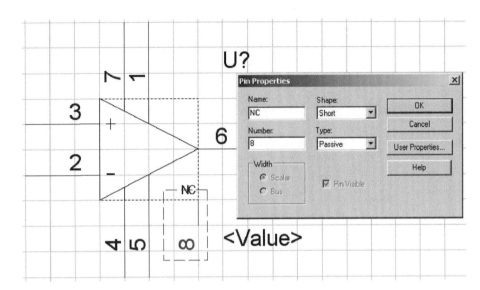

FIGURE 9-39
Adding a no-connect pin
to a part in Capture.

more than one of this type of part) when Capture asks Would you like to update only the part instance being currently edited, or all part instances in the design?

Make sure that you place a no-connect symbol on the unused pin on the schematic.

PERFORMING AN ENGINEERING CHANGE ORDER (ECO)

Once the part has been changed on the schematic, the changes need to be updated in the board design. In PCB Editor, an ECO is used to update the board design with changes that were made to the schematic in Capture, whereas an update from PCB Editor to Capture is called *back annotation*. Performing an ECO is a two step process: (1) recreate the netlist in Capture and (2) import the changes into the board in PCB Editor.

To perform an ECO, begin in Capture by making the Project Manager window active. Highlight the Design icon and then select Tools → Create Netlist from the menu bar to display the Create Netlist dialog box (Figure 9-40). So far this is the same process used to create the original netlist, but the next step is slightly different. In the Options section below Create or Update PCB Editor Board (netrev), use the Input Board File: Browse... button, ▥, to locate the current board you are working on (the one you want updated). Then use the Output Board File: Browse... button to choose the location and name of the new updated board. You can update a board into itself, so the input board and the output board are the same, as shown in the figure. In the Board Launching Option section, select Do not open board file if your board file is already open, otherwise select Open board in PCB Editor to open your board file. Click OK.

Browse... button

If you selected Open board in PCB Editor, the changes will be made automatically, but if you selected Do not open board file, then the board design must be updated with the changes.

FIGURE 9-40
Setting up for an ECO.

If you relaunched PCB Editor from Capture, the board will be automatically updated with the new information. If you selected Do not open board file, then you need to import the changes to update the board. *To update the design in PCB Editor*, select File → Import → Logic from the menu bar. The Import Logic dialog box will be displayed. Select Design entry CIS (Capture) in the Import logic type section. Select the desired option in the Place changed component section. Use the Browse... button to locate the directory listed in the Output Board File: (see Figure 9-41) if it is not already displayed. Click the Import Cadence button to finish the ECO process. Your board should now have the updated information.

Note: Depending on which version of the software you have, you may experience difficulties running the autorouter if you performed the ECO using the import logic method. If that occurs, save your design and close PCB Editor. Go back to Capture and perform another ECO, but launch PCB Editor from the Create Netlist dialog box. You will not need to import logic again and the autorouter should work properly. If you are using the demo version, the autorouter is nonfunctional.

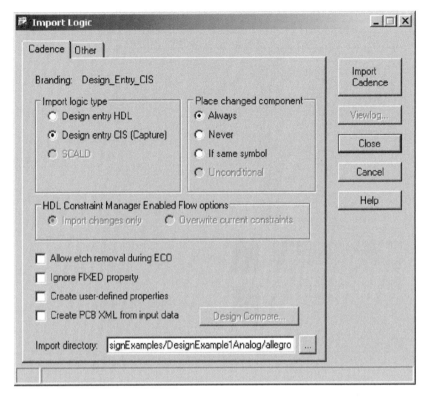

FIGURE 9-41
Import ECO information with the Import Logic dialog box.

If you were not able to place the op amp, U1, you should be able to place it now. The quickest way to place the part is to select Place → Manually from the menu (or click the Place Manual button on the toolbar). U1 should be on the list; check it and place the part on the board; click OK when you are finished.

Once the parts have been placed on the board, you need to position them. If all of the layers and nets are visible, it can be difficult to identify which parts are which. You can do a couple of things: turn off all or most of the nets and make only the necessary layers visible (e.g., silk-screen layers and pins).

FIGURE 9-42
Use the Options pane to control layer visibility.

CONTROLLING LAYER VISIBILITY

There are two methods for controlling layers (class/subclass) visibility. The first method is the Options tab (Figure 9-42) and the second is the Color dialog box (Figure 9-43). *To change the visibility of a layer using the* Options *tab,* move your mouse pointer over the Options tab (if the pane is not displayed) and select the class and subclass of interest. Toggle the colored button to the left of the subclass to show or hide that subclass.

FIGURE 9-43
Using the Color dialog box to control layer visibility.

Color button

To change the visibility of a layer using the Color *dialog box*, click the Color button, , on the toolbar. Most of the layers are common to both the Options tab and the Color dialog box, but the Color dialog box gives you access to more objects. Besides letting you change the visibility of objects and layers, it lets you change their colors. *To hide (or show) a layer*, find the class (the list of folders) and subclass you are interested in. Toggle the X beside the object or layer of interest (X means it is on, blank means it is off). *To change the color of an object or layer*, find and click on the color of choice in the color palette, then click the colored box for the object of choice. The box will change to the desired color. Click Apply to make the changes take effect, then click OK to dismiss the dialog box.

You can use the Color dialog box to change the visibility of one subclass or object at a time, an entire class, or the entire project simultaneously. *To change an entire subclass*, toggle the X on or off in the All row or column as indicated by points 1 and 2 in Figure 9-44. *To change the visibility of all classes and subclasses simultaneously*, click the Global visibility buttons (point 3 in Figure 9-44). Make sure you click the Apply and then OK buttons to make the changes take effect.

When placing parts and routing traces, it is sometimes easier to turn off all but a few essential subclasses. For example, when you first start to move parts around, you may want only the board outline and the component reference designators and outlines visible. In that case it is easier to turn off all classes,

FIGURE 9-44
Controlling the visibility
of multiple subclasses
simultaneously.

using the Global Off button, then turn on only the ones you want to be able to see. Some layers and classes/subclasses that you will likely need are

Board Geometry/Outline
Areas/Package Keep-In
Areas/Route Keep-In
Package Geometry/Silkscreen_Top (for component outline)
Components/RefDes/Silkscreen_Top (for component reference designator, R1, U1 etc.)

Optional ones are

Board Geometry/Dimensions
Board Geometry/x_Room (where x is Top or Bottom, etc.)
Areas/Package Keep-Out
Areas/Route Keep-Out

CONTROLLING NET VISIBILITY

There are three ways to turn nets on and off. The first method controls global rats visibility and the others control individual rat visibility.

To simultaneously make all rats invisible, click the Unrats All button, , and *to make them all visible* again, click the Rats All button, .

Unrats   Rats All
All button button

There are two methods to control individual rat visibility (or the visibility of specific groups of rats). One way to control the visibility of individual nets is to

Cmgr button

use the Constraint Manager. To launch the Constraint Manager, click the Cmgr button on the toolbar, ▦.

Note: The Constraint Manager takes a second to start, but if it doesn't seem to do anything for a long while, check the command window at the bottom. If it says Cannot launch Constraint Manager while a command is active, right click and select Done from the pop-up menu to end the current command. Then relaunch the Constraint Manager.

The Constraint Manager is shown in Figure 9-45. Select the Properties tab, select the General Properties icon under the Net folder. Locate the net or nets of interest and click inside the cell under the No Rat column. Select the On choice to apply the no rat property and make the net invisible. You can either close the Constraint Manager or leave it open. Changes made in the Constraint Manager are effective immediately.

FIGURE 9-45
The Constraint Manager.

The last method for controlling the visibility of specific nets is through the use of the Find filter pane. First click the Unrats All button to turn off all the rats. Next select Display → Show Rats → Net from the menu. Display the Find pane and check only the Nets box. In the Find By Name list, select Net and click the More... button. In the Find by Name or Property dialog box (see Figure 9-46), left click the desired nets in the left box to move them to the right box. When you finish selecting the nets, click the Apply button to make the changes take effect. Click the OK button to dismiss the dialog box. Only the selected nets will be visible.

CROSS-HIGHLIGHTING BETWEEN CAPTURE AND PCB EDITOR

When working with large projects, it can be difficult to find specific nets or parts in your board design, or you might want to know where a particular trace

FIGURE 9-46
Using the Find filter to
select specific nets.

belongs in the circuit. Cross-highlighting allows you to select a part in one application and see it highlighted in the other application.

To enable cross-highlighting, you need Intertool Communication enabled in Capture and you have to be in highlight mode in PCB Editor. *To enable* Intertool Communication *in Capture*, select Options → Preferences from the Capture menu bar. In the Preferences dialog box (Figure 9-47), select the Miscellaneous tab and check Enable Intertool Communication. Click OK.

To enter the highlight mode in PCB Editor, click the Highlight button, , on the PCB Editor toolbar (or select it from the Display menu on the menu bar).

Highlight button

Now if you select a net, a part, or a pin on a part in Capture, the corresponding object will be highlighted in PCB Editor. And, if you highlight a net, a part, or a pin on a part in PCB Editor, the corresponding object will be highlighted in Capture.

To dehighlight an object in Capture, simply click in a blank area in the schematic page, or click the Dehighlight button, , in PCB Editor then click that object.

Dehighlight button

To dehighlight an object in PCB Editor, you have to click the Dehighlight button and select that object. Deselecting an object in Capture does not dehighlight the object in PCB Editor. Also, selecting the Dehighlight button does not dehighlight all components; you have to click on each one individually or use the Options filter tab to select certain objects (see Figure 9-48). Once you select the Dehighlight button, cross-highlighting ceases. To restart cross-highlighting, you have to select the Highlight button again.

FIGURE 9-47
Enable Intertool
Communication in
Capture.

FIGURE 9-48
Use the Options pane to
select types of objects
to dehighlight.

Note: If you reopened a board during an ECO process, Intertool Communication may not work properly at first. If so, save the board, close it, then relaunch PCB Editor and reopen the board design normally (not through another ECO process).

FINDING PARTS USING THE FIND FILTER PANE

Once the parts are in their relative positions, you can begin to carefully and precisely place the components. It is often a shuffle game at first, until you get a feel for the parts and the space you have to work with. Besides the cross-selection and cross-highlighting tools, another method for finding parts or nets is to use the Find filter tab. To use the Find filter to highlight a part, display the Find pane (see Figure 9-49), select the object(s) you want to find (point 2), click the More... button (point 3) to display the Find by name or Property dialog box. In the list of components (or whatever you are looking for), select the item of interest (point 4) to move it to the Selected objects window and click Apply (point 5). The object will be centered in the screen and highlighted (as shown in the background of Figure 9-49).

FIGURE 9-49
Using the Find filter to highlight a part.

Once all the parts have been placed, your design might look something like that in Figure 9-50.

FIGURE 9-50
The placed components.

Some of the reference designators on the Silkscreen layer might be upside down or in an undesirable location. *To rotate the reference designators*, select the Move button, check the Text box in the Find filter tab. Move your mouse to the Options tab and select the desired angle from the list. Move your mouse back over to the board area and select the reference designator you want to rotate, right click and select Rotate from the pop-up menu. Left click to place the text, right click and select Done from the pop-up menu. Note: See also the Autosilk tool described in Chapter 10.

Design Rule Check and Status

The next steps in the design process are to define the layer stack-up and begin routing. Before we do that, though, it is a good idea to perform a design rule check (DRC) to check for any problems with the board and how the components are placed. By default the design rule checker is always on, so you do not usually have to specifically perform one unless it is out of date. There are two ways to update the DRC. The first way is to click the Update DRC button, , on the toolbar (if that toolbar is displayed). The second way is to use the Status dialog box.

Update DRC button

USING THE STATUS DIALOG BOX

To check for DRC errors through the Status tool, select Display → Status from the menu bar. The Status dialog box (see Figure 9-51) shows information about how many nets are routed, status of shapes, and if there are any DRC errors. If the DRC Errors box is green, then there are no errors. If the DRC Errors box is red, then there are errors in the design, but the box does not indicate what they are. If the DRC Errors box is yellow, then either the DRC is out of date (click the Update DRC to update it) or there are DRC *warnings.* Warnings may be flagged rather than errors, depending on what the problem is and how the DRC parameters are set in the Constraint Manager (described shortly). To actually view what DRC errors are, you need to generate a DRC report.

FIGURE 9-51
The Status dialog box.

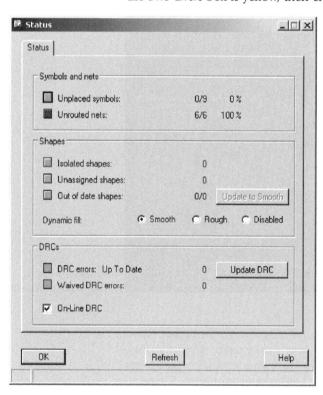

GENERATING DRC REPORTS

A DRC report is also used to find out the location of DRC errors and warnings as flagged by the Status tool. *To generate a DRC report,* select Tools → Quick reports → Design Rules Check Report from the menu bar. As shown in Figure 9-52, a DRC report tells you how many errors there are, where they are located, what type of errors they are, and who the offending parties are.

You can also generate a DRC report by selecting Tools → Reports from the menu bar. A Reports dialog box will be displayed as shown in Figure 9-53. From the Reports

Design Rules Check Report

Search: ☐ Match word ☐ Match case

Design Name C:/ORCAD/ORCAD_16.0/TOOLS/PROJECTS/DESIGNEXAMPLES/DesignExample1Analog/allegro/analog_ex4.brd
Date Sat Aug 16 13:57:56 2008

DRC Error Count Summary

DRC Error Type	DRC Error Count
Package to Room	1
Total DRC Errors	1

Detailed DRC Errors

Constraint Name	DRC Marker Location	Required Value	Actual Value	Constraint Source	Constraint Source Type	Element 1	Element 2
Package to Room Spacing	(3375.00 3500.00)	0 MIL	55 MIL	NONE	DESIGN	Filled Rectangle "Package Geometry/Place_Bound_Top"	Shape "Board Geometry/Top_Room"

FIGURE 9-52
A DRC report.

dialog box, you can select the type of report you want to see. It allows you to select additional options, such as generating multiple reports simultaneously and writing reports to files. To select a report, double click on the desired report in the Available Reports section and then click the Report button. A DRC report is the same with either method.

You can create custom reports using the Reports dialog box. An example of how to create a pick and place file is given in Chapter 10.

DRC ERROR MARKERS

If you click on the DRC marker location in the DRC report, the error marker will be centered in your display. The DRC marker is bowtie shaped, as shown in Figure 9-54. You can change the size of DRC markers to make them easier to see. *To change the size of the DRC markers*, select Setup → Design Parameters from the menu bar. In the Design Parameter Editor dialog box, select the Display tab and change the marker size to the desired value. Click Apply to apply the change and OK to dismiss the dialog box.

FIGURE 9-53
Selecting a report from the Reports dialog box.

You can find out additional information about a particular error with the Show Element window. *To display additional information on a DRC marker*, select the

(a)   (b)

FIGURE 9-54
A DRC marker (a) and its properties (b).

Show Element button

FIGURE 9-55
Reviewing a DRC error with the Show Element tool.

Show Element button, ⓘ, on the toolbar, select DRC Errors in the Find filter tab, and then left click the DRC marker. A text box will be displayed that provides in-depth information about the error, as shown in Figure 9-55.

Once all DRC errors related to placement are corrected, the layer stack-up can be defined if it has not already been done.

```
Show Element                                        _|□|×|
File  Close  Help
LISTING: 1 element(s)                                     ▲

         < DRC ERROR >

 Class:            DRC ERROR CLASS
 Subclass:         PACKAGE_TOP
 Origin xy:        (3375.00 3500.00)
 Constraint:       Package to Room Spacing
 Constraint Set:   NONE
 Constraint Type:  DESIGN

 Constraint value: 0 MIL
 Actual value:     55 MIL

 - - - - - - - - - - - - - - - - - -
 Element type:     FILLED RECTANGLE
 Class:            PACKAGE GEOMETRY
 Subclass:         PLACE_BOUND_TOP

 RefDes:           C4
 Room:             amp_stuff

                   xy-lower         xy-upper
 Rectangle Top:    (3500.00 3455.00) (3250.00 3455.00)
 Rectangle Bottom: (3500.00 3545.00) (3250.00 3545.00)

 Height: 90.00 mils
 Width:  250.00 mils
 Area:   0.0225  (sq in)
                                                         ▼
```

Defining the Layer Stack-Up

The op amp used in this example requires a dual rail supply and at least one Ground plane. Two Routing layers also are needed. Therefore a minimum of five layers is required. Since most multilayer boards are made up of double-sided cores (and therefore usually have an even number of layers), we use the extra layer as an additional ground plane. The layers could be stacked up in several ways (see Chapter 6 for other examples). The stack-up shown Figure 9-56 is used in this example because it provides the opportunity to demonstrate two methods for defining Plane layers.

SETTING UP THE LAYOUT CROSS SECTION

Once you have the parts in place, the next step is to set up the layers. In this design

Layer Function		Layer Name
Routing		TOP
GND Plane		GND_T
+ Pwr Plane		VPOS
– Pwr Plane		VNEG
GND Plane		GND_B
Routing		BOTTOM

FIGURE 9-56
Layer stack-up for Example 1.

we said that we needed six layers: two power planes, two ground planes, and two routing layers.

First, let's take an inventory of what we have. *To review or modify the layer stack-up*, open the stack-up editor (Layout Cross Section dialog box) by selecting the Xsection button, , on the toolbar. The Layout Cross Section dialog box is shown in Figure 9-57.

Xsection button

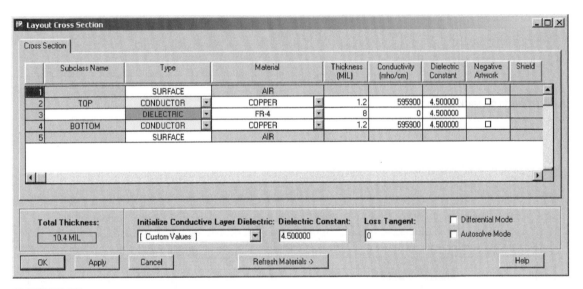

FIGURE 9-57
The layer stack-up as shown in the Layout Cross Section dialog box.

The Layout Cross Section dialog box lists the layers in the stack-up, their type, material properties, electrical properties, and image polarity.

By default the stack-up is defined by the outer surfaces (air), two Routing layers (top and bottom copper), and a dielectric layer (FR-4) between the copper layers. We now add four Plane layers and four Dielectric layers.

To add a Plane layer to the stack-up, right click on any one of the existing layers and select Add Layer Above (or Add Layer Below as appropriate) from the pop-up menu (see Figure 9-58). Do this eight times.

When you are finished, you will have eight dielectric layers. Convert every other layer to a Plane layer by selecting Plane from the Type list and Negative from the Negative Artwork column. The finished stack-up is shown in Figure 9-59.

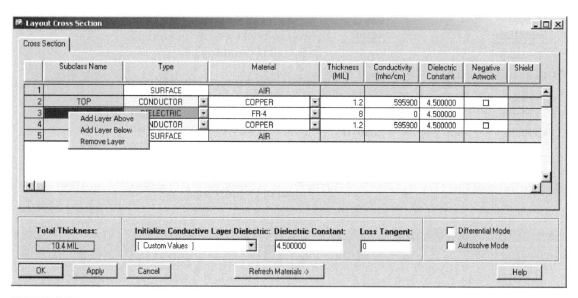

FIGURE 9-58
Adding layers to the stack-up.

FIGURE 9-59
The finished layer stack-up.

Click Apply and then OK. The Shield property is related to analyses that are not enabled in the OrCAD version of PCB Editor, so it doesn't matter if the option is checked or not.

Now if you display the Options pane, the new layers will be listed too. They all are given a default color, so you can customize your stack-up by changing the colors. *To change the color of a layer*, select the Colors button on the toolbar to display the Colors dialog box. Select the Stackup folder icon; the colors of all conductor and nonconductor layers will be shown. If you want to see only the conductor layers, select the Conductor folder icon to narrow the view to just the Routing and Plane layers. Select a color from the pallet then select the box in the Subclasses matrix to change its color. Figure 9-60 shows an example of a custom color scheme for the stack-up. Generally it is a good idea to select light, neutral, shades for Plane layers, so that they are not too distracting, since they typically cover a large area. When you make changes to the color or visibility

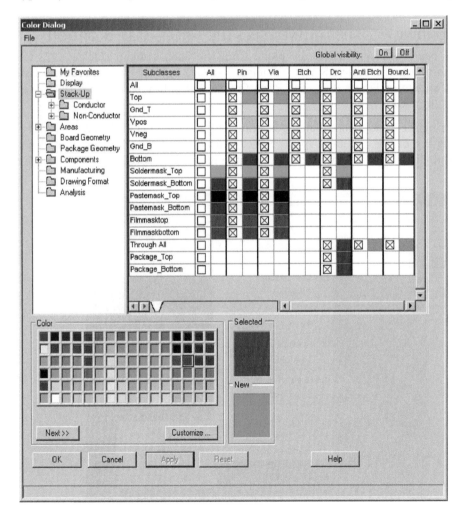

FIGURE 9-60
Setting custom colors for the stack-up.

remember to click the Apply button before clicking the OK button and dismissing the color dialog box.

Once the stack-up is defined, the copper planes can be poured onto the proper layers.

Pouring Copper Planes

The two ways to pour a copper plane are to use the Setup Plane Outline tool or the Add Shape tool. We demonstrate both, beginning with the Add Shape tool. Afterward how padstacks are connected to the planes using the Setup Plane Outline tool will be demonstrated.

Before pouring the copper, set the etch grid to the proper resolution (100 mils is suggested for this exercise). As you are placing pick points, you can pan the view around by using the arrow keys on the keyboard and the mouse wheel if you have one (highly recommended).

POURING A COPPER PLANE USING THE Setup → Outlines OPTION

To pour the copper, select the plane from the Options menu (Etch/GND_T, for example). Then select Setup → Outlines → Plane Outline... from the menu bar, as shown in Figure 9-61. The Plane Outline dialog box will be displayed (see Figure 9-62).

Using the Plane Outline dialog box, connect the proper net to the layer by selecting the Browse... button. In the Select a Net dialog box, select the desired net and click OK. If you are making a simple rectangle or box, you can select the

FIGURE 9-61
Starting a copper pour on a Plane layer.

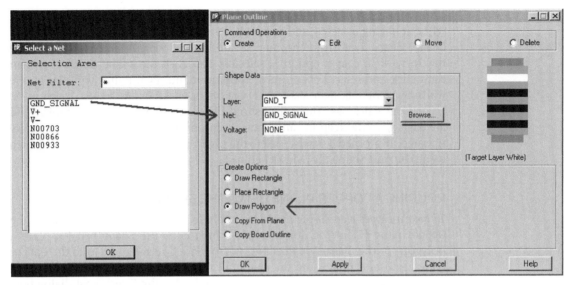

FIGURE 9-62
Drawing a plane area and assigning a net.

Draw or Place Rectangle radio buttons in the Create Options section. Otherwise select the Draw Polygon radio button (do not click OK or Apply yet).

Move your mouse pointer to the design area, left click and release to place the first vertex of the copper pour plane. If using the Polygon tool, place vertices (corners) to define the outline of the plane. Left click on the beginning point to complete the plane. If you are just drawing a rectangle, then left click at the corner opposite the starting point.

Note: In older versions of PCB Editor and newer versions that do not have the latest updates, *a copper area on a plane layer has to be dynamic copper*. Filled rectangles and static shapes may not produce proper connectivity to component pins or may violate route keep-in and route keep-out areas. In newer versions with the latest updates, connectivity can be made with filled rectangles and static shapes. Verifying connectivity between pins and planes is discussed in the next section.

Before pouring copper areas on the next plane layers, check to see what type of shape was placed on the plane. *To determine what type of shape the copper is,* hover your pointer over the plane; it should become highlighted and a data tips box should pop up with a description of the shape. If it does not, make sure that the Shapes option is checked in the Find filter pane. The poured copper will be one of three types and displayed as (1) Filled Rectangle "Netname, Class/ Subclass", (2) Shape, "Netname, Class/Subclass", or (3) Shape(auto-generated) "Netname, Class/Subclass". For example if a copper pour area is dynamic copper on the Vpos layer and is connected +V net, the data tips pop-up box should say Shape(auto-generated) "V+ , Etch/Vpos".

If the copper pour object is the wrong type of shape, you can usually change it without having to delete and redraw it. To change a shape's type, select Shape → Change Shape Type from the menu. Check the Options menu and make sure the To dynamic copper option is selected. Left click the shape to select it then right click and select Done from the pop-up menu. PCB Editor will display the warning Converting from static to dynamic shapes results in the loss shape voids. —Continue? Click Yes. The object should be converted to dynamic copper and all flashes, route keep-in, and route keep-out areas should be automatically updated. If problems persist, delete the object and redraw the copper pour area using the Add Shape tool, described next.

POURING A COPPER PLANE USING THE SHAPE ADD TOOL

To pour a copper plane using the Shape tool, select the target plane layer from the Options pane (Etch/Vpos, for example). Then click the Shape Add (to place a polygon) or the Shape Add Rect button (to place a rectangle). Go to the Options pane again and make sure the shape fill type is Dynamic copper. In the Assign net name: area, click the Browse... button to display the Select a net dialog box, as shown in Figure 9-63. Left click the net to select it then click the OK button to dismiss the dialog box. Left click in the work area to place each vertex of the plane outline. When you reach the last vertex, right click and select Complete from the pop-up menu. The outline will be drawn but still highlighted with the editing handles (boxes) visible. You can modify the shape if necessary or right click and select Done from the pop-up menu. Display the data tips box again to make sure the correct shape was drawn.

FIGURE 9-63
Drawing a copper pour shape and assigning a net.

Once the first plane outline (copper pour area) is drawn, you can copy and paste it onto the other layers rather than drawing each one individually. *To copy a plane outline*, start a new outline (Setup → Outline → Plane Outlines... from the menu). As described previously, select the target layer and assign a net to the plane using the Browse... button. In the Create Options area, select the Copy from Plane radio button, as shown in Figure 9-64. Select the existing plane from the list. At the command prompt it will say select a plane to copy. Select the plane on its edge. This is done since there could be more than one plane on a layer and PCB Editor needs to know specifically which plane to copy. Click OK when you have selected the plane. You should now have two Routing layers (top and bottom) and four Plane layers (two Ground and two Power planes). Again, make sure that the new copper pour areas are dynamic copper shapes, using the data tips box, and that connectivity is made from each plane to the correct pins as described next.

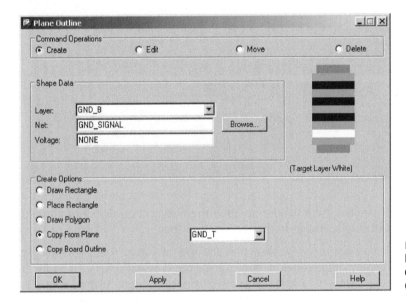

FIGURE 9-64
Drawing a new plane outline by copying an existing one.

Verifying Connectivity between Pins and Planes

A padstack that connects to a plane usually does so through a thermal relief, while padstacks that are not connected to a plane are isolated from the plane with a clearance area around the padstack. Now that the planes are in place from the previous step, we need to make sure that connections to the plane and isolations from it are done properly. If you are using a version of PCB Editor that has had its libraries modified, then some of the following discussion may not apply. The original library that comes with PCB Editor contains some parts that lack the proper thermal reliefs and clearances assigned to the padstacks. If you zoom into J1 and turn off all the layers except for the top Ground plane, you should see one of the four views shown in Figure 9-65.

FIGURE 9-65
Four possible views
after placing a new
plane outline.

If you see the bottom one, then the next section does not apply to you, so you can skip it and go to the section after that.

If you see the view shown at the top of Figure 9-65, you need to change the display options before you can see the thermal reliefs. To show thermal reliefs, select Setup → Design Parameters from the menu. At the Design Parameter Editor, select the Display tab. In the Enhanced Display Modes area, check the Thermal pads option (see Figure 9-66). Click Apply and then OK.

When viewing thermal pads is enabled in the Design Parameters Editor, you should see either the third or fourth image in Figure 9-65. If you see the second image, then the plane outline is the wrong type. See the previous section to determine the type of outline and how to change it if it is wrong.

If you see the third image, then no thermal flashes have been assigned to the padstack. The next section explains how to modify your design to use the thermal reliefs and clearances.

If you see the bottom view, then your pads have been set up with thermal flashes and clearances, so you can skip down to the Defining Trace Width and Spacing Rules section.

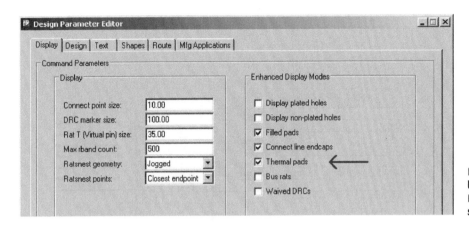

FIGURE 9-66
Use the Design
Parameters Editor to
show thermal reliefs.

ADDING THERMAL RELIEFS AND CLEARANCES TO PADSTACKS

To add the thermal reliefs to or change clearances on a padstack, you need to modify the padstack design using the Padstack Designer. You can modify just the padstack in the design or you can modify the padstack definition located in the symbols library. We will modify just the design padstack in this example. *To launch the Padstack Designer from PCB Editor*, select Tools → Padstack → Modify Design Padstack... from the menu bar. Move your pointer to connector J1, right click on pin 1 and select Edit from the pop-up menu. The Padstack Designer dialog box will be displayed. Select the Layers tab as shown in Figure 9-67.

On each of the layers, the thermal relief and clearance needs to be changed. Select a layer by clicking on it. Select Flash from the list of thermal reliefs and either type TR_180_150 in the flash name box or use the Browse... button beside it to use the Select flash symbol dialog box to choose from a list as shown in Figure 9-68.

Note: To do this you must have the TR_180_150 flash symbol copied to the PCB Editor symbol folder with the other footprints or in the working allegro folder. You can copy the one in the ...share/pcb/examples folder in the OrCAD file path or from the Web site for this book.

Assign the flash symbol and change the clearance to 188 on each layer. Once the changes have been made, select File → Update to Design... from the menu bar. Close the Padstack Designer to return to the board design. Right click and select Done from the pop-up menu. You should see the thermal reliefs now (make sure you have the thermals box checked in the Design Parameters dialog box).

Editing one pin actually edits all pins with the same name (even ones on different parts), so as you make each plane layer active, you should see the various clearances and thermals for the pins connected to that layer. If you want to change only one pin when there are many of that type of pin, select Tools → Padstack → Modify Design Padstack... from the menu bar as before. Left click on the pin and in the Options pane select the Instance radio button. PCB Editor will

FIGURE 9-67
Use Padstack Designer to modify a padstack definition.

FIGURE 9-68
Selecting a flash symbol for a pin.

automatically append a number to the name of the padstack to differentiate it from the others. Click the Edit button at the bottom of the pane to display the Padstack Designer.

You will probably need to change the padstacks on part symbols C1 and C2, which use flash symbol TR_110_93. The remaining components in this design are surface mount and have no internal layers. And as we will see shortly, the vias are full contact type and do not use thermal reliefs.

Unless you modify the actual library parts, you will have to do this every time you start a new design and use the parts in the library. Instead of selecting the Update design option, you can select the Save to file option prior to closing the Padstack Designer. This saves a copy of the padstack (*name*.pad) in the working directory. You can copy the padstack and put it in the symbols library, so that it can be used in other projects or you can select Tools → Padstack → Modify Library Padstack... from the menu bar to modify it directly in the symbols library. You need to know the name of the padstack(s) that you want to modify to use this option. Over time many of the parts in the library will be modified and you should rarely have to do these modifications. However, you may want to back up the original library before you make a bunch of changes to it, and it is a good idea to make a backup of custom padstacks, flash symbols, and footprints that you develop as well.

Once the plane layers are properly set up you can fan out the surface-mount components to their respective Plane layers. To do this however, we need to assign or verify which via will be used for fan-outs and routing and that trace width and spacing requirements have been defined, as described in the next section.

Defining Trace Width and Spacing Rules
DETERMINING TRACE WIDTH

The trace width depends on three design considerations. The first consideration is the capabilities of your board manufacturer. The traces need to be wider than their minimum fabrication capability. The second consideration is the required current handling capability, and the third is the impedance. According to the data sheet, the short circuit output current for the 741 op amp is 64 mA. If we use a safety margin of, say, 100 mA and we are using a board with 1 oz copper, then per Eq. (6.17) in Chapter 6 (or the graph from Figure 6-38), the minimum trace width is 1.3 mils for inner layers and 0.5 mils for outer layers. From Table 6-8 in Chapter 6, you can see that, with 6-mil traces, you can run up to about 300 mA on inner traces and about 600 mA on outer traces. So the default value of 5 mils is adequate for most small-signal applications. However, the chosen width must also be manufacturable by your board house. Five mils is fairly narrow for many board houses, so we will change the widths.

For this example we specify that all signals be of very low frequency (<20 kHz) so that trace impedance is not a concern. An example of designing controlled-impedance transmission lines is given later in the digital design example (Example 4).

Cmgr button

Trace width and spacing rules are set using the Constraint Manager. *To set trace width rules*, open the Constraint Manager by clicking the Cmgr button, 🖳, on the toolbar. Select the Physical window and select the All Layers icon in the Net folder, as shown in Figure 9-69. The default minimum width for all nets is 5.00 mils and maximum is set at 0.00 (which means there is no maximum limit). Set the line widths to different values and note the differences when the board is routed later. Changes made in the Constraint Manager are effective immediately.

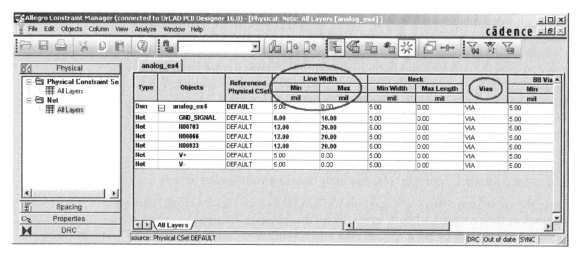

FIGURE 9-69
Use the Constraint Manager to set trace width rules.

ASSIGNING VIAS

Notice also in Figure 9-69 that you can *specify particular vias for each net* on the Physical tab of the Constraint Manager. To select a different via, left click inside the cell for the via you want to change. In the Edit Via List dialog box shown in Figure 9-70, select the via that you want to use on that net and remove unwanted via(s). Click OK when your selection is complete. If you need a via that is not listed, you can make a custom via using the Padstack Designer, which is discussed in Chapter 8.

DETERMINING TRACE SPACING REQUIREMENTS

The op amp used in this example requires a dual rail supply, which can be up to ±15 V (and the input and output voltages must be between the rails). The greatest voltage difference possible between any two traces then is 30 V peak to peak, so the spacing between the traces must be able to support this. If we decide to play it safe and use the 31–50-V range and we plan on using a soldermask, then, using Table 6-8 in Chapter 6, the route spacing should be greater than 4 mils on internal layers and greater than 5 mils on external layers. Again the selected spacing rules must fall within your board manufacturer's capabilities also.

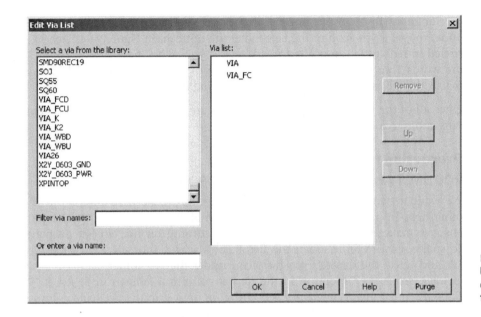

FIGURE 9-70
Use the Edit Via List dialog box to establish the allowed vias.

To set the spacing rules, select the Spacing tab in the Constraint Manager. You can set spacing rules by layers or by nets. *To set the rules by layers*, select the All Layers icon in the Spacing Constraint folder. In the Objects column, you can click the + sign to expand and view the entire list of objects. *To set the rules by net*, select the All Layers icon in the Net folder, as shown in Figure 9-71. To set

FIGURE 9-71
Setting trace spacing rules in the Constraint Manager.

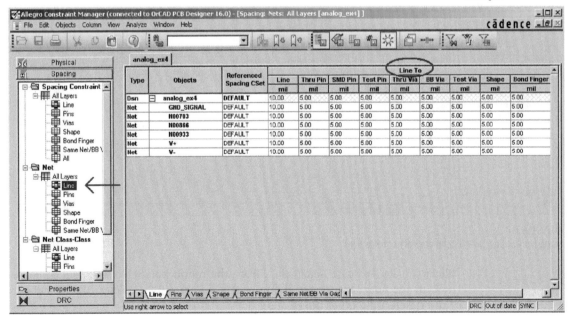

specific rules (for example, line-to-line spacing, as shown), select the Line icon under the All Layers icon or the Line tab at the bottom. In Figure 9-71, Line To is circled in red at the top of the spreadsheet. If you select the Pins icon (or tab), the words Thru Pin To will be listed in place of Line To. The spreadsheet is a matrix that gives you control over all the spacing rules.

Once all of the trace width and spacing rules are set we can begin routing fan-outs and traces.

Prerouting the Board

ROUTING FAN-OUTS FOR POWER AND GROUND

We want to fan out only power and ground, so we need to disable all the other nets. In particular we start with the Signal_RTN net on the GND layers. First we make the general nets invisible by turning on the No Rat property. To make nets invisible, open the Constraint Manager and select the Properties tab and select the General Properties icon under the Net folder. *To make a net invisible*, select the On option in the No Rat column as shown in Figure 9-72. *To disable a net from being routed*, select the On option in the No Route column. You can set a property on many nets simultaneously by dragging a box across the nets then setting the property. For this example make all nets invisible except for the SIGNAL_RTN net. For demonstration purposes we also want to prevent the power nets from being fanned out, so the No Route restriction is applied. Note that the signal nets will not be fanned out because we can control that by configuring the autorouter as shown next.

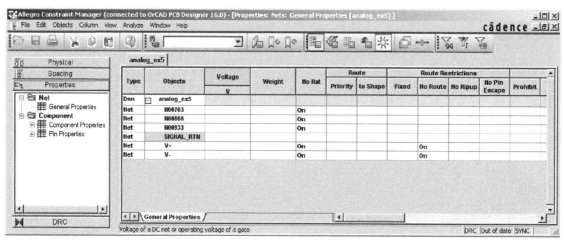

FIGURE 9-72
Using No Rat and No Route to control net routability.

Select Route → Route Automatic... from the menu bar. Select the Routing Passes tab in the Automatic Router dialog box (see Figure 9-73). Select the Fanout box and deselect the other boxes.

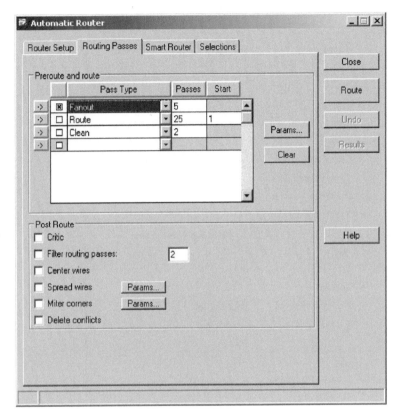

FIGURE 9-73
Using the autorouter to perform fan-outs.

Click the Params... button to display the SPECCTRA Automatic Router Parameters dialog box (see Figure 9-74). Here we can control what type of nets get fanned out, in what direction the fan-outs are routed, and how pins and vias are shared. Note that not all options are functional in the OrCAD version of PCB Editor. Under Pin Types check the Specified radio button and the Power Nets box. This allows only nets assigned to Plane layers to be fanned out and, because the V+ and V− nets have a No-Route restriction (in Constraint Manager), only the SIGNAL_RTN net will be fanned out to the ground planes. Click OK to dismiss the Router Parameters dialog box. Click the Route button on the Automatic Router dialog box to begin the fan-out. Only the SIGNAL_RTN nets should be fanned out.

To complete the fan-outs for the power nets, go back to the Constraint Manager and clear the no-route restriction; repeat the fan-out procedure for the V+ and V+ nets.

If you look closely at the vias used for the fan-outs (and routing by default), you will notice that they have no thermal reliefs to the Plane layers. If you want thermal reliefs for the vias, you can add them using same procedure as described previously to add thermals to the component pins. However, since leads or wires are typically not soldered into vias, many designers do not use

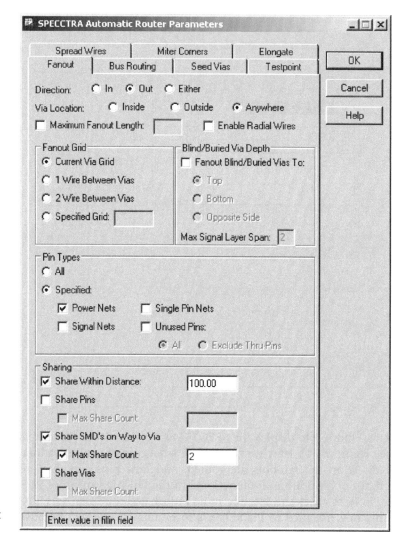

FIGURE 9-74
Setting up fan-out
options.

thermal reliefs on routing and fan-out vias. There are arguments for both sides, but thermals are not used on the vias in this example.

MOVING AND DELETING FAN-OUTS

Slide tool

If the location of a fan-out is a problem, you can easily move it. *To move a fan-out*, select the Slide tool, ![], left click on a via to select it. Move the pointer to where you want the via and left click to place the via.

Delete tool

If for some reason you want to delete a fan-out and route it manually, you can do that too. *To rip up a fanout*, select the Delete tool, ![], from the toolbar, select only the Clines and Cline Segs in the Find filter, and select only Clines and Vias in the Options tab. Double click the trace and via you want to rip up. Right click and select Done from the pop-up menu when you are finished.

To manually redo a fanout, use the Add Connect tool, , to route a trace and fan-out via from the component pin (on the top or bottom layer) to a Plane layer. Make sure that the net is visible and does not have a No-Route restriction on it. Select the Add Connect tool, and left click the rat's nest near the component's pin. Left click at the location where you would like the via. Right click and select Add Via from the pop-up menu. Right click and select Done from the pop-up menu.

Add Connect tool

Figure 9-75 (left) shows two shared via techniques, where the variation is a result of the via location. Figure 9-75 (right) shows using individual vias instead of shared vias, where each part has its own vias. This approach makes routing a little easier, since it leaves an open path between the components on the top routing layer. As described in Chapter 6, there are arguments for all three types of fan-outs.

FIGURE 9-75
Three fan-out methodologies.

At this point it is a good idea to update the DRC. After all errors are fixed (there should be none), we can route the rest of the board.

Manually Routing Traces

Traces (clines) can be routed manually or by the autorouter. We begin by looking at manual routing.

TO BEGIN MANUALLY ROUTING TRACES

To manually route a trace, select the Add Connect tool, , left click on the desired trace, and left click again to place a vertex and route the trace. The default trace

Add Connect tool

characteristics will be determined by the Constraint Manager. You can make changes to the trace as you are routing by using the Options filter tab (see Figure 9-76), where you can change routing layers, line width, and other properties as shown in the figure.

LOCKING (FIXING) TRACES

Unfix Fix button button

Show Element button

Once the fan-outs and specially routed traces are completed, you can lock (fix) them to prevent them from being inadvertently unrouted or shoved by the autorouter. *To fix traces*, select the Fix button, 🔍, from the toolbar and left click the desired trace. *To unlock (unfix) a trace*, select the Unfix button, 📌, and left click a trace to unfix it. To find out if a trace is fixed or not, use the Show Element button, ⓘ, and left click on a trace to find out the status. You can also use the Constraint Manager to fix traces (select the Properties tab and set the Route Restrictions properties).

RUNNING THE AUTOROUTER

Once the fan-outs and manual routing tasks are completed, you can quickly route the rest of the board using the autorouter. *To start the autorouter*, select Route → Route Automatic... from the menu bar. As shown in Figure 9-77 (Route Setup tab, Strategy options area), there are three ways to use the autorouter: (1) specify routing passes, (2) use the smart route, and (3) set up a SPECCTRA (Allegro) DO file. You use the Route Setup tab on the Automatic Router dialog box to select which method to use. Methods 1 and 2 are both easy to use, method 3 is not discussed here. The smart router has fewer options to set than the routing-passes method, but the routing-passes method gives you more control over the auto-router. Methods 1 and 2 are described next. Click the radio button for the method you want to use then click the corresponding tab to change the route settings for that method.

FIGURE 9-76
Using the Options tab to control manual routing properties.

Use the Smart Router tab to set up the autorouter for that method. As shown in Figure 9-78, you need to specify a minimum set of options, and they are pretty straightforward.

With the routing-passes method, you have greater control over the routing process. In addition to the settings on the Router Setup tab and the Routing Passes tab, you also have access to the Params... dialog box, as was shown during the fan-out description. Some of the features available in the routing-passes method are used in later examples. For this example just uncheck the Fanout box (since we have already done that) and check the Route and Clean boxes. Click the Route button to finish the board (Figure 9-79).

FIGURE 9-77
Setting up the auto-router strategy and options.

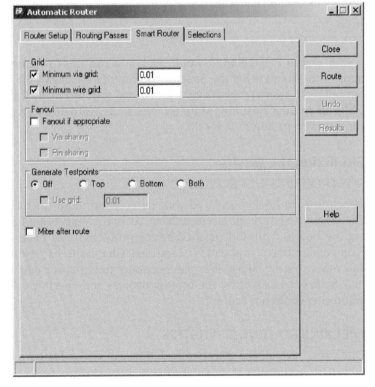

FIGURE 9-78
Setting up the Smart Router.

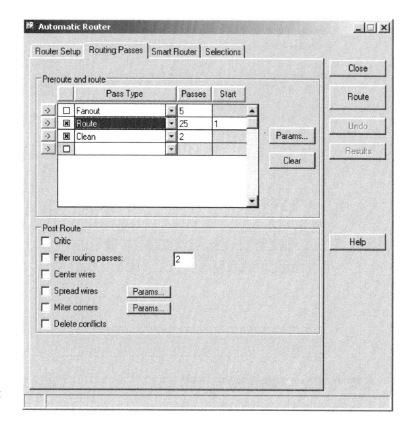

FIGURE 9-79
Setting up the Routing Passes options.

If your design is large and complicated, you can limit routing to specific sections of the board using the Selection tab filter (see Figure 9-80). As shown, you can select (or restrict) specific nets for (from) routing. Using the Selection tab in addition to the Constraint Manager. No-Route restrictions gives you extensive control over the autorouter.

Finalizing the Design

POSTROUTING INSPECTION

Depending on how you placed your parts, your board might look something like Figure 9-81 (not all layers shown). After the autorouter has finished routing the board, it should be checked for problem areas, such as acute angles, long parallel traces, and bad via locations. Chapter 6 discusses several of these and other routing issues. It is also necessary to check for new DRC errors that may have been caused by the routing process and check for any nets that the autorouter could not route.

CHECKING ROUTING STATISTICS

Once you have all the traces routed, you can check for new DRC errors and unrouted nets using the Status dialog box (Figure 9-82). *To view routing*

FIGURE 9-80
Selecting which traces to route.

FIGURE 9-81
Board design after final parts placement and autorouting.

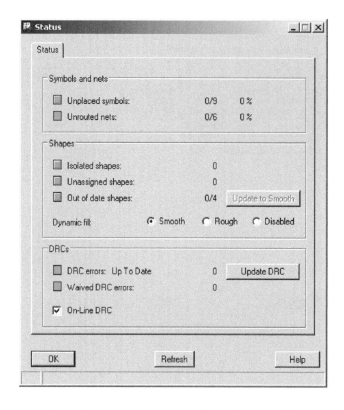

FIGURE 9-82
Use the Status
dialog box to check for
unrouted nets and DRC
errors.

statistics, select Display → Status from the menu bar. The Status dialog box provides information about the number of unrouted nets as well as unplaced components.

Rats All button

If the routing statistics are less than 100% but you cannot easily locate unrouted nets, use the Visibility filter tab to turn off all Etch layers and make sure that the Rats All button is active, . Use the Find filter (look for nets only) and the Show Element tool and drag a box across the board area. Any unrouted nets should be displayed in the Show Element text box with information about the net's name, to which pins the net belongs, and its location. An example is shown in Figure 9-83.

CLEANING UP A DESIGN BY GLOSSING

After the autorouter has finished routing the board (or after performing any last-minute manual routing tasks), you can have PCB Editor go over the design and perform a cleanup using the Gloss command. *To clean up (gloss) a board design*, select Route → Gloss... from the menu. At the Line Smoothing dialog box (see Figure 9-84), check the options you would like PCB Editor look for and fix. When you have made your selections, click the Gloss button.

You can attach either the NO_GLOSS property or the FIXED property to the nets so that gloss does not modify the routing of these nets. Use the Constraint

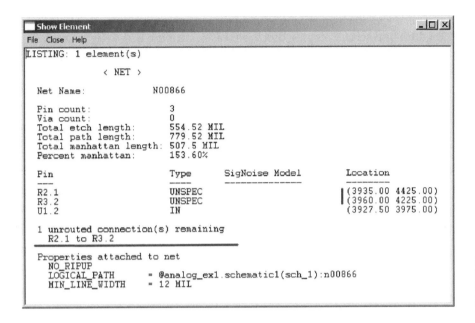

FIGURE 9-83
Using the Find filter and Show Element tool to identify unrouted nets.

FIGURE 9-84
Use the Gloss command to add finishing touches to the board design.

Manager to apply the NO_GLOSS property to a net. Select the Properties tab and the Net/General Properties icon then set the NO_GLOSS property to On. To apply the FIXED property, you can also use the Constraint Manager or use the Fix button on the toolbar.

SYNCHRONIZING THE DESIGN WITH CAPTURE (BACK ANNOTATION)

Once the board design is finished, you can export information from your PCB design to your schematic by performing a back annotation. Back annotation in PCB Editor is called feedback (search for it in fcoms.pdf). Also see *Backannotating to Allegro Design Entry HDL or System Connectivity Manager* in *Transferring Logic Design Data* (algrologic.pdf), and the *Capture User's Guide* (cap_ug.pdf). See *Back annotation from PCB Editor*, p. 508.

Information that is back annotated has to be common to PCB Editor and Capture. To see some examples of these properties, go to the schematic page and double click a part to display the properties spreadsheet. Select the Cadence-Allegro filter. Some examples are HARD_LOCATION, NO_SWAP_GATE, NO_SWAP_PIN. The configuration file allegro.cfg, which is located in the C:\OrCAD\OrCAD_16.0\tools\capture\path by default, lists all of the properties included during back annotation.

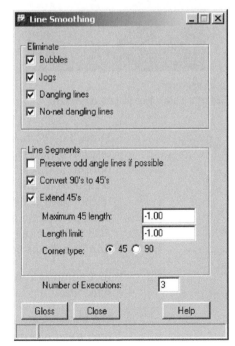

To perform a back annotation from PCB Editor to Capture, go to the PCB design in PCB Editor and select Setup → User Preferences.... In the User Preferences dialog box, left click the Logic folder once, and select capture in the schematic_editor: list, as shown in Figure 9-85.

FIGURE 9-85
Setting the Logic type for back annotation.

From the File menu, select Export Logic...; the Export Logic dialog box will be displayed (see Figure 9-86). Select the Cadence tab, select the Design entry CIS option, and select the destination path from Export to directory: area (use the Browse... button). The default path is your current working directory, displayed in the Export to directory field as a period (unless another location was specified and applied in a previous session). Click the Export Cadence button. That's it for PCB Editor. Now go back to Capture to import the back-annotation information into the project design.

In Capture go to the Project Manager pane and select the Project icon. Then select Back Annotate... from the Tools menu, as shown in Figure 9-87.

The Backannotate dialog box will be displayed. Select the PCB Editor tab and select the proper files and paths using the Browse... buttons. You can change what gets

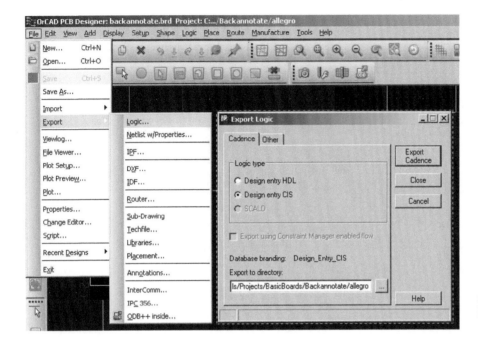

FIGURE 9-86
Displaying the Export
Logic dialog box.

imported by clicking the Setup... button, as shown in Figure 9-88. The allegro.cfg file lists the parameters that will be imported. To edit the file, click the Edit button on the Setup dialog box. If you edit the file, remember to save the changes. Dismiss the .cfg file and Click OK three times to dismiss the dialog boxes and perform the back annotation.

To see what changes were made, double click a part in the schematic to view the properties sheet. Select Cadence-Allegro from the Filter by: list. Figure 9-89 shows the trace width properties that were set in the Constraint Manager in PCB Editor (a) are imported into the Edit Properties spreadsheet in Capture (b).

This completes the first example. At this point you would produce the manufacturing files for the board and send them off to a board house for manufacturing. We look at this in greater detail in Chapter 10.

FIGURE 9-87
Performing back annotation in Capture.

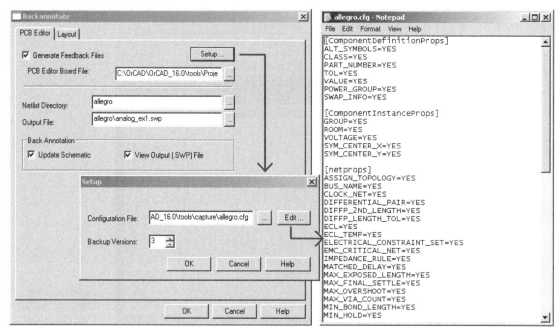

FIGURE 9-88
Setting up the back-annotation path and allegro.cfg file.

(a)   (b)

FIGURE 9-89
Net properties back annotated to (a) PCB Editor (b) Capture from.

EXAMPLE 2   MIXED ANALOG/DIGITAL DESIGN USING SPLIT POWER, GROUND PLANES

This example shows how to design a circuit with a mixture of analog and digital parts, multiple power planes, and a single Ground plane split into analog and digital sections that have a common reference point.

Mixed-Signal Circuit Design in Capture

Figure 9-90 shows the circuit design example. The circuit consists of a mixture of analog and digital parts, multiple power sources, and separate analog/digital

FIGURE 9-90
Mixed-signal schematic design in Capture.

Ground planes. It has an off-board connector, an analog signal conditioning circuit (the op amp U1), an analog-to-digital converter (U2), a digital microcontroller (U3), and a digital serial-to-parallel shift register (U4). Note that the bypass capacitors have been omitted to accommodate the part count limitations of the demo version of the software.

The circuit has five global power nets, which include analog power (V+ , V−) and analog ground (AGND) for U1 and U2, and 5-V digital power (VCC) and digital ground (GND) for U3 and U4. Analog and digital grounds are often kept separate to help prevent digital switching noise from affecting the analog ground. But, for the circuit to work (especially when using analog-to-digital and digital-to-analog converters), the two grounds must have a common reference point. The two grounds are tied together through a low-impedance, single-point connection at the connector.

The procedures and tools described in the first example are used to start a new project; place the parts onto the schematic page, connect the parts, and assign footprints. Table 9-4 lists the parts used and the footprint assignments. The conn14 footprint symbol is a modified version of the one included with the software (it is not included with the demo version); a copy of the modified symbol is included on the Web site that accompanies this book.

Table 9-4	Design Parts List and Schematic and Footprint Libraries for Design Example 2		
Reference	**Part**	**Source library**	**PCB footprint**
J1	CON14	CAPTURE/LIBRARY/CONNECTOR.OLB	conn14
R1	1 k	CAPTURE/LIBRARY/DISCRETE.OLB	Smdres
R2, R3, R4	2 k	CAPTURE/LIBRARY/DISCRETE.OLB	Smdres
U1	LM741	CD...ANALOG_EX1.DSN	soic8
U2	TLC548	CD...CHAPTER9.OLB	soic8
U3	PIC16C505	CD...CHAPTER9.OLB	dip14_3
U4	CD74HC164	CD...CHAPTER9.OLB	soic14

There are a couple of significant differences between this and the last example. The differences include

1. All analog parts and some digital parts have explicit connections to power and ground nets, while the shift register is connected to the digital power and ground globally, as described later.
2. Two different ground symbols (CAPSYMs) are connected to one net.
3. Some of the connections are made with a bus rather than wires.

POWER AND GROUND CONNECTIONS TO DIGITAL AND ANALOG PARTS

Digital gates from the Capture libraries have nonvisible (invisible) power and ground pins. Their power pins are named *VCC* and their ground pins are named *GND*. Connecting a wire to an invisible power pin does nothing. Connections from nets to invisible power pins can be made only through global net names (made global by CAPSYM power, ground symbols, or off-page connectors). So the power pin VCC on U4 (CD74HC164) is connected to the schematic's VCC symbol by the name *VCC* and the ground pin is connected to the digital ground symbol by the name *GND*.

To connect a digital gate's ground pin to a ground net in your schematic, you have to either name the digital ground net *GND* or change the name of the digital gate's ground pin to match the name you assigned to the digital ground net (e.g., *DGND* instead of *GND*)—this is required for all digital gates. It is easiest to use the default *GND* name for the digital ground.

Most other Capture parts use either visible power pins or input or passive pins for power and ground. And instead of using *VCC* and *GND* as names of the power pins, they often use *V+* and *V−* or *VDD* and *VSS*. When parts use visible power and ground pins, you can either take advantage of their global nature (and simply use their names to make connections to global nets) or make direct connections to the pins with wires. When parts use input or passive pins for power and ground, you have to make explicit connections (using wires) between the pins and the power or ground nets. Said another way, you cannot make connections to invisible power pins, and you have to make connections to pins that are not power pins; but visible power pins may have either wired or global connections to them. When you make an explicit connection to a visible power pin with a wire, the pin becomes connected to that net (the name of the wire) regardless of the name of the pin.

CONNECTING SEPARATE ANALOG AND DIGITAL GROUNDS TO A SPLIT PLANE

The CAPSYM power and ground symbols are global in nature and share characteristics of nonvisible power pins (i.e., connections to nets can be made by name or by physical connections). Figure 9-91 shows how the analog and digital grounds are named and displayed. Two CAPSYM symbols are used with distinct names to differentiate the two ground systems. The analog parts of the circuit are referenced to the analog ground (AGND) symbol and the digital components are referenced to the digital ground (GND) symbol. At some point the two grounds need to be connected together for the analog-to-digital converter to work correctly (particularly with single-ended inputs). At the connector the two grounds may leave the board separately and be connected at the power supply, or they may be connected at a pin (or pins) and leave the board together as one common ground. In this example the two grounds are connected at pin 5.

FIGURE 9-91
Connection of the analog and digital grounds and bus connections.

Place Ground tool

To place the analog and digital ground symbols, use the Place Ground tool, ⬇, and select the desired ground symbols. To change the symbol's name, select the ground symbol (for example, from GND_SIGNAL to AGND), select the symbol on the schematic page to highlight it, right click, and select Properties from the pop-up menu to display the Properties spreadsheet (see Figure 9-92). Select the Name row, right click, and select Edit from the pop-up menu to display the Edit Property Values dialog box. Enter the new name of the ground symbol and click OK. To display the name of the symbol on the schematic page, click the Display button on the top of the spreadsheet and select the appropriate radio button on the Display Properties dialog box (Figure 9-92).

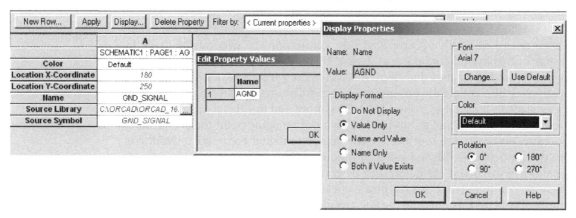

FIGURE 9-92
Changing a ground/power symbol's name and displaying it.

Place the second ground symbol and rename it (if applicable). Use the Wire tool to connect the two ground symbols together then to the connector pin. The two ground symbols were distinct global nets until they were connected by the wire (net). Nets typically have names such as N04432 unless they are connected to a global net or symbol, then they take on the global name. A net can have only one name, and since the net just placed is connected to two global symbols with different names, it has to pick one of them (it decides alphabetically).

FIGURE 9-93
Selecting the ground net name.

To select the name of a global net, double click the net (or select the net and right click and select Properties) to display the Schematic Net Properties spreadsheet (Figure 9-93), select the Name row, right click, and select Edit to display the Edit Property Values dialog box. Use the dropdown selection list to choose which name the net should go by. Click OK and then close the spreadsheet.

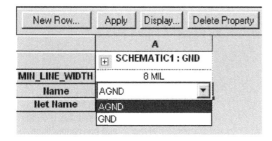

It appears on the schematic that the analog and digital grounds are separate ground systems, connected only at J1. As far as Capture is concerned, they are actually the same net—the name of which was decided in the dropdown list in Figure 9-93. Once we get to PCB Editor, we will separate the ground systems using a split plane.

USING BUSES FOR DIGITAL NETS

Digital circuits often have multiple, related input/output wires. The serial-to-parallel shift register in Figure 9-90 is an example, in which Q0 through Q7 are data lines that represent the analog voltage sampled by the analog-to-digital converter as an 8-bit word. Rather than draw all eight lines throughout your schematic, you can group the eight members of the word into a single bus (the blue line in Figure 9-90 and Figure 9-91).

A bus is a solid line that represents a group of wires or signals. Buses are not used to make connections to pins, but buses contain the signals connected to pins. Signal lines are added to a bus through aliases. Nets that are contained in buses are named, for example, A0, A1, A2, and A3, and the bus that contains them is named A[0..3].

Bus_Button tool

Buses are placed on the schematic using the Place Bus tool, [Bus_Button]. After you have placed the bus, you define (name) it with an alias using the Place Net Alias tool, . Buses are named by a specific convention, busname[n..m], in which busname is the alias (a name you give it) and [n..m] is a member list made up of integers. The integer n in busname[n..m] is the first member of the list and integer m is the last member of the bus list. The integers within the list can be separated with two periods, [n..m]; a colon, [n:m]; or a dash, [n–m]. The nets (wires) that make up the members are placed on the schematic using the Place Wire tool, . The bus members are placed on the schematic as you would place any other wire, but they are added to the bus by giving the wire a name (a member alias) that associates it with the bus. The members are named as busnamen through busnamem with the Place Net Alias tool. So in this example, the bus alias is named Q[0..7], and the member aliases are named Q0, Q1, to Q7. The short, slanted lines that are used to connect the wires to the bus are placed on the schematic using the Place Bus Entry tool, . Or you can *add slanted lines* yourself: Select the Place Wire tool, hold down the Shift key on your keyboard, left click to start the wire, release the Shift key, and left click to end the wire.

Net Alias tool

Place Wire tool

Place Bus Entry tool

The microcontroller (U3) has a couple of unused pins. To indicate that the pin is a "no-connect," place a no-connect symbol on the pins by selecting the Place no connect button, , from the toolbar and place no-connect symbols on the unused pins.

Place No Connect button

Note: Any pin in a schematic used to make a PCB can have a no-connect symbol, but when PSpice simulations are performed, some pins are not allowed to "float" and the no-connect symbol cannot be used. In that case enter RtoGND in the FLOAT property in the Properties spreadsheet. For more information see the *Capture User Guide*, cap_ug.pdf, and search for RtoGND.

Once the design is completed, it is a good idea to do some housecleaning and design checking before making the netlist. Such tasks (as described in the previous example) include

1. Perform an annotation to clean up part numbering.
2. Make sure that global power/ground nets are properly named.
3. Make sure all parts have the correct footprints assigned to them (use the BOM to assist).
4. Perform a DRC to verify that the circuit design has no issues.
5. Correct any errors and reperform the DRC as needed.

Once the design has been second-checked and is free of errors, create the PCB Editor netlist as described in the first example (from the Project Manager go to Tools → Create Netlist and select the PCB Editor tab).

Defining the Layer Stack-Up for Split planes

As discussed in the first example, it is a good idea to have the board constraints defined prior to starting the board layout. Items to consider are the board technology, board size constraints, noise and shielding requirements, part placement restrictions, the number of Power and Ground planes and Signal layers, and trace width and spacing requirements.

The voltage and current constraints are similar to those of the analog design, except that the analog parts should be segregated from the digital parts on the board. In the first analog design example, we used four planes for V+, V−, and ground. A similar approach could be taken here, but in this example we use one plane for V+, V−, and VCC, and one plane for both the analog and the digital grounds, as shown in Figure 9-94, to demonstrate the process of

FIGURE 9-94
A mixed-signal power and ground system.

working with split planes. The default PWR layer is divided into three completely separate Power-plane areas, using three copper shapes to supply analog V+ (PA+), analog V− (PA−), and digital VCC (PD), as indicated in the figure. The Ground plane is partially divided into two Ground planes for analog ground (AGND) and digital ground (GND) but left connected by a small section of copper near the connector's ground pin. This arrangement segregates the analog components from the digital components and therefore reduces switching noise in the analog circuitry. To accomplish this, the board requires two Plane layers and at least two Routing layers. Table 9-5 lists the design constraints.

Table 9-5	Mixed-Signal PCB Design Constraints: Component and Board Requirements
Component mounting type	**Mixed SMDs and THDs**
Number of sides with parts	1
Number of plane layers	2
Number of routing layers	2
Total number of layers (min)	4
Smallest leads Pad spacing (min) Pad width (max)	SOIC 50 mils 25 mils
Maximum current Inner trace width (min) Outer trace width (min)	0.1 A 1.3 mils 0.5 mils
Maximum voltage Inner trace spacing Outer trace spacing	10 V 4 mils 5 mils

To begin the board layout, we follow the same procedure as outlined at the beginning of these examples and discussed in detail in the analog design example.

From Capture, perform the Create Netlist procedure and launch PCB Editor.

As described in the first example, draw the board outline, use one of the place tools to find and place parts into the board outline, and perform a DRC to check for footprint problems prior to doing anything else. Figure 9-95 shows an example of how the parts might be placed. Notice that the analog parts are segregated from the digital parts.

FIGURE 9-95
Initial parts placement for the mixed analog and digital design.

DEFINING THE LAYER STACK-UP

When the board constraints were defined earlier, we said that we wanted two Plane layers and two Routing layers, so we need to add the Plane layers. Follow the procedure from the first example (right click and add layers) and add the two plane layers, as shown in Figure 9-96.

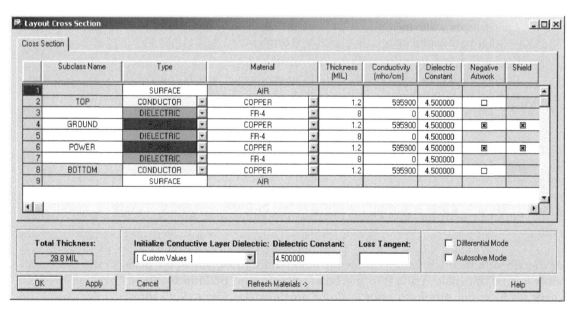

FIGURE 9-96
Layer stack-up for the mixed-signal board design.

SETTING UP A SPLIT GND PLANE

The objective in splitting a Plane layer into analog and digital sections is to remove copper from the plane using a shape void. Figure 9-97 shows the desired split Ground plane. Electrically the AGND and GND nets are the same but physically we want to separate them. To do this, a copper shape, which is attached to the GND net, is added to the Ground layer. This makes them electrically connected. To physically separate the analog from digital circuitry, we create a void strip between the analog and digital areas of the board.

FIGURE 9-97
A split place for analog and digital circuits.

Before placing the plane, you may need to adjust your route grid. Set it to about 5 mils, because you need the resolution to properly place the vertices (corners). *To add the GND shape to the ground plane,* use the procedure used in the first example. In short, select Setup → Outlines → Plane Outline... from the menu bar (or use the Shape Add tool and select Dynamic copper in the Options pane). Select the Ground plane, the GND net, and the Draw Rectangle option in the Place Outline dialog box. Draw the Ground plane then click OK to dismiss the dialog box. To view the thermal reliefs, follow the procedure used in the first example (View thermal pads in the Design Parameters dialog box and assign flashes to the padstacks as necessary). Next we split the plane into analog and digital areas.

Note: Ground planes can be drawn with filled rectangles, static shapes, or dynamic copper shapes. For best results use dynamic copper shapes. To identify the shape type of existing shapes, hover your mouse over the shape edge (make sure the Shapes option is selected in the Find filter pane). To change a shape to dynamic copper, select Shape → Change Shape Type from the menu then left click the shape you want to change.

Shape Void Polygon
button

To split a plane, select the plane layer from the Options tab by selecting the Etch class and Ground subclass. Select the Shape Void Polygon button, 🔳, on the toolbar. At the command prompt, PCB Editor asks Pick shape or void to edit. Left click the GND plane. PCB Editor then asks Pick void coordinates. Left click at the vertices of the void area you want to define, as shown by the arrows in Figure 9-97. Draw the void so that the analog and digital areas are separate but leave a connected area by the connector, J1, so that the two Ground planes remain electrically connected. The void between the ground areas should be about 50 mils wide.

U2 is the analog-to-digital converter (ADC) and has both analog and digital properties. Some ADC manufacturers specify that the analog and digital grounds should be common near the IC. Right now the analog and digital grounds are separate, but we will connect them directly under the IC by placing a small patch of copper, which will be connected to the GND net, to bridge the void separating the analog and digital planes, then merge the patch with the main Ground plane.

Shape Add Rectangle
button

To create a copper bridge, select the Shape Add Rectangle button, 🔳, and use the Options tab to define the shape on the Etch Ground plane and connect it to the GND net, as shown in Figure 9-98. Note that, if the Ground plane was made with a dynamic copper shape, then the bridge also needs to be dynamic copper rather than static filled.

Place the bridge as shown in Figure 9-99. Left click and release at the upper left and lower right corners to draw the shape. Make sure to overlap the shape and the Ground plane to allow the merge process to work properly.

To merge copper shapes, select Shape → Merge Shapes from the menu bar. At the command line, PCB Editor will ask you to Pick primary shape to merge to. Select the Ground plane. Then PCB Editor will ask Pick shapes to be merged. Select the bridge, right click, and select Done from the popup menu. The bridge is now part of the Ground plane.

Notice that at the J1 connector we have the "common ground area" in Figure 9-97, but that the analog and digital ground areas are separated throughout the rest of the board. The area under the ADC does not defeat the "split" plane. Recall from Chapter 6 that return currents follow the path of least impedance, so return currents related to U3 and U4 do not go out of their way to pass through the area under the ADC.

FIGURE 9-98
Setting the properties of the copper bridge.

FIGURE 9-99
Placing a copper shape
to bridge the two ground
planes.

SETTING UP SEPARATE PLANES ON A SINGLE PLANE LAYER

Unlike the ground system, in which there is really only one ground net and one Ground plane, the power system has three distinct power nets (V+, V−, and VCC) that need to share the Power plane. To accomplish this, distinct copper areas are drawn for each net on the one Power plane. The goal is shown in Figure 9-100.

We begin by placing the VCC Power plane in the digital section of the board. As described in previous sections, we start by selecting Setup → Outlines → Plane Outline from the menu bar. At the Place Outline dialog box, select the Power plane and select the VCC net. Select Draw Polygon and left click to place the eight vertices that make up the VCC plane area. Click OK to complete the process and close the dialog box. Keep the VCC plane above the digital Ground plane and away from the void area. Once you have the VCC plane area drawn, there should be a thermal relief around the connector pin attached to VCC (pin 4 in this example).

Repeat this process for the V+ and V− plane areas. Note that the area between the digital and analog plane areas should match up with the void that was

FIGURE 9-100
Setting up multiple power areas on a single plane.

placed in the GND plane until you get close to the connector. Also note that we need not place voids on the Plane layer to split the plane, since the copper areas are distinct objects and do not need to be split. When drawing the V+ copper pour, make sure to include J1 pin 1 and the V+ fan-outs for U1 pin 7 and U2 pin 8 and make sure the copper pour stays on the analog side of the split Ground plane. When drawing the V− copper pour, make sure to include J1 pin 2 within the boundary and stay on the analog side of the Ground plane, as shown in Figure 9-100.

Setting Up Routing Constraints

Cmgr button

Before fanning out or routing the board, we need to set up the design constraints using the Constraint Manager. Launch the Constraint Manager by selecting the Cmgr button, ▦, on the toolbar. *To change the spacing constraint*, select the Spacing tab and the Line icon under the Net folder. Change some of the nets to whatever you like, just to see what happens. Figure 9-101 shows the line-to-line spacing of all the nonbus nets set to 15 mils.

To change the line width constraint, select the Physical tab and the All Layers icon under the Net folder. Change a couple of the minimum line widths to see the effect. The GND and a couple of the analog nets were changed, as shown in Figure 9-102. Note that changing the GND line thickness changes the width of the fan-out traces (the short trace going from an SMD pad to the via to GND) and thermal spokes on positive Plane layers, as we will see.

Next we begin the process of fanning out power and ground. We could disable all the nets except for the power and ground nets to do the fan-out, as we did in the last example, but we will force the autorouter to ignore the signal nets, so that we need not do that extra step. Perform the fan-out as described in the first couple of steps of the first example. Select Route → Route Automatic...

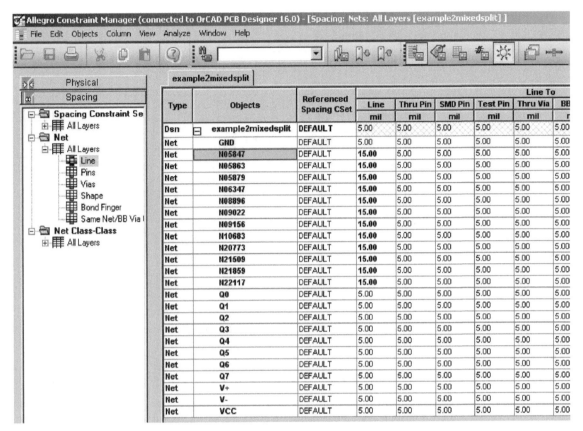

FIGURE 9-101
Setting the trace spacing rules.

from the menu. In the Automatic Router dialog box, select the Router Setup tab and select Specify routing passes in the Strategy section. Select the Routing Passes tab, select Fanout option, and deselect the Route and Clean options. Click the Params… button. In the SPECCTRA Automatic Router Parameters dialog box, select only the Power Nets option under the Pin Types section. Click OK then click Route to generate the fan-outs.

Check the fan-outs to make sure the router did its job and check that there are no DRC errors by checking the Status dialog box (Display → Status…) or generating a DRC report (Tools menu). If everything looks OK, then we can route the rest of the board.

To make sure that the fan-outs or any manually routed traces remain intact, they should be locked in place (fixed). To fix traces, display the Constraint Manager and select the Properties tab. Under the Net folder, select the General Properties icon. Set the On flag in the Fixed column for any trace you want to protect. Alternately you can use the Fix button, , and the Unfix button, , on the toolbar to select traces and components to fix or unfix.

Fix button   Unfix button

Allegro Constraint Manager (connected to OrCAD PCB Designer 16.0) - [Physical: Nets: All Layers [example2mixedsplit]]

File   Edit   Objects   Column   View   Analyze   Window   Help

				Line Width		Neck	
				Min	Max	Min Width	M
Type	Objects		Referenced Physical CSet	mil	mil	mil	
Dsn	⊟	example2mixed	DEFAULT	5.00	0.00	5.00	0.0
Net		GND	DEFAULT	8.00	0.00	5.00	0.0
Net		N05847	DEFAULT	12.00	0.00	5.00	0.0
Net		N05863	DEFAULT	12.00	0.00	5.00	0.0
Net		N05879	DEFAULT	12.00	0.00	5.00	0.0
Net		N06347	DEFAULT	5.00	0.00	5.00	0.0
Net		N08896	DEFAULT	5.00	0.00	5.00	0.0
Net		N09022	DEFAULT	5.00	0.00	5.00	0.0
Net		N09156	DEFAULT	5.00	0.00	5.00	0.0
Net		N10683	DEFAULT	5.00	0.00	5.00	0.0
Net		N20773	DEFAULT	5.00	0.00	5.00	0.0
Net		N21509	DEFAULT	5.00	0.00	5.00	0.0
Net		N21859	DEFAULT	5.00	0.00	5.00	0.0
Net		N22117	DEFAULT	5.00	0.00	5.00	0.0
Net		Q0	DEFAULT	5.00	0.00	5.00	0.0
Net		Q1	DEFAULT	5.00	0.00	5.00	0.0

Physical

Physical Constraint Se
  All Layers
Net
  All Layers

example2mixedsplit

FIGURE 9-102
Setting the trace width rules.

After performing the fan-out and fixing routed traces to protect them, we can *set up to route the signal nets*. Select Route → Route Automatic... from the menu. In the Automatic Router dialog box, select the Router Setup tab and select Specify routing passes in the Strategy section. Select the Routing Passes tab and select the Route and Clean options and deselect the Fanout option; click the Route button to route the rest of the board. If you want to restrict which signal nets get routed, use the Constraint Manager to set routing constraints. Use the Route Restrictions on the Properties tab of the Constraint Manager.

An example of a routed board is shown in Figure 9-103.

As in the earlier example, a postrouting inspection is performed after autorouting to look for problem areas, and offending traces are unrouted then rerouted. Figure 9-103 shows an area that could use some work (Figure 9-104 shows a close up view of the area). The red arrow points to a trace that was routed over the void in the GND plane.

Slide tool

As described in Chapter 6, we do not want this to occur, as it increases the loop inductance of the trace (an impedance-matching issue), which can lead to ringing problems and EMI. In this case the easiest fix is to simply move the trace down. *To move a trace*, select the Slide tool, ⌐, left click and release on the trace to select it. Move the mouse to the desired location, left click and release to place the trace. To prevent this from happening on subsequent routing actions,

FIGURE 9-103
The circuit after fan-out
and routing.

we can also place a Route Keep-out area in the void to keep the autorouter
from routing other traces over the void.

FIGURE 9-104
Look for problem areas
after autorouting.

ADDING ROUTE KEEP-OUT AREAS

To place a route keep-out shape area, select the Route
Keepout class and Top subclass from the Options tab.
Select the Shape Add tool and place a polygon on
the top layer that overlaps the void by 10 to 20 mils,
as shown in Figure 9-105. This will keep traces away
from the void.

Another potential problem area is the traces that
run between the microcontroller (U3) and the ADC
(U2). The traces run side by side in close proximity
for a relatively long distance. The traces should be
moved farther apart (use the Slide tool as described
previously) or guard traces can be added to provide isolation between the
traces (discussed later and in the next example). The traces that run from U4 to
J1 could also be a problem. A simple solution there would be to move some of
the traces to the bottom layer, as described next.

MOVING A ROUTED TRACE TO A DIFFERENT LAYER

As long as a trace has a through hole or via on each of its ends, you can easily
change which layer the trace is on without having to rip it up. *To move a routed*

FIGURE 9-105
Add a route keep-out
shape to control routing.

General edit Button

trace to another layer, select the General Edit button, ▦, select the trace you want to move (use the Find filter if necessary), right click, and select Change to layer from the pop-up menu then select the layer to which you want the trace to move, as shown in Figure 9-106.

FIGURE 9-106
Moving a trace to a
different layer.

ADDING GROUND PLANES AND GUARD TRACES TO ROUTING LAYERS

As discussed at the beginning of these examples and in Chapter 6, you can reduce noise levels on signal lines by surrounding traces with copper planes and guard traces. Not everyone agrees with this practice, but it is demonstrated here in the interest of completeness. The following procedures demonstrate how to use the Add Connect tool to add ground nets and use shapes to add copper planes to routing layers.

ROUTING GUARD TRACES AND RINGS

If you have one or two traces that could have cross-talk problems, you can add guard traces between them that can be attached to the Ground plane with vias. You can also add guard rings around component pins. Like the guard traces, the guard rings are attached to the Ground plane with vias. It should be noted though that their usefulness is debated in the literature, because they can cause more problems than they fix if not applied correctly (see Appendix E for references). Figure 9-107 shows examples of guard traces between the control and the data lines and guard rings around one of the micro-controller pins.

FIGURE 9-107
Add guard traces to help minimize cross talk.

To place guard traces that are attached to the Ground plane, select the Ground plane on the Options tab and select the Add Connect tool. Left click on the Ground plane where you want to begin the guard trace. Right click and select Add Via from the pop-up menu. The Add Connect tool should still be active and you should now be on the Top layer (but check the Options tab to make sure). Draw the trace where you want the guards to be. At the end of the trace, right click and select Add Via. At various intervals along the trace, add vias to securely stitch the guard trace to the ground plane. To add the vias, select the Add Connect tool, left click on the trace where you want the via, right click and select Add Via. Repeat this where ever a via is needed.

To add a guard ring around a component pin that is attached to the Ground plane, start by placing a via attached to the Ground plane near where the guard ring will be. The via is used to start and end the ring and can be moved to a different location after the ring is made. To begin with, the via should be placed across from (or above) one of the vertical (or horizontal) quadrant points, indicated by the numbered squares in Figure 9-108. To route a guard ring as shown in the figure, select the Ground plane from the Options tab, select the Add Connect tool and left click at the square marked 1, and immediately right click and select Add Via. Check the Options tab to make sure you are on the layer you want the ring to be, then left click at the 2 position. Right click and navigate through the pop-up menu to Options → Line Types → Arc, or select the Arc option on the Options tab. Route a curved trace around the pin and left click at point 3 continuing to points 4 and 5. At point 6, left click to tack the trace then right click and select Done from the pop-up menu. Once the

FIGURE 9-108
Adding a guard ring around a pin.

ring is in place, you can use the Slide tool to move the via to another location. In doing so you may need to temporarily delete the straight part of the trace, move the via, then reroute the trace to connect the ring to the via again. To delete just the one segment of a trace, select the Delete tool, go to the Options tab and select Clines, go to the Find filter tab and select Cline Segs only. Left click on the straight segment, make sure it is the only thing that gets highlighted, then right click and select Done from the pop-up menu. After you move the trace, use the Add Connect tool to reconnect the ring to the via.

Adding Ground Planes to Routing Layers

ShapeAdd button

ShapeAddRect button

Next we add Ground planes to the top and bottom Routing layers. *To add Ground plane areas to a routing layer,* make the Top (or Bottom) Etch layer active and select either the ShapeAdd button, 🔲, or the ShapeAddRect button, 🔲. Display the Options tab and select Dynamic copper and assign the GND net to the shape, as shown in Figure 9-109.

Left click to place the first vertex. If placing a rectangle, left click the opposite corner to complete the plane. If you are placing a polygon, left click to place other vertices to define the boundaries of the plane. To finish the polygon, right click and select Complete from the pop-up menu.

An example of a copper plane is shown in Figure 9-110. Notice that thermal reliefs are automatically placed on pins and pads. The spacing of the plane to the pad (pin) and the spoke width are set using the rules in the Constraint Manager.

FIGURE 9-109
Selecting Etch properties for a ground area on a routing layer.

Thermal reliefs are set to orthogonal (+) by default, but you can change them to diagonal (×) if you prefer. *To change the direction of thermal reliefs on a positive Plane* (Routing) layer, select Setup → Design Parameters from the menu bar. In the Design Parameter Editor dialog box, select the Shapes tab then select the Edit Global Dynamic Shape Parameters... button. In the Global Dynamic Shape Parameters dialog box (see Figure 9-111), select the Thermal relief connects tab and choose the desired direction.

To change the spacing of the Ground plane to pins on a Routing layer, launch the Constraint Manager and select the Spacing tab (Figure 9-112). Select the Shape icon under the Spacing Constraint Set folder and select the TOP layer (or whichever layer you want to change). Change the Shape to Line, Shape to Thru Pin, and so forth settings to the desired width. Changes are effective immediately; however, if your plane areas were not drawn using dynamic copper, your design may look as if something has gone horribly wrong.

If the ground plane on the Top layer suddenly takes over, the design is actually OK, the shapes just need to be updated. *To update the Ground plane shape,* select Display → Status from the

FIGURE 9-110
An example of a poured ground plane on a routing layer.

FIGURE 9-111
Setting thermal relief properties for a positive plane.

menu bar. In the Status dialog box (Figure 9-113), select the Update to Smooth button. The Ground plane should now look correct again and have the proper spacing.

To change the spoke width of thermal reliefs on a routing layer, select the Physical tab in the Constraint Manager (see Figure 9-114). Select the All Layers icon

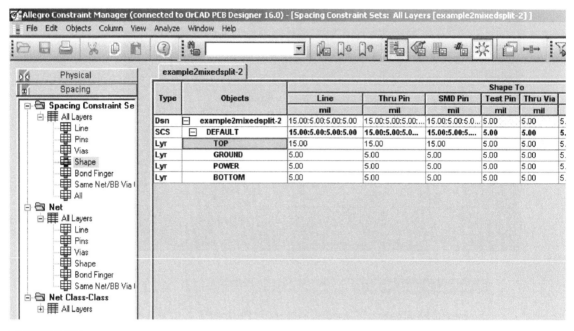

FIGURE 9-112
Setting trace to copper pour spacing rules.

FIGURE 9-113
Use the Status dialog
box to update shapes.

Type	Objects	Referenced Physical CSet	Line Width		
			Min	Max	
			mil	mil	
Dsn	⊟ example2mixed	DEFAULT	5.00	0.00	5
Net	GND	DEFAULT	8.00	0.00	5
Net	N05847	DEFAULT	12.00	0.00	5
Net	N05863	DEFAULT	12.00	0.00	5
Net	N05879	DEFAULT	12.00	0.00	5
Net	N06347	DEFAULT	5.00	0.00	5
Net	N08896	DEFAULT	5.00	0.00	5

FIGURE 9-114
Use the Constraint Manager to control thermal relief spoke width on routing layers by changing the trace line width.

under the Net folder. Change the line width of the net that has the thermal reliefs you are interested in. The spokes and traces will now be the same for the nets you modify. Perform the Update Shapes step as described previously to update your board design. If you need to have certain trace widths for specific segments on a net and you do not want to have these changes affect them, you must go back and change those segments back to the way you want them using the Change command (in the Edit menu) and the Options and Find filter tabs. Once the changes have been made, you can then Fix them so that the changes are not undone.

Once your plane is established, remember to place voids and merge copper areas and the like on the new plane to match the inner Ground plane as appropriate. Update the DRC to make sure the copper pour did not cause any errors.

You can add a Ground plane to the Bottom layer (or any other Routing layer for that matter) using the same procedures used for the Top layer. When you have several Ground planes, it is important that they have many low-impedance connections to help keep the planes at the same potential throughout the board area. This is done by placing vias, which are connected to the net for those planes, at various places. These vias are sometimes called *stitching vias*. *To place stitching vias on a board* to connect multilayer Ground planes, select the Add Connection tool, left click on the plane that needs the via (a trace will be started), right click and select Add Via from the pop-up menu. *To place multiple vias*, use the Copy Button, , to copy the via and left click wherever you want a via placed (right click and select Done to quit). An example is shown in Figure 9-115. It is also a good idea to place a couple of vias underneath the ADC (U2).

Copy Button

If you look closely at the ground strips between some of the traces and between some of the pins on the connector, J1, in Figure 9-110 and Figure 9-115, you will notice that some areas are isolated from the rest of the Ground plane.

FIGURE 9-115
Use stitching vias to provide good connections to ground planes.

These isolated strips are called *islands*. Islands can act like antennas, which can pick up high-frequency noise (EMI) and cause problems for the rest of the circuit. To solve this problem, the strips and islands need to be either tacked down to the Ground planes using the stitching vias, trimmed, or removed altogether.

Use vias as just described to tack the strips to the underlying Ground plane. For parts of strips or islands that you want to trim from larger, stitched sections, use one of the Shape Void tools to remove unwanted sections of the strip. *To trim islands and isolated strips*, select one of the Shape Void tools and use the Options tab to select the correct layer and net to void. Select the shape, then draw the void shape over the area you want removed.

Island_Delete tool

To completely remove strips or islands, use the Island_Delete tool, ⬆. By default all islands on a layer associated with a particular net will be deleted. Use the Options tab to select whether to delete all the islands on that layer or just specific ones. *To delete only specific islands*, select the First button on the Options tab then left click on the island you want to delete. Right click and select Done when finished.

This completes Design Example 2.

EXAMPLE 3   MULTIPAGE, MULTIPOWER, AND MULTIGROUND MIXED A/D PCB DESIGN WITH PSPICE

Introduction

Multipage schematics can be used to organize and simplify large circuit designs and to incorporate PSpice simulations prior to laying out the board design.

The mixed analog/digital circuit from the last example (see the schematic in Figure 9-90) is reused in this example but is modified to demonstrate how to route a single PCB from a multipage schematic project and add PSpice simulation capabilities. The example also demonstrates two methods used to establish isolated Ground planes using blind vias and a buried chassis shield. The technique allows quiet circuits to be placed on one side of the board and shielded from noisy digital circuits, which are placed on the other side of the board.

Multiplane Layer Methodologies

In the previous example, a single Plane layer was used to produce a digital ground and an analog ground even though there was really only one ground net. The two ground systems were produced by physically segregating the parts on the board and removing a strip of copper from the one plane (creating a split plane) between the two circuit areas.

In high-density, high-frequency digital designs, multiple Ground planes are often used even when there is only one type of circuit ground (as demonstrated in Example 4). This helps reduce loop inductance when using multiple Routing layers and control trace impedance (see Chapter 6 for details). When two Plane layers are used for a single net, connections are made to both planes simultaneously through plated through holes (whether for through-hole leads or fanouts from surface-mounted devices) anytime a connection is made to the net. This is shown in Figure 9-116.

FIGURE 9-116
Multiple Ground planes and multiple connections for one ground net.

In this example continuous Plane layers are desired for both the analog and the digital parts of the circuit. As described in Chapter 6, significant cross talk can occur between adjacent planes if there is any overlap between the two plane areas. Since both the Analog and Digital planes extend out to the limits of the board, there is complete overlap. One way to minimize cross talk is to insert a shield between them that carries no signal currents.

Figure 9-117 shows the system design concept for this example. This is just one of many possible types of PCB power distribution schemes. The system uses a dual power supply for ± analog power for op amps and a single power supply (VCC) for digital circuits. The analog and digital systems each has its own ground system on the PCB; however, a common reference voltage is required for the analog-to-digital converter. To facilitate both requirements, the grounds are tied together at a single point on the PCB before returning to the power supply. To keep the two ground systems from experiencing cross talk on the PCB, they are separated by a Shield layer buried inside the PCB, which is connected to the chassis ground and the shielded wire bundle. Extensive coverage of noise reduction and shielding is provided in the literature (see Ott 1988 especially for detailed coverage of this topic).

The two grounds are connected at a controlled point (or points) in several ways. The simplest is to use a jumper wire at the connector. However, the analog-to-digital converter in this example also requires a common ground area under the IC package. A jumper wire is not practical in this case.

The challenge in setting up a ground system like this is that the two ground systems must be kept separate everywhere except at the tie point. This is not possible in Capture (at least not in a straightforward manner). As soon as the two ground nets are connected on the schematic (even if only at one point), the two nets become one everywhere in the netlist and cannot be separated in PCB Editor, since it is a single net. Therefore the two distinct ground nets (or three, counting the shield) are kept separate in Capture and only made to appear connected in the schematic (for documentation purposes). The individual nets are then tied together at the common reference point in the board layout.

Two methods can be used in PCB Editor to tie the distinct grounds together at the common reference point. The first method uses a plane-to-plane connector (a shorting strip), as shown in Figure 9-118. The shorting strip is a footprint with two padstacks that can be shorted together with a wire jumper or a copper strip (a line or shape on the Top Etch class) placed in the footprint symbol or on the board design. To use the shorting strip, a Capture part must be made that has two pins but no electrical connection between them, then the PCB

FIGURE 9-118
A shorting strip or jumper wire can be used to make plane-to-plane connections.

Editor footprint is assigned to the Capture part. This method is demonstrated in Example 4.

Note: The shorting strip can be replaced with a jumper wire or ferrite bead soldered into the footprint padstacks. The copper strip method is demonstrated in the example, since it lowers cost and simplifies assembly.

The second method uses a specialized padstack that forces the planes to be shorted together by the padstack definition, as shown in Figure 9-119. Normally clearances are specified on Plane layers when there is no connection to the layer and if the netlist specifies a connection to the Plane layer, PCB Editor knows to insert a thermal relief (if a flash is assigned) in place of the clearance. However, if you explicitly specify that the clearances on Plane layers are smaller than the drill diameter, then the clearance will be drilled out and shorted to the padstack barrel during the plating process. If you force a

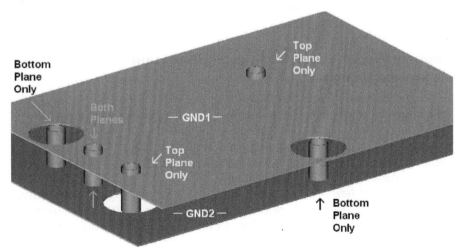

FIGURE 9-119
A specialized shorting padstack used as a plane-to-plane connector.

connection to a Plane layer, PCB Editor will generate a DRC error, which can be waived. This method also is demonstrated in Example 4.

The stack-up in this example demonstrates how to use PCB Editor to implement one method of EMI and cross-talk reduction. A 10-layer board and blind vias are required. The layer stack-up is shown in Figure 9-120. Since two methods can be used to construct blind/buried vias, both are discussed. The first method is demonstrated on the analog half of the board using a blind via named BBANLG and the second method (which makes several blind/buried vias simultaneously) is used on the digital side of the board. Since the second method produces multiple vias, the names of the vias have a base name (BBDIG in this example) and a suffix that describes which planes are connected (e.g., BBDIG-x-y). So the middle blind via on the bottom of Figure 9-120 is BBDIG-VCC-BOTTOM. Both methods are explained in greater detail.

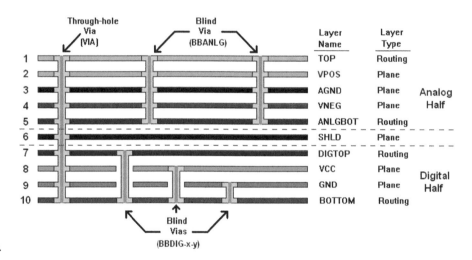

FIGURE 9-120
Layer stack-up for the shielded dual plane example with blind vias.

Capture Project Setup for PSpice Simulation and Board Design

Setting up a Capture project for PCB design and PSpice simulations at the same time has not been widely covered in the literature, so this example addresses that point. Extensive coverage of PSpice simulations in general is not provided in the example, but the process of setting up a project that allows PSpice simulations and the result of a basic simulation are included. To begin a PCB design project that can be simulated with PSpice, start Capture, and from the File menu, select New → Project. When the New Project dialog box is displayed, select the Analog or Mixed A/D option as shown in Figure 9-121. Enter a name for the project, use the Browse... button to set up a new folder for the project, and click OK.

FIGURE 9-121
Beginning a new project
for PCB design plus
PSpice simulation.

When the Create PSpice Project dialog box is displayed, select the Create based upon an existing project radio button and select the empty.opj project template as shown in Figure 9-122. Click OK.

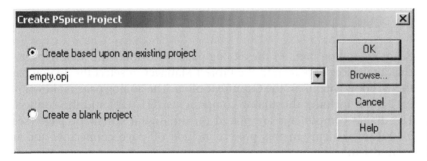

FIGURE 9-122
Selecting a PSpice
project template.

ADDING SCHEMATIC PAGES TO THE DESIGN

Three schematic pages are needed for this project: one for analog circuitry, one for digital circuitry, and one for PSpice simulation sources. A new project initially contains one schematic page, PAGE1. This page will be renamed, and two more pages will be added. To change the name of an existing schematic page, select the Schematic Page icon, right click, and select Rename from the pop-up menu. Enter the name Analog in the dialog box and click OK. To add a schematic page to a schematic folder, select the SCHEMATIC1 folder, right click, and select New Page from the pop-up menu, see Figure 9-123(a). Enter a name for the schematic (e.g., Digital) and click OK. Add another schematic page to the SCHEMATIC1 folder and name it PSpice. The final schematic page structure is shown in Figure 9-123(b).

(a)　　　　　　　　　　　　　　　　　　　　(b)

FIGURE 9-123
Setting up a multipage project in Capture.

Begin by adding parts to the analog page. To display the analog page, double click the Analog Page icon in the Project Manager or select the analog page from the Windows menu on the Capture toolbar. The analog page is shown in Figure 9-124 and includes the analog components (U1, U2) and the connector (J1). The digital components are placed on the digital schematic page (see Figure 9-100). New items in this project include off-page connectors and multiple ground symbols.

USING OFF-PAGE CONNECTORS WITH WIRES

Generally speaking, off-page connectors are used to continue signal nets across page boundaries. Off-page connectors are used in this example to connect signal lines between the ADC (on the analog page) and the microcontroller (on the digital page) and from the shift register (on the digital page) back to the connector (on the analog page). The off-page connectors are also used here to

FIGURE 9-124
The analog schematic page.

inject a PSpice signal originating from the PSpice page onto the analog input line (the Input net is shown in Figure 9-124 and described later).

To place off-page connectors, select the Place Off-Page Connector tool button, , on the schematic page toolbar or select Off-Page Connector… from the Place menu. In the Place Off-Page Connector… dialog box, select one of the off-page connector symbols and enter the name to which the connector will be attached in the Name: text box; click OK and place the off-page connector on the schematic page. You can change the name of an off-page connector after it has been placed on the schematic. *To change the name of an off-page connector*, double click the name; enter the new name in the Display Properties dialog box.

Note that the off-page connector to pin 5 on U2 (chip select) does not contain the overbar (e.g., $\overline{CS}$, indicating an active low line), as does the pin name. Overbars can be generated on Capture schematic parts for nonpower-type pins, but do not use overbars on power symbols, as this will produce invalid netlist names. With PCB Editor you can use overbars on off-page connectors not used for power nets without producing invalid netlist names. However, overbars on nets do not display the same as pin names. For example, when you make a Capture part, you can put an overbar above CS by typing the pin name as C\S\ and it will be displayed as $\overline{CS}$. But if you type C\S\ for a net alias name or an off-page connector, it will simply be displayed as C\S\, which will follow through and be displayed the same way in PCB Editor.

Place Off-Page
Connector tool button

Off-page connectors cannot be used with power symbols, but they are not required, since they are already global and known by all pages within the design.

USING OFF-PAGE CONNECTORS WITH BUSES

Both nets and buses can be connected across page boundaries with off-page connectors. If nets belonging to buses cross page boundaries by off-page connectors, net aliases (using the ▣ button) are not required on the nets because the off-page connectors produce the aliases, otherwise net aliases are required to connect nets to the bus. For example, the nets connected to board connector J1 require aliases to be connected to the bus.

SETTING UP MULTIPLE-GROUND SYSTEMS ON THE SCHEMATIC

Another difference between this design example and the previous one is the way the ground connections are made. In the previous example, there were two ground symbols (AGND and GND) but only one actual ground net (GND). In that design the grounds were physically separated on the board using a split plane, even though they were still of the same net. In this example there are three ground symbols and three distinct ground nets (AGND, GND, and SHLD). The shield ground is indicated by a GND_EARTH symbol (renamed SHLD); it is connected to J1.6 by itself, and it will be the only connection to the plane layer called *SHLD*. In Figure 9-124 AGND and GND appear to be jointly connected to J1.5 but are actually separated by the special Capture part symbol G1 (shown in Figure 9-125 along with the pin properties), which uses the PCB Editor footprint cap300. The footprint can be any two-pin footprint and cap300 is used for convenience. Capture part G1 is a custom part and is not included with the standard OrCAD libraries but is included in the

FIGURE 9-125
The Capture part G1 used as a multiground net connector.

GNDJCT.OLB on the Web site if you care to see it. Part G1 contains two pins that are graphically connected in the part but are not connected as far as the netlist is concerned. The purpose of G1 is to indicate on the schematic that the grounds are connected, while allowing the grounds to remain separate nets in the netlist.

After the connector has been placed on the board and wired to the appropriate ground nets, you can turn off the part reference (G1) by selecting Do Not Display in the Display Properties dialog box (double click the part to show the dialog box).

You can also make the pin names invisible to make the part look like a wire. *To turn off the pin names*, select the part, right click, and select Edit Part from the pop-up menu. From the Part Editor menu bar, select Part Properties from the Options menu. At the User Properties dialog box (Figure 9-126), select the Pin Names (Numbers) Visible option(s) and select False from the selection list.

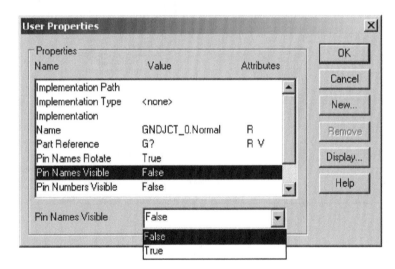

FIGURE 9-126
Making pin names and numbers invisible.

Note from Figure 9-124 that a connector, J2, is used with the chassis ground symbol. J2 is a single pin connector with a single padstack footprint that is used to connect the buried shield to the chassis enclosure. When we begin working in PCB Editor, you will see how the three ground connections will be made on the board.

Figure 9-127 shows the digital schematic page. Off-page connectors are used as described earlier. Unlike on the analog page, where each net in the bus had its own off-page connector, here the bus itself (and all the nets it contains) is attached to a single off-page connector that has the same name as the bus (e.g., Q[0..7]). Note also that off-page connectors are not used with the power and ground symbols, as they are global and known by all schematic pages in the design.

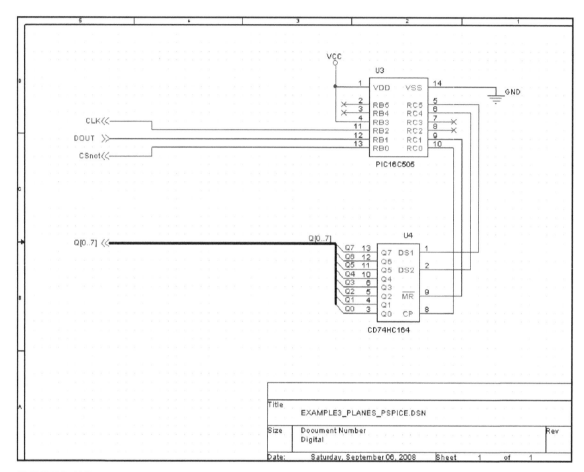

FIGURE 9-127
The digital schematic page.

SETTING UP PSPICE SOURCES

Figure 9-128 shows the PSpice page. Sources are VDC and VSIN, which can be found in the SOURCES.OLB library located in the Capture/Library/PSpice folder. Set the VDC and the VSIN source values as shown in the figure.

All PSpice simulations require a 0/GND symbol to which all sources can be referenced. The 0/GND symbol is included with the other GND symbols in the CAPSYM library. It is connected to both the analog and the digital grounds only during the simulation. After the circuit has been satisfactorily simulated, the 0/GND symbol must be deleted so that the different ground nets remain separate.

So that no footprints are added for the PSpice parts, make sure all PSpice parts are PSpiceOnly = TRUE and that the PCB Footprint cell is blank. To check these features double click a part to display the Part Properties spreadsheet. A partial spreadsheet is shown in Figure 9-128.

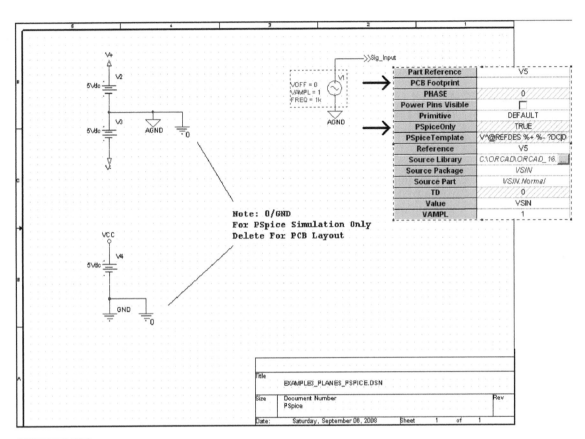

Note: 0/GND
For PSpice Simulation Only
Delete For PCB Layout

Part Reference	V5
PCB Footprint	
PHASE	0
Power Pins Visible	☐
Primitive	DEFAULT
PSpiceOnly	TRUE
PSpiceTemplate	V^@REFDES %+ %- ?DC|D
Reference	V5
Source Library	C:\ORCAD\ORCAD_16.
Source Package	VSIN
Source Part	VSIN.Normal
TD	0
Value	VSIN
VAMPL	1

FIGURE 9-128
The PSpice simulation page.

Although the voltage and signal sources are marked as PSpiceOnly = TRUE and PCB Editor will not attempt to make them available for placement, they are still a part of the netlist and the netlister will review them for netlist errors. One error that it checks for is pin numbering, and pin numbers less than 1 will be flagged as an error. If this occurs the netlist will fail and you will not be able to place any parts in PCB Editor. When fatal errors occur, you can open the netlist. log file to identify the errors that caused the netlist failure. The netlist.log file is located in the same file folder as your board file (the allegro folder by default). So, along with making sure the voltage and signal sources are marked as PSpiceOnly = TRUE, you must also check their pin numbers and change any zeros to numbers greater than or equal to 1. Voltage and signal sources usually have pin numbers starting with 0, so the ones in this example must be changed or deleted. To change the pin numbers, left click the part to select it, right click and select Edit Part from the pop-up menu. From the Part Editor, double click a pin to display the Pin Properties dialog box to determine the pin's name and number. If the pin has 0 as its number, change it to the next available number (probably

2 since the other pin should be 1). Changing the pin number will not affect PSpice simulations, since it matches the pin *names* to the PSpice template. When you change the pin numbers and close the Part Editor, the changes will be saved to your design only and the original library part will be unaffected.

Another issue can crop up when using PSpice model parts rather than non-modeled parts. One example is the resistor. Resistor can be found in both the PSpice/Analog library and the Discrete library. The PSpice resistor contains a variable, VOLTAGE, which is set to RVMAX by default. PCB Editor also has a variable called VOLTAGE it expects to be blank or a number. The value RVMAX will be imported into PCB Editor during netlisting and cause an error. So, if you are using the PSpice part models for simulation and PCB design, some variable settings may need to be changed.

Any parts that do not have PSpice templates will not be simulated. Parts U3 and U4 and all the connectors have no PSpice templates. When the simulation is run, they will be marked with green dots and ignored.

PERFORMING PSPICE SIMULATIONS

Once the circuit is made, a PSpice simulation profile needs to be established. A default simulation profile is included with the project, because the Analog or Mixed A/D... radio button was chosen during the project setup. All that needs to be done is to edit it for this design. *To edit the PSpice simulation profile*, select Edit Simulation Profile from the PSpice menu as shown in Figure 9-129.

FIGURE 9-129
Editing the PSpice simulation profile.

At the Simulation Settings dialog box (Figure 9-130), select Time Domain (Transient) from the Analysis type: list. Enter a value in the Run to time: box to display three or so complete cycles of the waveform (3–5 ms for a 1-kHz signal).

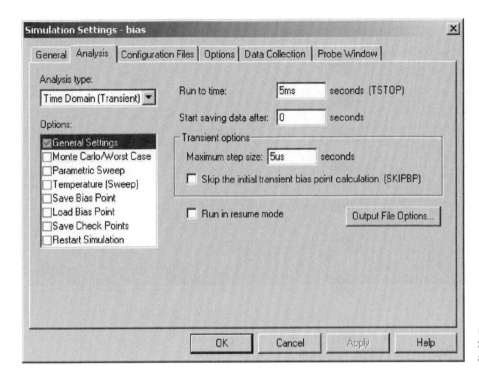

FIGURE 9-130
Setup for time domain analysis.

You can specify a value in the Maximum step size: box also, but this is optional. A value that is about 1/1000 of the run time produces very smooth waveforms but takes longer to run. Click OK when you are finished.

To run the PSpice simulation, click the Run PSpice button (blue triangle button) located on the schematic page toolbar (shown in Figure 9-131).

FIGURE 9-131
Run PSpice button.

The PSpice results are shown in Figure 9-132. Three voltage markers (probes) were placed on the design, but only two waveforms are displayed in the probe window, because not all the parts in the design had PSpice models (PSpice Templates) attached to them. PSpice will inform you that not all data were displayed by telling you, No simulation data for marker 'V(DOUT)', as is indicated by the blue arrow in Figure 9-132.

FIGURE 9-132
PSpice simulation results using a VSIN stimulus in the time domain.

To perform time domain simulations, use the VSIN source; to perform frequency domain simulations, use the VAC source and select AC Sweep/Noise in the Simulation Settings dialog box (Figure 9-130). Many other types of sources can be used to perform simulations. You can even create specialized stimulus files (including noise signals) using one of the VPWL_FILE sources. To see how to use these other sources, see the *PSpice User's Guide* (pspug.pdf, search for "stimulus files") included in the OrCAD documents folder, or see one of the many texts that specialize in PSpice simulations.

Designing the Board with PCB Editor

Once the PSpice simulations are complete and the circuit has been verified, the design is ready to be prepared for PCB Editor. One of the first tasks is to assign (or verify) footprint assignments for all parts. As described in the previous examples, a custom BOM can be generated to list the footprints to make it easier to identify missing or incorrect footprints (see the previous examples and Chapter 10 for details). A sample BOM for this design is shown in Table 9-6. Custom parts and footprints are available from the Web site for this book.

Table 9-6	BOM Footprint List for the Dual-Page Plane Example		
Reference	**Part**	**Part library**	**Footprint**
G1	GNDJCT	GNDJCT.OLB	cap300
J1	CON14	ORCAD/.../CONNECTOR.OLB	conn14
J2	CON1	ORCAD/.../CONNECTOR.OLB	Con1
R1	1 k	ORCAD/.../DISCRETE.OLB	Smdres
R2	2 k	ORCAD/.../DISCRETE.OLB	Smdres
R3	2 k	ORCAD/.../DISCRETE.OLB	smdres
R4	2 k	ORCAD/.../DISCRETE.OLB	smdres
U2	TLC548	.../USERLIB/CHAPTER9.OLB	soic8
U3	PIC16C505	.../USERLIB/CHAPTER9.OLB	soic14
U4	CD74HC164	.../USERLIB/CHAPTER9.OLB	soic14
U5	OP-27/LT	.../EXAMPLE3_PLANES_PSPICE. DSN	soic8

The remaining tasks were described in the preceding text or earlier examples and are listed here without details:

- Remove 0/GND symbols used for PSpice simulations, and change voltage and signal source pin numbers to something other than 0.
- Perform an annotation to clean up part numbering (optional).
- Make sure that global power nets are properly utilized.
- Perform a DRC in Capture to verify that the circuit design has no issues. Correct any errors and reperform the DRC as needed.
- Use Capture to generate the netlist and launch PCB Editor.

CREATE THE BOARD OUTLINE

As described in earlier examples, set the design size to a size A sheet (Setup → Design Parameters... → Design → Size), and draw a board outline (Setup → Outlines → Board outline...). A 3.00 × 2.00-in. or larger board is sufficient for this design. Use one of the Place Part tools to place parts inside the board outline. Place digital parts on the bottom side of the board, as shown in Figure 9-133.

PLACING PARTS ON THE BOTTOM (BACK) OF A BOARD

To place parts on the bottom side of a board, select the Generaledit button, ▨, select Symbols in the Find filter tab, select the component, right click and select Mirror from the pop-up menu. Left click off to the side to deselect the part. The part should now be on the bottom of the board. If you do not see the part, make sure the bottom Silk-Screen and associated layers are visible. Check the DRC for footprint and placement problems prior to doing anything else.

Generaledit button

FIGURE 9-133
Initial part placement for
Example 3.

LAYER STACK-UP FOR A MULTIGROUND SYSTEM

The layer stack-up shown in Figure 9-120 is used in this design. Set up Power and Ground planes, the Shield plane, and the analog and digital Routing layers as shown in the Layout Cross Section dialog box in Figure 9-134 (click the Xsection button,). To add a layer, right click and select Add Layer Above (or

Xsection button

Below) from the pop-up menu. Add 16 layers, then name and change the types as shown in the figure (eight Dielectric layers, two Routing layers and six Plane layers). Select Negative Artwork for all the Plane layers, click the Apply button, then the OK button to complete the setup and dismiss the dialog box. Select distinctive colors to differentiate the planes using the Color tool as described in the earlier examples.

Once the parts are in place and the stack-up is defined, we can pour copper on the Plane layers, perform fan-outs, and begin routing traces on the board. We begin by pouring the copper on the planes.

ADDING COPPER TO THE PLANES

As described in the previous examples, use the Setup Plane Outlines tool (Setup → Outlines → Plane outline... from the menu bar) or one of the Shape Add tools (on the toolbar) to add the copper pours to the Plane layers. Remember that dynamic copper shapes are preferred over filled rectangles or static shapes and to assign each shape to its correct net.

ESTABLISHING NET, PLANE, AND CONSTRAINT RELATIONSHIPS

The next thing we want to do is to fan out power and ground, but before we can perform fan-outs or route traces, we need to set up blind via definitions

FIGURE 9-134
The board stack-up in the Layout Cross Section dialog box.

and custom routing constraints to limit the layers on which certain nets can be routed. Table 9-7 shows a summary of which layers, vias, and physical constraint sets will be assigned to each type of net. Blind and buried via definitions are described in the next section, and physical constraint definitions are described in the following section.

DEFINING BLIND VIAS

To use a blind or buried via in a design, an appropriate padstack must exist so that it can be used. If one does not exist, you need to make one. There are two ways to do this, as described next.

Padstacks used as vias exist as library padstacks and board design (layout) padstacks. A *library padstack* is a padstack definition contained in the symbols library. A *layout padstack* is a padstack definition associated with a pin or via in a board design. However, when a padstack is used in a board design, its definition (as used) is stored in the board layout file itself and not in any library. So there are two ways to work with via padstacks that are used as blind/buried vias, the first is through the Padstack Designer application and the second is

	Table 9-7	Relationships among Nets, Layers, Vias and Physical Constraints			
Net	**Routing**	**Plane**	**Via**		**Physical Cset**
Analog nets	TOP, ANLGBOT	—	BBANLG		PCSanalog
V+	Top	VPOS	BBANLG		—
AGND	Top	AGND	BBANLG		—
V−	Top	VNEG	BBANLG		—
SHLD	—	SHIELD	VIA (modified)		—
VCC	BOTTOM	VCC	BBDIG-VCC-BOTTOM		—
GND	BOTTOM	DGND	BBDIG-GND-BOTTOM		—
Digital bus nets	DIGTOP, BOTTOM	—	BBDIG-DIGTOP-BOTTOM		PCS1
Digital ADC nets	TOP, DIGTOP, BOTTOM	—	VIA		PCS2

through PCB Editor itself, which is the preferred method in most cases. Only a basic overview of the method using the Padstack Designer is given here. Using Padstack Designer is covered in detail in Chapter 8. Via definitions for this design are made using the second method from within PCB Editor.

BASIC OVERVIEW OF USING PADSTACK DESIGNER

To define and store a padstack in the PCB Editor library, you can create one from scratch or you can copy an existing padstack, modify it, then save it with a new name in the library using the Padstack Designer. You can launch it from within PCB Editor or independent of PCB Editor (called *stand-alone mode*).

To launch Padstack Designer from within PCB Editor select Tools → Padstack and select one of the Modify padstack options.

To make a padstack from scratch using the Padstack Designer in stand-alone mode, go to the Windows Start button on your tool tray and from the All Programs option, select OrCAD 16.0 → OrCAD PCB Editor Utilities → Pad Designer, as shown in Figure 9-135.

The Pad Designer dialog box is shown in Figure 9-136 opened in stand-alone mode. Using the Pad Designer, you can construct through padstacks (for through-hole pins on components or for vias), blind/buried padstacks (for vias), or single-layer padstacks (for surface-mount component pins).

Padstack design is covered in detail in Chapter 8 and mentioned only briefly here. In general you set the drill diameter, pad shapes and sizes, and thermal relief and anti pad shapes and sizes. In the Layers tab, you can specify layers in addition to the default ones. By selecting specific layers and certain types of connections, you can use this tool to construct blind/buried vias.

FIGURE 9-135
Launching Padstack
Designer in stand-alone
mode.

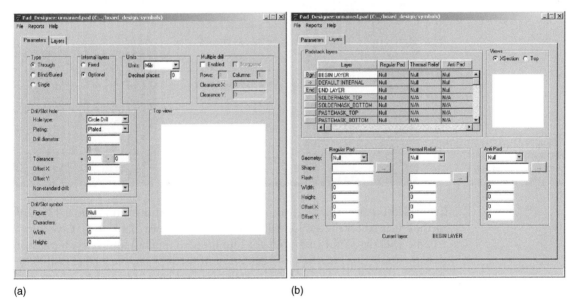

(a)   (b)

FIGURE 9-136
Padstack Designer: (a) Parameters tab, (b) Layers tab.

You can also launch the Padstack Designer from PCB Editor by selecting Tools →
Padstack → Modify Padstack from the menu. You can modify library padstacks
or padstacks associated with the design only.

To use the Padstack Designer, you need to duplicate the layer stack-up in your
board design and specify a specific layer/pad combination that satisfies the via
requirements. The via padstack is saved to the library or the design and assigned to
specific nets in the board design. This approach can be cumbersome and the via is
easily reusable in future designs only if they have an identical layer stack-up.

The other method is to set up vias interactively from the design using the BBvia tool. This is the easiest way to make a blind or buried via, because all the layer definitions are known, and set up, by the BBvia tool. Interactive design of blind and buried vias is initiated by selecting Setup B/B Via Definitions, as shown in Figure 9-137, where you can select from two different methods of interactively designing custom blind and buried vias.

FIGURE 9-137
Two ways to interactively set up blind and buried vias in a design.

If you select Define B/B via…, you get the dialog box shown in Figure 9-138. This method is used for the analog half of the board (see also Table 9-7). Give the via a name (BBanlg in this example, for blind/buried analog) and select a padstack to copy from. This sets the basic definition of the new padstack (e.g., drill diameter and pad shapes and sizes). The Start and End layer entries define how "tall"

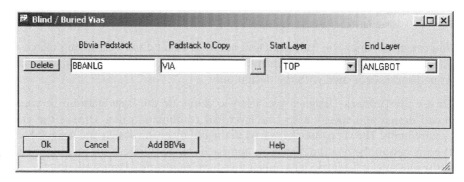

FIGURE 9-138
Blind/Buried Vias setup dialog box (option 1).

the padstack will be (see Figure 9-120). So in this example, if you choose VIA as the padstack to copy (which of course is a through-hole padstack), the new padstack will be identical to it but will not go all the way through the board (it will go between only the Start and End layers). This method creates one padstack that goes from the Start Layer to the End Layer and includes all the layers between them. Click OK when finished or AddBBVia to make another one.

If you select Auto Define B/B via..., then you get the dialog box shown in Figure 9-139. We use this method to make the blind vias for the digital half of the board. Again select a padstack on which to base the new one. Check the Add prefix box and select a prefix name. Select the Start and End layers as before. You can select which layers will be used in setting up the via; we use all the layers in this example. Once the settings are entered, click Generate to start the process. Click Close when finished.

FIGURE 9-139
Blind/Buried Via setup dialog box (option 2).

This method actually makes several vias. Essentially it makes as many blind and buried vias as necessary to connect each of the adjacent planes (two layers at a time) and to connect the Start and End layers to each other and to each of the internal layers. You can view the bbvia.log file, and in this example, the following padstack vias were created:

    BBDIG-DIGTOP-VCC
    BBDIG-DIGTOP-GND

BBDIG-DIGTOP-BOTTOM
BBDIG-VCC-GND
BBDIG-VCC-BOTTOM
BBDIG-GND-BOTTOM

So any via that has BOTTOM in its name is a blind via, and the others where BOTTOM is not included are actually buried vias. One of the vias will never be used in this example, VCC-GND, because it would short the Power plane to GND and result in a scrapped board. Once the planes and vias are defined, we will assign the proper via(s) to each of the nets.

Cmgr button

Assigning Vias to Nets

Use the Constraint Manager to assign vias to the nets. Open Constraint Manager by clicking the Cmgr button, ▦, on the toolbar. Select the Physical tab, and select All Layers under the Net folder. In the Vias column, you will see a listing of all vias in the design (including the default, Via). Select a net to modify by selecting the cell in the Vias column for that net. The Edit via list dialog box will be displayed, as shown in Figure 9-140.

FIGURE 9-140
Using the Constraint Manager to assign vias to nets.

Note that the BBAnlg is in the "available" list but not in the "used" list. Add the BBAnlg via to all of the analog-related nets (including V+, V−, and AGND). Remove any vias that are prohibited or unnecessary. The final setup is shown in Figure 9-141. Notice that the CLK, CSNOT, and DOUT nets use the default via, VIA, since it is a through-hole padstack and these nets need to pass through the Shield layer (to connect the ADC on the analog side to the microcontroller on the digital side). Although the SHLD net does not need a via, the default via, VIA, is assigned to it because each net must have at least one via assigned to it, so the default via was assigned. Next we see how to restrict the routing of certain nets to specific layers.

| Type | Objects | Line Width | | Neck | | Vias |
| | | Min | Max | Min Width | Max | |
		mil	mil	mil	mil	
Dsn	⊟ example3_plane	5.00	0.00	5.00	0.00	VIA:BBDIG-DIGTOP-VCC:BB...
Net	AGND	8.00	0.00	5.00	0.00	BBANLG
Net	AIN	12.00	0.00	5.00	0.00	BBANLG
Net	CLK	5.00	0.00	5.00	0.00	VIA
Net	CSNOT	5.00	0.00	5.00	0.00	VIA
Net	DOUT	5.00	0.00	5.00	0.00	VIA
Net	GND	8.00	0.00	5.00	0.00	BBDIG-GND-BOTTOM
Net	N14049	12.00	0.00	5.00	0.00	BBANLG
Net	N14065	12.00	0.00	5.00	0.00	BBANLG
Net	N25559	12.00	0.00	5.00	0.00	BBDIG-DIGTOP-BOTTOM
Net	N25591	12.00	0.00	5.00	0.00	BBDIG-DIGTOP-BOTTOM
Net	N25595	12.00	0.00	5.00	0.00	BBDIG-DIGTOP-BOTTOM
Net	N25631	12.00	0.00	5.00	0.00	BBDIG-DIGTOP-BOTTOM
Net	Q0	5.00	0.00	5.00	0.00	BBDIG-DIGTOP-BOTTOM
Net	Q1	5.00	0.00	5.00	0.00	BBDIG-DIGTOP-BOTTOM
Net	Q2	5.00	0.00	5.00	0.00	BBDIG-DIGTOP-BOTTOM
Net	Q3	5.00	0.00	5.00	0.00	BBDIG-DIGTOP-BOTTOM
Net	Q4	5.00	0.00	5.00	0.00	BBDIG-DIGTOP-BOTTOM
Net	Q5	5.00	0.00	5.00	0.00	BBDIG-DIGTOP-BOTTOM
Net	Q6	5.00	0.00	5.00	0.00	BBDIG-DIGTOP-BOTTOM
Net	Q7	5.00	0.00	5.00	0.00	BBDIG-DIGTOP-BOTTOM
Net	SHLD	10.00	0.00	5.00	0.00	VIA
Net	SIG_INPUT	12.00	0.00	5.00	0.00	BBANLG
Net	V+	8.00	0.00	5.00	0.00	BBANLG
Net	V-	8.00	0.00	5.00	0.00	BBANLG
Net	VCC	8.00	0.00	5.00	0.00	BBDIG-VCC-BOTTOM

FIGURE 9-141
Via-net assignments.

ASSIGNING NETS TO LAYERS USING CUSTOM, PHYSICAL CONSTRAINTS

The next step is to assign nets to the proper Routing and Plane layers. The net layer assignments are shown in Table 9-7. To make net layer assignments, we need to set up new physical constraint sets (CSet). To set up a new CSet, open the Constraint Manager and select the Physical tab. Select Objects → Create → Physical CSet from the menu, as shown in Figure 9-142.

At the Create Physical CSet dialog box (see Figure 9-143), enter a name for the new constraint and click OK. If a net or layer is selected in the spreadsheet, the dialog box shows a Copy constraints from: area. To avoid this, deselect the cell by left clicking on a column header and restart the process.

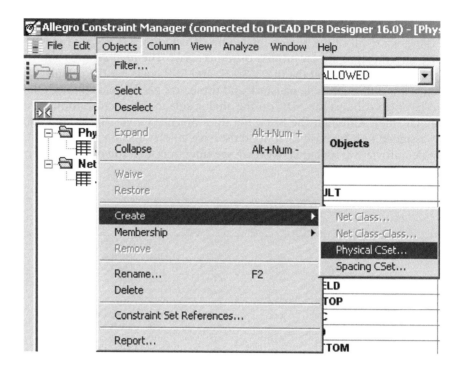

FIGURE 9-142
Setting up a new
constraint with the
Constraint Manager.

FIGURE 9-143
The Create Physical
CSet dialog box.

To set up the new constraint, select the Physical tab and select the All Layers icon under the Physical Constraint Set folder. The new constraint will be listed beneath the DEFAULT constraint, as shown in Figure 9-144. In this example we use the constraint to limit routing to specific layers, so the appropriate flag is set under the Allow Etch column, as shown in the figure.

Once the constraint set is defined, assign it to proper nets (see Table 9-7). *To assign a constraint set to a net*, select the All Layers icon under the Net folder (see Figure 9-145). Select the cell or cells in the Referenced Physical CSet column for the nets to which you want to assign the new constraint. When you select the cell(s), a dropdown list will be displayed. Select the desired constraint set from the dropdown list. As shown in the figure, the new constraint, PCS1, was selected for the digital bus traces. With this constraint the autorouter will route only these traces on the DIGTOP and BOTTOM layers. If these nets are routed

FIGURE 9-144
The new physical constraint listed in the Constraint Manager.

FIGURE 9-145
Assigning constraints to nets.

manually on a layer not included in the constraint set, DRC errors will be issued (see Figure 9-146 as an example).

Continue naming and assigning constraint sets to the nets as outlined in Table 9-7.

Once vias have been assigned to nets and nets have been assigned to layers, the next step is to begin the fan-out process.

FAN-OUTS USING BLIND VIAS

Once the vias are set up and assigned to the nets, we can begin performing the fan-outs. We do the first couple of ones by hand so that you can see how it works. We begin by doing the first fan-out for VCC. Make all the Plane layers

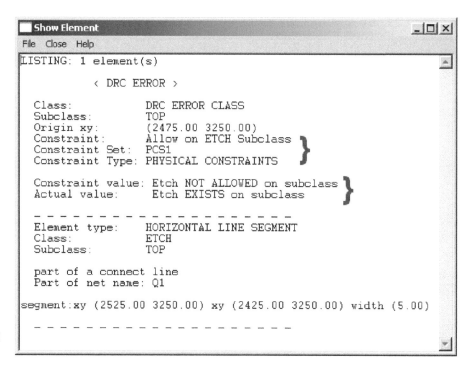

FIGURE 9-146
DRC errors resulting from violating a physical constraint set.

and nets invisible except VCC. Set the Etch grid to 25 mils (All). Zoom to the area around U3 (PIC16C505) and locate pin U3.1 (check the Pins option in the Find tab if necessary). Select Add Connect... and click on the net (make sure Nets is checked in Find tab). Route a small section out from the pad and left click to place a vertex. Display the Options tab (see Figure 9-147) to make sure that the two layers involved are Bottom and Vcc and that the BBDIG-VCC-BOTTOM via is available. Then right click and select Add via from the pop-up menu. A bbvia should be inserted. Right click and select Done. Note that the correct Alt layer has to be displayed in the Options tab or you may not be able to place a via if the via assigned in the Constraint Manager cannot physically make a connection from the Act. (active) layer to the Alt. (alternate) layer displayed in the Options tab.

If you toggle back and forth between the Bottom, Vcc, and GND layers, you should see that the via connects the trace and pad on the bottom to the VCC plane and a clearance area around the via is on the GND layer. If you toggle through the analog planes, you should see no evidence of the trace or via at all.

Next we try one of the analog fan-outs. Locate resistor R3 and its pin R3.2 We will route a fan-out from this pin to the analog Ground plane. In the Constraint Manager, make sure that the AGND net (and layer) is visible and able to be routed. Select the Add Connect... tool, click on the net to start a trace, and left click a short distance (about 50 mils) to place a vertex. Check the Options tab and make sure that the active layer is Top and the alternate layer is AGND,

FIGURE 9-147
Routing a blind via to Vcc.

and that the BBANLG via is available. Right click in the work space and select Add Via from the pop-up menu.

Note that the process of using the blind via fan-outs is the same no matter which method was used to generate them. Now we use the autorouter to place the rest of the fan-outs.

We fan out the remaining AGND and Vcc nets and complete all the V+ , V− , and GND fan-outs using the autorouter. Open the Constraint Manager and make sure that all the power and ground nets are visible and no routing restrictions have been placed on them. Select Route → Route Automatic... from the menu. As described in detail in the previous examples, use the Automatic router dialog box to set the fan-out parameters. As a summary select the Router Setup tab, and in the Strategy group box, select the Specify routing passes radio button. Select the Routing Passes tab and select only the Fanout option in the Pass Type column. Click the Params... button to display the router parameters dialog box. Select the desired options and click OK to dismiss the dialog box. Click the Route button to perform the fan-out. You can also use the Create fanout tool to fan out pins and components.

An example of a completed fan-out is shown in Figure 9-148. Notice where a via goes from Bottom to GND (digital GND) on U3 right under the corner of the pad on R4. Note also that this does not cause a DRC error, because the via under the pad of R4 does not go all the way through the board and therefore does not touch R4. Other occurrences of this type are shown on the pads for the ICs.

FIGURE 9-148
Fan-outs completed using blind vias.

Once all the fan-outs are complete, the rest of the board can be routed. Set up the autorouter as described in the previous examples and route the rest of the board. As a quick overview, select Route → Route Automatic... from the menu. At the Automatic router dialog box, select the Router Setup tab, and in the Strategy group box, select the Specify routing passes radio button. Select the Routing Passes tab, uncheck the Fanout option in the Pass Type column, and select the Route and Clean options.

Click OK to dismiss the dialog box. Click the Route button to route the board.

Note: Designs like this can be tricky. If the autorouter has difficulty routing the board or performing the fan-outs, make sure that all the Plane layers and design constraints are set up properly, the proper vias are assigned to the correct nets, and the appropriate nets are enabled.

An example of the fully routed board is shown in Figure 9-149. The grid and all the Plane layers are turned off so that it is easier to see the traces.

Alternate Methods of Connecting Separate Ground Planes

In this example the two Ground planes are electrically separate and the component G1 in the schematic is used to connect the planes on the board by soldering a wire jumper, an inductor, or ferrite bead in the G1 footprint. In the following sections, alternatives to inserting and soldering the wires or components

FIGURE 9-149
The final board design
for Example 3.

are discussed. The first alternative is to short the G1 padstacks with copper, and the second alternative is to short the planes with a shorting padstack.

SHORTING THE PLANES WITH COPPER ETCH

To reduce assembly complexity, a copper etch object can be used to short G1's padstacks rather than soldering a wire or installing a component into the footprint. The copper etch can be a trace or a copper area. *To add a thick trace* across the padstacks, select the Etch/Top classes in the Options pane. Select the Add Line tool, and from the Options pane, set the line width to 30 mils or so. Left click on the first padstack of G1 and draw a line (trace) to the second padstack. The copper etch line is shown in Figure 9-150(a).

Another way to make the connection is to use a copper area. To do so, select the Shape Add Rect tool. In the Options pane, select Static solid as the Shape Fill Type. You can leave the Dummy Net assigned to the shape. The static solid is shown in Figure 9-150(b).

FIGURE 9-150
Adding copper etch,
(a) a trace or (b) a shape,
to short the planes.

(a)   (b)

When using either of these etch objects, DRC errors will occur because the shapes violate pad spacing rules. Since this is what we want, we can override the DRC errors. To do so, click the Waive DRC button on the toolbar. Make sure that the DRC Errors option is checked in the Find filter pane then left click on the DRC markers to override them.

SHORTING THE PLANES WITH A PADSTACK

Rather than take up space on the board using a dedicated footprint, you can use a via attached to one of the Ground planes and modify it so that it is attached to both Ground planes at the same time. We place a VIA on the board, attach it to the GND plane, and modify it so that it is connected to the AGND plane too.

When using this approach, the special part (G1) in Capture is not used. Instead a small segment of a graphical line—using the Place Line tool instead of the Place Wire tool—is used to indicate that the two grounds are connected at the header pin. This eliminates the footprint in PCB Editor.

Go back the schematic page and delete part G1 in Capture. Use the Place Line tool to make the AGND and GND nets look like they are connected as shown in Figure 9-151. The Line Tool is graphical only and does not create a connection in the netlist.

FIGURE 9-151
Placing a "virtual" ground connection using the Line tool.

Perform an ECO to forward annotate the changes to PCB Editor as described in the earlier examples. In short, save the board design and close PCB Editor. In Capture select the project icon, select Tools → Create netlist... from the menu. Select the PCB Editor tab, check the Create or Update PCB Editor Board box and relaunch PCB Editor. When the board design is reopened, the footprint for G1 should be gone.

The next step is to place a via on one of the planes and modify the via to connect it to the other plane. To do so, select the Add Connect tool and from the Options pane, make the AGND plane (class) as the active class and the Top layer as the alternate. Left click on the AGND plane near the connector J1 to place a vertex and start a trace. Immediately right click and select Add Via from the pop-up menu; right click again and select Done from the pop-up menu. The via (which should be padstack BBANLG) is now connected to the AGND plane but nothing else.

To modify the via, select Tools → Padstack → Modify Design Padstack... from the menu. Left click the via to select it. In the Options pane, select the Edit Instance button (see Figure 9-152). The original BBANLG name will be listed and a new via BBANLG-1 will also be listed. Click the Edit... button to display Padstack Designer.

The Padstack Designer, Layers tab, is shown in Figure 9-153 with the new padstack settings. Note that the Flashes and Anti Pads are set to 5 mils, which is much smaller than the drill diameter. So, when the holes are drilled into the board during the manufacturing process, the small clearances will be drilled away and the copper on the planes will butt up against the drill hole. Then, when the hole is plated, the plating will short the planes together. Remember to add clearances to the other plane layers (e.g., Vcc and SHLD) or they will be shorted to the planes too.

To save the changes, select File → Update to Design... from the menu. A warning box will be displayed telling you that a pad will be drilled away (see Figure 9-154). That is what we are after, so close the warning box. Another warning will be displayed (see Figure 9-155). Click Yes to complete saving the changes and quit editing.

FIGURE 9-152
Modifying a via to short the two ground planes.

FIGURE 9-153
Use Padstack Designer to modify a padstack definition.

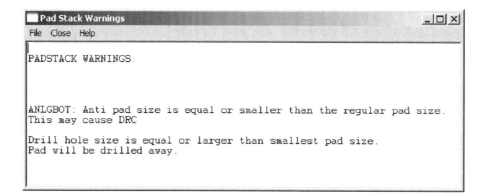

FIGURE 9-154
Pads drilled away
warning.

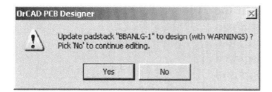

FIGURE 9-155
Save padstack modifications with warnings.

Close Padstack Designer then right click and select Done from the pop-up menu to complete the changes. Now if you toggle between the two planes you will notice a small hole in the AGND and DGND planes. If you toggle the Display plated holes option on and off (Setup → Designer parameters... → Display tab), you can see that the drill hole is larger than the clearances on the two Plane layers, so the small clearances will in fact be drilled away. Check the other planes to make sure that the clearances on those planes are larger than the drill hole, so that they will not be exposed to the plating process and shorted to the Ground planes.

The modified padstack will cause a DRC error, because the padstack, which is connected to the AGND net, is touching the DGND plane (GND net). This is exactly what we want. Use the Waive DRC tool to override the error.

Note: This method is not necessarily recommended, since connection between the two Ground planes is not automatically "documented" by the software. The process should be manually documented on the schematic and some type of marker should be placed in silk screen on the board, indicating which via is shorting the two planes together. In the event that some problem occurs and the planes need to be separated, it would be a simple matter to drill out the via. If the via is not marked somehow, it would be impossible for someone not familiar with the board design to know where or how the planes were shorted together. Even if a person were to look at the design in PCB Editor without some markings in the design, the only way to tell which via is shorting the planes is to look at all of the padstack definitions or create a Waived Design Rules Check Report (if the person happened to think of it).

This concludes the third design example.

EXAMPLE 4   HIGH-SPEED DIGITAL DESIGN

This example demonstrates how to stack up layers and design transmission lines for a high-speed digital PCB. The example also demonstrates how to create a moated ground area with a bridge around a high-frequency crystal oscillator, how to perform pin/gate swapping, and how to create a heat spreader using vias to the Ground plane. The example circuit is shown in Figure 9-156.

FIGURE 9-156
High-speed digital circuit schematic.

The BOM for this example is shown in Table 9-8. The circuit consists of a (fictional) high-speed, low-pin-count microcontroller/digital signal processor (uP-EXD10) driven by a 66-MHz clock (X1), a digital to fiber optic interface IC (FO-TX, which mimics an ADN2530 but with fewer pins), a fiber-optic laser

Table 9-8		Bill of Materials for the Digital Design Example			
Item	Quantity	Reference	Part	Nomenclature	Footprint
1	2	C1, C2	1 μF	Bypass capacitor	smdcap
2	2	C3, C4	27 pF	XTAL shunt capacitor	smdcap
3	1	J1	CON10	10-pin board connector	conn10xx100tr
4	1	LD1	LD/NX8311	Fiber-optic laser diode	to18-4
5	1	U1	uP-EXD10	Microcontroller/DSP IC	soic14
6	1	U2	FO-TX	Digital-to-fiber interface IC	plcc12
7	1	U3	54ALS00A	NAND logic gate	soic14
8	1	X1	66 MHz	Crystal oscillator	XTAL2smd

diode (LD1), and a couple of 54ALS00 NAND gates used for I/O decoding. The digital signals have rise and fall times from 200 ps to 1.9 ns and require controlled impedance traces (see the Analog Devices ADN2530 data sheet for an example application). In a real design, more bypass capacitors would be used on the circuit, but the design is scaled down to keep the design simple. The parts and footprints are located on the Web site for this book.

Using the procedures described in the earlier examples, start a new Capture project, and place and connect the parts as shown in Figure 9-156. After completing the schematic, make sure all footprints are assigned to the parts and create the netlist for and launch PCB Editor.

Start a new board project using the procedures described in the previous examples.

As in the previous examples, the first step is to make a board outline and place the parts inside the boundary. The initial board layout is shown in Figure 9-157. Signal flow is from left to right, with the highest-frequency components located close together near the laser diode connector on the right side of the board.

FIGURE 9-157
Initial board layout for the digital design example.

The next few steps were covered in detail in the previous examples. The following tables and figures show the design parameters for this example, but step-by-step instructions are not repeated here. The required steps are (1) define the layer stack-up and enable the appropriate layers using the Layout Cross Section dialog box, (2) define two vias in addition to the default VIA using the Padstack Designer and Constraint Manager, and (3) fan out power and ground for the surface-mounted components.

Layer Setup for Microstrip Transmission Lines

Since there are so few parts, a simple four-layer board design is used. The layer stack-up and net assignments are shown in Figure 9-158. The layer thicknesses

Layer Name	Layer type	Used by nets:
Top	Routing	Any
Ground	Plane	GND
Power	Plane	PWR
Bottom	Routing	Any

FIGURE 9-158
Layer stack-up for the
digital example.

depend on the board manufacturer; the values (units in mils) shown in the figure are typical. The Top layer and Ground plane will be used to route surface-type microstrip transmission lines and most of the lower-speed digital traces. Only low-speed traces that cannot be routed on the top will be routed on the bottom layer.

Figure 9-159 shows the layer stack-up as defined in the Layout Cross Section dialog box. Notice that the material thicknesses have been added for documentation purposes. High-speed signal analysis tools are not functional in the OrCAD version of PCB Editor.

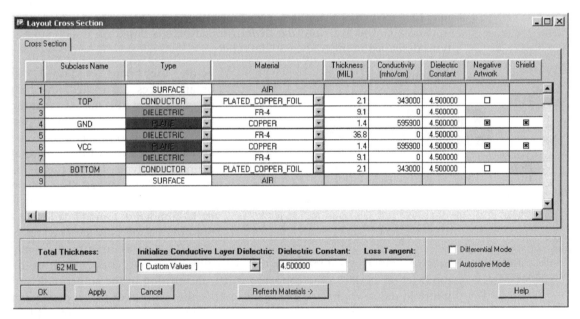

FIGURE 9-159
The layer stack-up defined in the Layout Cross Section dialog box.

Three via types are used in this example (see Table 9-9). VIA is the default via included in the symbols library folder, while VIATENT and VIAHEAT are custom vias (included in the design folder on the book's Web site). VIATENT is similar to the default VIA but is tented (i.e., the padstack contains no soldermask definitions) and used here for the fan-outs as a demonstration. Multiple copies of VIAHEAT will be used as heat pipes to connect a thermal pad (a copper

Table 9-9	List of Vias Used in the High-Speed Digital Example					
Via name	Function	Drill dia.	Pad dia.	Clearance dia.	Connection to plane	Soldermask
Via	Default	13	24	30	Full	Yes
ViaTent	Fan-outs	13	24	30	Full	No
ViaHeat	Heat pipes	10	20	26	Full	Yes

pour area) beneath U2 to the Ground plane and function as a heat spreader. VIAHEAT has smaller dimensions so that they can be placed close together to provide low resistance to heat flow to the ground plane.

Constructing a Heat Spreader with Copper Pours and Vias

The design of heat spreaders on PCBs depends significantly on the type of device and how it is attached to the board. For design examples and thermal management calculation, see the application note references listed in Appendix E. The heat spreader demonstrated here is based on design suggestions described in the ADN2530 data sheet.

FIGURE 9-160
Functional diagram of a heat spreader.

Before the board is fanned out or any traces are routed, the heat spreader is put into place so that the router avoids that area, thereby preventing having to rip up and reroute fan-outs or traces. A functional diagram of one type of heat spreader is shown in Figure 9-160. The silicon die inside the component is thermally bonded to a metal pad on the bottom of a specially designed package. The pad is in turn thermally bonded (either by soldering or thermal compound) to a copper area on the top layer of the PCB. The copper area has an opening in the soldermask and multiple vias to connect it to a Plane layer (either ground or power depending on the chip design). The vias function as thermal conductors (heat pipes) that allow heat to flow away from the component. If a component dissipates excessive heat, the Plane layer can be mechanically (and thermally) connected to a larger heat sink or other mounting hardware to help dissipate the heat.

VIA DESIGN FOR HEAT SPREADERS

Before we begin to construct the heat spreader, we need to define the VIAHEAT padstack. To make the heat pipe efficient at conducting heat, solid connections to the plane are used rather than thermal reliefs. To define the new via (VIAHEAT.pad), we begin with an existing padstack, save it with a new name, modify it to our specifications, then save it. If you have your board design open, launch the Padstack Designer by selecting Tools → Padstack → Modify Design Padstack... from the menu bar. Display the Options tab, select Via from the list, then click the Edit... button.

Before making any changes to the padstack choose File → Save as... from the menu bar. An information window will be displayed with the following warning:

PADSTACK WARNINGS:

Drill hole size is equal or larger than smallest pad size.

Pad will be drilled away.

Click the X button to dismiss the window (this will be explained shortly). When asked Save padstack with WARNINGS? click Yes. Save the padstack as VIAHEAT.pad in either the working directory or the symbols library.

Using the Parameters and Layers tabs as shown in Figure 9-161 modify the padstack per the specifications in Table 9-9.

The reason that the "drilled out pad" warning was given is because of the thermal flash symbol, AB00. If you open the AB00 drawing (ab00.dra in the symbol library) you will notice that it is a simple copper dot with a diameter of 5 mils. Remember that a copper dot on a negative plane actually defines a spot where the copper is removed. Since this opening's diameter is much smaller than the drill diameter, the drill bit will remove the opening and contact the copper plane. When the board's holes are plated, the plane layer will be connected directly to the plating (no thermal relief), which is exactly what we want for a heat pipe.

Since the top and bottom pads in this padstack are smaller than the drill diameter, they will also be drilled out. Normally we would not want this, but in this example many vias will be placed side by side and a copper rectangle will be placed over the whole group and will act as one big pad for the entire group of vias. Top and bottom pads can be left in place, but it is visually more appealing in the design if the pads are not shown.

Note also that the soldermask is left rather large but could have been removed because, as will be shown later, one large soldermask opening will be placed on the board to expose the heat spreader and heat pipes and will overwrite the individual soldermask openings.

After the VIAHEAT padstack is finished and saved, repeat the process to create the tented via, VIATENT, which will be used as the default fan-out via. The VIATENT padstack is identical to the default VIA padstack, except that no soldermask openings are defined, as shown in Figure 9-162.

Once the VIAHEAT padstack is finished, use the Constraint Manager to assign it to the GND net, as shown in Figure 9-163. *To assign a via to a net*, left click in the cell to display the Edit Via List dialog box, as shown in Figure 9-164. Left click a via in the left-hand box to add it to the via list box.

(a)

(b)

FIGURE 9-161

Use the Padstack Designer to define the heat pipes: (a) Parameters tab, (b) Layers tab.

FIGURE 9-162
Layer definitions for the tented via.

FIGURE 9-163
Use the Constraint Manager to assign vias to nets.

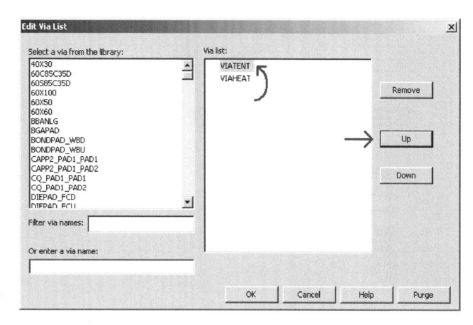

You can also remove vias from the list by left clicking the one you want to remove, then clicking the Remove button. The via at the top of the list will be the default via. To change the priority of use, select the desired via and click the Up or Down button to move the via within the list. When you are finished, click OK. The ground layer will need the VIATENT via for the fan-outs and the VIAHEAT via for the heat spreader, so assign both vias to the GND net. All other nets get only the VIATENT via, and no nets should be allowed to use the VIA via (select VIA and click the Remove button).

FIGURE 9-165
Options settings for the
heat spreader plate.

Now that the heat pipes have been defined, we can build the heat spreader. We begin by placing a copper plate on the Top layer to define the boundaries of the heat spreader. To do this click the Shape Add Rect button on the toolbar, display the Options pane, and select the options shown in Figure 9-165.

Draw a copper pad in the center of the footprint, as shown in Figure 9-166. Note: You may need to adjust the grid settings (Setup → Grids) to achieve the necessary drawing resolution.

The next step is to place the heat pipes. To do this, begin by clicking the Add Connect button. Display the Options pane and select GND as the active layer (Act) and Top as the alternate layer (Alt). Left click on the design at the insertion point. Display the Options pane again and select VIAHEAT from the via list (see Figure 9-167). Move your mouse back over to the design area, right click and select Add via from the pop-up menu. A VIAHEAT

FIGURE 9-166
The heat spreader placed within the component footprint.

FIGURE 9-167
Select the tented via with the Options pane.

via should be placed. Repeat this process for each pipe. See Figure 9-168 for reference.

Note: If you are using the Allegro PCB Editor software, you can easily place many vias at once using the via array generator. Although OrCAD PCB Editor is based on Allegro PCB Editor, the via array generator is not included, so the vias must be placed manually, one at a time. As an alternative to placing the heat pipes in the board design, you can make a custom package that includes the heat pipes and copper plate.

For a good thermal connection to occur between the package and the heat spreader, an opening needs to be made in the soldermask. *To make an opening in the soldermask*, select the Shape Add Rect button on the toolbar then select the Board Geometry class and Soldermask_Top subclass (or Package Geometry class and Soldermask_Top

FIGURE 9-168
The final heat spreader design.

subclass) from the Options pane. Draw a filled rectangle over the copper pad that is 5 mils larger than the pad on all sides. If the thermal bond will be made by soldering, repeat the preceding steps to place a filled rectangle on the Package Geometry class and Pastemask_Top subclass.

Note: There is no Pastemask subclass in the Board Geometry class and many of the footprints included in the library do not have a pastemask defined in the Package Geometry class. So, if you are planning to use pick and place assembly, you must modify some of the footprints to include the pastemask definitions.

The finished heat spreader is shown in Figure 9-168. At this point click the Fix button (the green thumb tack) and fix all the vias and the dynamic copper plate so that they are not removed by the autorouter or any routing or cleanup processes.

The next step is to fan out the board. Earlier we used the Constraint Manager to establish *net* default vias, but when the autorouter is used to fan out a board, it uses the *design* default. *To change the default design via*, open the Constraint Manager, select the Physical tab and the All Layers icon under the Physical Constraint Set folder. Left click the Default cell in the Vias column. In the Edit Via List dialog box, select the VIATENT via to add it to the list and remove the VIA via; click OK.

Now you can set the fan-out parameters using VIATENT. To set fan-out parameters, select Setup → Design Parameters from the menu bar. In the Design Parameter Editor, select the Route tab then click the Create Fanout Parameters button, as shown in Figure 9-169. In the Create Fanout Parameters dialog box, you can now select VIATENT whereas it would normally list only VIA.

Once these settings are correct, perform the fan-out as described in detail in the previous examples. In summary, select Route → Route Automatic... from the menu bar. At the Automatic Router dialog box, select the Router Setup tab, and select the Specify routing passes radio button. Next, select the Routing Passes tab, select the Fanout and deselect the Route and Clean options, click the Params... button. In the SPECCTRA Automatic Router Parameters dialog box, select the Fanout tab. In the Pin Types section, select the Specified: button, select the Power Nets and deselect all others. Click OK, then click the Route button.

Once the router has finished, it is a good idea to check the vias to make sure PCB Editor did what you asked and performed the fan-out using the right via. To check the fan-out vias, display the Find pane and make sure that the Vias box is checked. Hover your pointer over a via and an information box will be displayed, which will tell you which via was used.

If for some reason the wrong via was used, you can change it. *To change placed via types*, select Tools → Padstack → Replace from the menu bar. Display the Options tab and, as shown in Figure 9-170, select the via to be replaced and the new via. Click the Replace button to make the changes take effect.

Once the fan-outs are complete, use the Constraint Manager to fix ground and power nets and enable all the other nets. Next we route critical traces before autorouting the board.

FIGURE 9-169
Set the default fan-out vias using the Create Fanout Parameters dialog box.

Determining Critical Trace Length of Transmission Lines

Since the controlled impedance traces are critical, they are routed next. The first step is to determine which traces need to be handled as transmission lines and which ones do not. As mentioned previously the digital-to-fiber interface IC, FO-TX (U2), was modeled after the Analog Devices ADN2530. In the data sheet, the digital signal lines going to the part and the modulation signals leaving the part (going to the laser diode) are to be handled as transmission lines. The digital control lines going to U2 need not be handled as transmission lines. The only traces left to consider are the ones related to the crystal oscillator and the NAND gates.

The literature states that the propagation time, PT, should be less than one half of the rise time, RT (or fall time, FT); that is, PT < ½RT. If possible it is better if PT < ¼RT (see Chapter 6 for more details). So we need to calculate PT

FIGURE 9-170
Use the Options pane to switch via types.

for this board layout and look up RT and FT for the oscillator and the NAND gates. Since the crystal is a fictional part here, let us assume that RT = FT = pulse width = ¼ the total period of a 66-MHz square wave. Under that assumption RT = 3.8 ns for the oscillator. The typical RT for ALS family logic is 1.9 ns.

The critical maximum length can be calculated using Eq. (9.1),

$$\text{Length}_{\text{trace}} < \frac{\text{RT}}{k \times t_{\text{PD}}} \tag{9.1}$$

where $\text{Length}_{\text{trace}}$ is the maximum allowed trace length in inches, RT is the signal rise time in picoseconds, k is the safety factor ($k = 2$ minimum), and t_{PD} is the propagation delay of the board material in picoseconds per inch.

The propagation delay for the surface microstrip (see Table 6-6 in Chapter 6) is

$$t_{\text{PD}} = 85\sqrt{0.457\varepsilon_r + 0.67} \tag{9.2}$$

using $\varepsilon_r = 4.2$ for FR4, $t_{\text{PD}} = 137$ ps/in., and the critical trace lengths are given in Table 9-10 for various values of k. As indicated, there is no way that the ADN2530 traces can be treated other than as transmission lines, but as long as none of the other traces is longer than 3.5 in., they need not be treated as transmission lines. Note that we are neglecting the length of the cables leaving the board through connector J1, but that is beyond the scope of the example.

Table 9-10	Maximum Safe Trace Lengths			
		Maximum Length$_{\text{trace}}$ (in.)		
	RT (ns)	**k = 2**	**k = 3**	**k = 4**
66-MHz OSC	3.8	13.9	9.26	6.95
ALS logic	1.9	6.95	4.63	3.47
ADN2530	0.026	0.095	0.063	0.048

ROUTING CONTROLLED IMPEDANCE TRACES

The objective is to design surface microstrip transmission lines with a characteristic impedance of $Z_0 = 50\,\Omega$. Using the design equations from Chapter 6, repeated here in Eq. (9.3), the width of the trace is calculated as

$$w = 7.47h \times e^{(-Z_0\sqrt{\varepsilon_r + 1.41})/k} - 1.25t \tag{9.3}$$

where, from Figure 9-158, $t = 1.35$ mils (1 oz copper), $h = 10$ mils, $k = 87$ for $15 < w < 25$ mils (most references use this number—87 is used here), or $k = 79$ for $5 < w < 15$ mils (Montrose offers this option), $Z_0 = 50\,\Omega$ (the design goal), and the desired trace width in mils $w = 17.5$ mils (17 mils $= 50.9\,\Omega$).

To specify the width of a net, open the Constraint Manager. Select the Physical tab and select the All Layers icon under the Nets folder. Set the Min trace width to 6 mils and the Max value to 17.5 mils in the Line Width columns for the ModN and ModP nets.

With these settings the traces will be 6 mils by default (good for connections to the small surface-mount pads) but can be as wide as 17.5 mils (which is needed for the transmission lines).

FIGURE 9-171
Use the Options pane to change the trace width during manual routing.

Next the transmission lines are routed manually. Choose the Add Connect tool. Select a net on U2 at a point close to the pad to begin routing. Place a vertex just outside the place outline by left clicking once (this short, narrow trace allows for a thermal relief during reflow but is too short to interfere with the trace impedance).

With the vertex in place, we now want the 17.5 mil trace. *To change the trace width,* display the Options pane and enter 17.5 in the Line width: box, as shown in Figure 9-171. The trace attached to your cursor should now be the correct width and you can continue routing to the laser diode pin. Repeat this process for the other net. Figure 9-172 shows the completed transmission line.

Note that, because of the pin-out of the component and the lead spacing of the diode, the lengths of the traces may not be equal (which is recommended in the data sheet). You can use the Show Element tool to measure the lengths of the traces. Figure 9-173 shows a comparison of the two traces and reveals that the top trace (MODP) is about 58 mils longer than the bottom trace (MODN).

FIGURE 9-172
The routed transmission line.

FIGURE 9-173
Use the Show Element tool to compare transmission line lengths.

If we want the trace lengths to be within a certain tolerance, then we need to reroute the shorter traces to include extra length (using trombones, accordions, or sawtooths) or change the position or orientation of either or both components. As an example Figure 9-174 shows U2 rotated 45° and relocated. The trace lengths in this configuration are equal to within 0.01 mils.

FIGURE 9-174
A component rotated 45°.

To rotate a part 45°, select the Move tool, display the Options pane, and select 45 in the Angle: selection list. Select the part (make sure the Symbols box is checked in the Find filter pane), right click, and select Rotate from the pop-up menu. Move the cursor around to rotate the part. When the part is in the correct rotation, left click, move the part to the correct place and left click again to place the part.

MAXIMUM NECK LENGTH

When routing traces with varying widths, DRC errors may result because a trace has been "necked down." To clear maximum-neck-length DRC errors, open the Constraint Manager, select the Physical tab, and select the All Layers icon under the Physical Constraint Set folder. Change the DEFAULT Maximum Neck Length as necessary (e.g., 40 mils) to eliminate the DRC errors.

Moated Ground Areas for Clock Circuits

The oscillator is routed next. In many applications a moated ground plane around the clock circuitry is recommended to prevent stray ground currents from affecting other circuits. Before adding the moated ground area around the oscillator, the traces should be routed so that the size of the required ground area is known. Begin by enabling all the nets associated with the clock circuitry (set Rats on). Route the traces manually. Figure 9-175 shows the routed, curved clock traces (along with the moat, which is described later).

FIGURE 9-175
Clock circuitry with curved traces and a moated plane area.

ROUTING CURVED TRACES

Notice that the traces between the crystal (X1) and U1 are curved. While they are not necessary, the curved corners are used here as a demonstration. *To route curved traces*, select the Add Connect tool, left click the net near one of the pads on X1, route the trace straight out from the pad about 100 mils or so, and place a vertex by left clicking. Display the Options pane and select Arc from the Line lock: list (Figure 9-176). As you move your mouse around, you will see that the trace is a curve instead of an angle. Place vertices to define the curves. You can switch between curves and lines as needed using the Options pane.

The next step is to etch a moat into the GND plane around the clock circuitry (see Figure 9-175). *To etch a moat into a plane*, make the GND plane visible and choose the Shape Void Polygon tool, . In the command window, PCB Editor

Shape Void Polygon tool

FIGURE 9-176
Use the Options pane to select between curved and straight traces.

will ask you to Pick shape or void to edit. Left click on the ground plane to select it (it will become highlighted). Create the void by picking void coordinates with the left mouse button. To define the shape shown in Figure 9-175, 32 pick points were required. When you get to the last pick point, right click and select Complete from the pop-up menu.

Make sure to leave a "bridge" attached to the main ground plane. The bridge should be wide enough to include the ground pin and the area under the clock traces on U1. The local ground area under the clock circuitry must be attached electrically to the rest of the ground system, but the moat is used to "corral" the ground currents back to the ground pin on the IC. Ground areas (and moats) will be placed on the Top and Bottom layers, and ground stitching will be used on all Ground planes; but before that is performed, the rest of the board needs to be routed.

Disable and lock all routed traces (use the Constraint Manager to turn on the Fixed parameter). Enable the remaining unrouted nets and set the routing grid to 25 mils (Setup → Grids, All Etch Spacing:). Autoroute the board (Route → Route Automatic..., Routing Passes: Route and Clean on, Fanout off). Figure 9-177 shows the result. Many of the traces have wandered around due to poor usage of the gates (as assigned on the schematic), particularly around the areas marked 1 and 2. Note

FIGURE 9-177
The design after autorouting.

also that, at area 3, two traces were routed over gaps (the moat) in the Ground plane. As described in Chapter 6, we do not want to allow this, as it will increase the loop inductance and introduce EMI issues. To fix these two problems, we now look at how to perform pin and gate swapping and how to define a route keep-out area to prevent the autorouter from routing traces over the moat.

Gate and Pin Swapping

Two methods can be used to swap gates and pins. The first is to swap the gates (or pins on a gate) on the schematic page and run an ECO to PCB Editor. The second method is to swap pins in PCB Editor and back annotate the changes to Capture to update the schematic.

The second method described is demonstrated here. Before doing the swap, save the design (or save as a new .brd file) so that the preswap design can be recovered if something goes wrong.

The first task is to unroute the gates. Make sure all gate nets are enabled and all other nets are disabled and fixed. Select the Delete tool and drag a box across multiple nets to rip up more than one trace at a time.

Before a swap can be made, you need make sure pins are swappable in Capture. To verify that they are swappable, go to the schematic page, double click on a pin, select <Current Properties> from the filter list; its Swap ID will be −1 if it is *not* swappable, and it will be 0 or a positive number if it *is* swappable. If pins are not swappable, you need to change the property of the pins in the part definition in Capture.

To make pins swappable, go to the schematic in Capture, left click on the logic gate you want to make swappable, go to the Edit menu and select Part. The part editor will be displayed. In the part editor select View → Package from the menu bar. You will see all the gates in the work space. Now select Edit → Properties from the menu bar to display the Package Properties spreadsheet, as shown in Figure 9-178. In each of the PinGroup cells, enter an integer number greater than 0 to make those pins swappable. You can enter specific numbers for specific gates to make them swappable with each other and different numbers to make them swappable with only those pins (gates) with the same number. This gives you control over which gates/pins can be swapped with certain other gates/pins. You need to do this only with the input pins, because Capture and PCB Editor are smart enough to know which output pins will go with their respective input pins. If you want all gates to be equally available for swapping with any other gate in that IC, then you can enter integers in just the two input pins on one gate then click the Update button and Capture will automatically put the same integer number in all of the input pins on all of the gates. Click OK to dismiss the spreadsheet.

To save the changes, click the X in the upper right-hand corner to close the part editor. When Capture asks if you want to change only the one or all of them, choose Update All. Choosing Update Current will cause problems with the

FIGURE 9-178
Making pins swappable in Capture.

netlister, because when you change a part with the part editor, the link between the part and the library from which it came will be broken and the part will be given a modified name with a suffix (e.g., 54ALS00A will be renamed to 54ALS00A_1). Every time you modify a gate, the suffix is incremented. So if you modify a gate and choose Update Current, only that gate will be modified and its suffix will be different from the other gates within the same IC package. You can check this by selecting the parts, right clicking, and selecting Edit Properties from the pop-up menu. Select <Current Properties> from the filter list and look at the Source Package cells for the individual gates. If gates from the same IC have different names (suffixes), the netlister will fail during an ECO (check the netlist.log file for details if this happens). If you choose Update All, then all gates in that IC will be given the same suffix. Gates contained in different ICs will not be changed (or renamed) by Update All, so if you want them to be swappable, you will need to modify them as well. If the parts within a package end up with different names, you cannot modify a single part to change its name to match the others, because Capture will increment its suffix to one greater than the highest suffix in the package, so they will always end up being different. The only way to fix it is to open one of the gates with part editor, as described previously, then immediately close it again (no changes are necessary) and select Update All when prompted. This will cause all parts to be renamed with the same name.

Repeat these steps for pins 10 and 11 on U1.

If you have had to modify parts a couple of times, a record of each change is maintained in the Design Cache folder in the Project Manger pane. This is not a problem, but you can remove the outdated parts from the Design Cache selecting (left click on) the Design Cache folder then selecting Design → Cleanup Cache from the menu bar.

After you have made gates/pins swappable, you need to perform an ECO to let PCB Editor know that the pins can be swapped.

Once the ECO is complete and the board design is open, swapping is accomplished through the Place menu. The Place menu offers two swap options, Swap and Autoswap, and in each of the two options are additional options. In short, the difference between the two options is that the Swap option gives *you* specific control over pin, function, and component swapping actions; while the Autoswap option allows *PCB Editor* control over swapping actions (with some input from you). We first look at the Swap option.

Using Swap Options

PIN SWAPPING

In this example we use the Swap option to manually uncross the two nets between U1 (pins 10 and 11) and U2 (pins 11 and 12). *To perform a manual pin swap*, select Place → Swap → Pins from the menu bar. In the command box, you will see Pick a pin you wish to swap. Pick pin 10 on U1. Then in the command box, you will see

last pick: 3500.00 3200.00
Pick a pin you wish to swap from those that are highlighted.

The package to which the pin belongs will have the other pins highlighted that were set in Capture with the same pin group number. Select pin 11 on U1. Then you should see

last pick: 3500.00 3200.00
Pick DONE/NEXT or pick the first pin of the next swap.
Command >

Right click and select Done from the pop-up menu. The nets should now be swapped between the two pins and no longer be crossed. A back annotation will be performed at the end of this section to show the effects of the pin swap in Capture.

You can perform pin swaps on logic gates too. When doing so, PCB Editor will allow you to swap pins only within a single gate and only pins that are of the same type. For example if you pick pin A on a NAND gate (an input pin), it will allow you to swap it only with pin B on the same gate. You cannot swap pin A with any pin on another gate (even if it is on the same IC), and you cannot swap it with pin Y (the output) on the same gate, because pins A and Y are different types of pins.

FUNCTION (GATE) SWAPPING

If you need to swap an entire gate (e.g., U3A for U3C) and not just pins within a gate, then you use the Swap → Function option. *To swap two gates* (and therefore their pins), select Place → Swap → Functions from the menu bar. In the command window, PCB Editor will instruct you to Pick a function you wish to swap.

In this example select pin 1 on U3 (this is input pin A on gate A). All the pins on U3 will become highlighted and the command window will display

last pick: 2400.00 2700.00
Pick a function to swap with from those highlighted …

Select pin U3.13 (this is input pin B on gate D); the command window displays:

last pick: 2600.00 2700.00
W- (SPMHA2-12): Pick NEXT/DONE or pick the first function of the next swap.

Right click and select Done from the pop-up menu. The rat's-nest lines that were connected to gate A will now be connected to gate D and the pins will be utilized in the best way to minimize length of the rat's nest (referred to as *virtual wire length* by the Autoswap tool).

Swap → Components does not perform gate or pin swapping, it just physically swaps (trades) the locations of two components on the board. The components do not even have to be the same type or be "swappable." *To perform a component swap*, select Place → Swap → Components from the menu bar. The command window will not ask you for anything, but left click one of the parts in the design (it will become highlighted) then left click another part. The two parts will immediately trade places. To undo the component swap, right click and select Oops or Cancel from the pop-up menu.

BACK ANNOTATING THE SWAP OPERATIONS TO CAPTURE

If you are working through these examples on your own board design, you can perform a back annotation to see the effects of the gate and pin swapping on the schematic. *To perform a back annotation*, select File → Export → Logic from the menu bar. At the Export Logic dialog box, select the Cadence tab and click the Browse… button at the bottom to select the file path where your board design is located. Click the Export Cadence button. After the back annotation is complete, close the dialog box. A swap file has now been generated, and we need to import it into Capture. *To import the swap information into Capture*, go to the Project Manager pane in Capture, select the Design icon and select Tools → Back Annotate... from the menu bar. In the Backannotate dialog box, select the PCB Editor tab and make sure that the correct folders and board and swap files are selected, then click the OK button. When you go back to your schematic, it should be updated with the new information, as shown in Figure 9-179 (compare this to Figure 9-156).

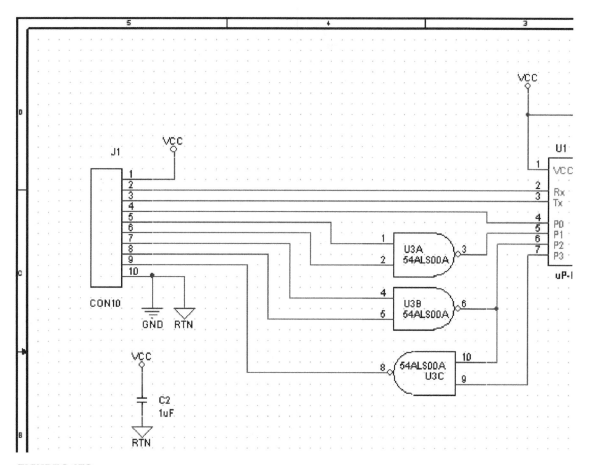

FIGURE 9-179
Logic gate connections in the schematic design in Capture after the pin swap.

In the case of the pin swap on the microcontroller (U1), pins 10 and 11 were actually moved on the part definition (and a suffix was added to the parts name, as described previously) while the nets were physically unaltered (see Figure 9-180). This is obviously different from the way pin swaps look in PCB Editor and can be easily missed. If you or another designer replaces this

FIGURE 9-180
Microcontroller connections in the schematic design after the pin swap.

modified part with the original from the library and an ECO is performed, the pin swap will be effectively undone. Another point to consider in an example like this is that any software written for the microcontroller would have to be changed to accommodate the pin swap.

Using the Autoswap Option

In using the Autoswap option, you basically let PCB Editor figure out the best gate usage and pin connections then back annotate the results to the schematic. You do have some control over the swapping action. From the Autoswap menu, you can have PCB Editor work the whole board design, a room, or a selection window. After you select the type of autoswap you want to do, you select Place → Autoswap → Parameters to execute the swap. Examples of each follow. Once you perform an autoswap, you *cannot* Undo it, so to try all the autoswap types on this small design, you can try a swap and close the PCB design without saving it (and without doing a back annotation) then reopen the board design to try a different one.

AUTOSWAPPING AN ENTIRE DESIGN

To have PCB Editor review and autoswap and entire board design, select Place → Autoswap → Design from the menu bar. Then select Place → Autoswap → Parameters... to display the Automatic Swap dialog box shown in Figure 9-181.

FIGURE 9-181
Setting up the Automatic Swap tool.

The numbers represent the time limit (in minutes) for each pass. After a pass PCB Editor records the virtual wire length for each net and compares the lengths to the previous pass. PCB Editor will continue trying to shorten the wire lengths until all passes have been completed or the wire lengths cannot be made any shorter. By default only two passes are enabled. Passes with a 0 time limit are disabled.

The interroom option allows or prohibits PCB Editor from attempting to shorten the wire lengths by swapping gates between different rooms.

You can specify more or fewer passes and change the time limit depending on the complexity of your design. You can see what PCB Editor attempted to do during each pass by reading the swap.log file that PCB Editor generates after a swap operation. The swap.log file is located in the same file folder as the board design.

To begin the autoswap, click the Swap button. PCB Editor will look over all the components and nets and automatically swap all gates and pins in a way that will result in the shortest traces. Note that gates that were not placed somewhere on the design will not be considered for swapping and neither will gates with "no-connect" markers on pins or pins connected to nets that are fixed or nets that have a No Ripup property assigned in Constraint Manager. If you want unused gates that were not placed on the schematic to be considered for gate swapping, place them on the schematic and leave the pins floating. After the swap and subsequent back annotation, you can apply no-connect pins and delete unused gates.

AUTOSWAPPING A ROOM

You can restrict how much the design PCB Editor is allowed to modify by defining rooms around a component or groups of components and telling PCB Editor

FIGURE 9-182
Use the Physical Room
Browser for swap
operations on rooms.

to work on only the rooms you want swapped. To use this option, you must have rooms defined (review Design Example 1 to see how to set up rooms). Once rooms are defined, select Place → Autoswap → Room from the menu bar. The Physical Room Browser will be displayed, as shown in Figure 9-182. Select the room you want PCB Editor to review and autoswap and click OK. Then select Place → Autoswap → Parameters... to display the Automatic Swap dialog box, and click the Swap button. PCB Editor will automatically swap all gates and pins in the selected room but will leave the rest of the design alone.

AUTOSWAPPING A WINDOW SELECTION

You can also restrict how much of the design PCB Editor is allowed to modify by defining a window around a component or groups of components. To use the Autoswap Window function, select Place → Autoswap → Window from the menu. Define the window by clicking your left mouse on the design to define a window corner then drag a box (window) around the components you want swapped; left click again to complete the window. Then select Place → Autoswap → Parameters... to display the Automatic Swap dialog box, and click the Swap button. PCB Editor will automatically swap all gates and pins within the selected window (if swapping results in an improvement) but will leave the rest of the design alone.

VIEWING THE SWAP LIST AND THE SWAP LOG

Once you have made a swap type selection, you can review what PCB Editor will be looking at with a swap list before you actually perform the swap. To display a swap list, select Place → Autoswap → List from the menu. The swap list shows you where PCB Editor will focus during the swapping operation, but it does not give you information about the swap itself. You can obtain detailed swap information by opening the swap.log file generated after a swap is performed. The swap.log file is located in the same file folder as the board design.

Once the swaps have been completed you can rerun the autorouter. Figure 9-183 shows the results of the autoswap and autoroute. The figure also shows the route keep-out area placed over the ground moat. Notice how the autorouter hugs the keep-out area but does not cross it. Defining a route keep-out area is described next.

DEFINING A ROUTE KEEP-OUT AREA

To define a route keep-out area on your board, begin by making the Route Keepout class active on the Options pane. In the subclass list, you can choose any of the layers. In this example select the All subclass to place the keep-out on all layers simultaneously. Next, select the Shape Add button, ⟳, on the toolbar. Draw a polygon on the Route Keepout / All class/subclass that matches the moat in the Ground plane. This will keep the router from routing any traces on Routing (conductor) layers and will also create a void in the VCC plane layer

Shape Add button

FIGURE 9-183
The board design after autoswap and autoroute.

similar to what the void shape did on the Ground plane (see Figure 9-175). The keep-out also works if you pour a copper ground area on the Top and Bottom layers, such as the one demonstrated in Example 3.

This completes the Example 4 design.

POSITIVE PLANES

As we saw in the previous design examples, keeping track of the necessary details to use negative Plane layers can be a bit of a task (making flash symbols, modifying padstacks, etc.). Negative planes are used because they have been used historically and are recommended in the OrCAD user's manuals. One of the reported advantages of using negative planes is that the artwork files are smaller than for positive artwork files. With the current PC speed and memory capacity, file size is less of an issue. Another argument for negative planes is a shorter processing time for negative planes at the manufacturer's end. That certainly was true when vector plotting with the flash lamps, but for current photoplotting equipment, that has also become less of an issue. Yet another argument for negative planes is that, during the board design, the planes can be distracting and result in slow drawing regeneration. But PCB Editor displays the negative planes in the positive view (WYSIWYG mode) anyway, and you can easily turn off the Plane layers. Additionally the author is aware of some designers who use positive planes for most of their work. That being the case and for the sake of completeness, we look at a very simple design to demonstrate how to use positive planes and take care of the corresponding details.

The circuit for this example is shown in Figure 9-184. As shown, the circuit consists of only two resistors. We focus on the Plane layers that will be used by the VCC and GND nests, so this simple circuit suffices.

As with the previous design examples, the netlist is created and PCB Editor is launched. A board outline is drawn and the parts placed. The next step is to define the layer stack-up.

FIGURE 9-184
Simple circuit to test positive planes.

In the previous design examples, the Ground and Power planes were defined as negative Plane layers. In this example, we use positive planes. The layer stack-up is shown in Figure 9-185. As indicated the layer stack-up is similar to the other design examples except that the Negative Artwork boxes are unchecked.

A positive plane layer should still be specified as PLANE rather than CONDUCTOR so that the autorouter does not try to route traces on that layer. Routing traces on Plane layers (particularly GND planes) results in slots in the plane, which

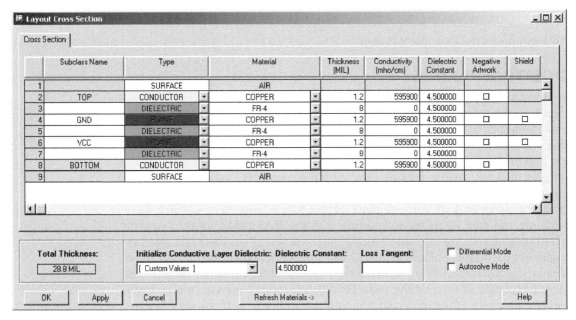

FIGURE 9-185
Layer stack-up for the positive plane circuit.

disrupts the return paths for signals on Routing layers and can lead to signal integrity issues. Setting the type as Plane prevents this.

Once the layer stack-up is specified, the next step is to draw the copper areas on the planes. The process is identical to the previous examples. Select the Shape Add tool, select the GND or VCC subclass in the Etch class, and draw a dynamic copper area attached to the appropriate net.

Unlike with negative planes, you do *not* have to enable the Thermal Pads setting in the Design Parameters dialog box to see the thermals. Figure 9-186 shows the two resistors and the VCC plane. Note that R1 is connected to the plane with a thermal relief that was automatically generated by PCB Editor. The padstacks used in this design are native to the symbols library and have no thermals assigned to them. This result is similar to the Ground plane placed on the Top layer in Example 3.

FIGURE 9-186
Thermal reliefs shown on a positive Plane layer with default settings.

The padstack definition is shown in Figure 9-187 using Padstack Designer. As indicated the inner pads are 55 mils in diameter and the clearances are 75 mils in diameter. This indicates that there should be 10 mils between the edge of a pad and the edge of a plane (if the pads are present).

A close-up inspection of the padstack reveals that the inner pad is indeed 55 mils but the diameter of the clearances (outer diameter of the thermal relief) is only 65 mils (leaving a space of only 5 mils between the pad and the copper area). So, on a positive plane, neither the thermal relief diameter nor the anti-pad diameter in the padstack definition determines the diameter of a clearance or the outer diameter of a thermal relief.

Since PCB Editor automatically generates the thermal reliefs, a question arises as to how it determines the thermal relief dimensions. The answer was described briefly in Example 2. The space between the copper area and the padstack is determined by the spacing constraints set in the Constraint Manager, and the spoke width is determined by the trace width setting in the Constraint Manager. If we change the shape-to-pin spacing constraint to 10 mils (see Figure 9-188), the thermal and clearance areas should increase. And if we change the trace width constraint to 10 mils (see Figure 9-189), the spoke width should increase.

Figure 9-190 shows the difference between (a) the default constraints and (b) the result of changing the trace width and spacing to 10 mils—the spoke widths and the void width are wider.

FIGURE 9-187
Padstack design for the positive Plane layers example.

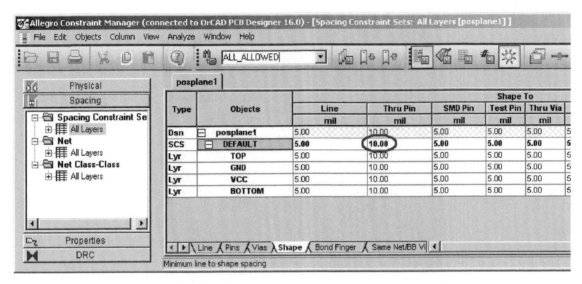

FIGURE 9-188
Changing the trace (and void) spacing constraint.

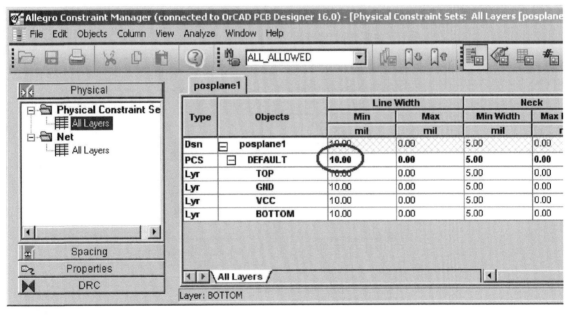

FIGURE 9-189
Setting the trace (and spoke) width constraint.

Besides the constraint settings in the Constraint Manager, you also have control over thermal reliefs on positive planes using the Design Parameter Editor. As shown in Figure 9-191, additional parameters can be set by selecting the Shapes tab and clicking the Edit global dynamic shape parameters button.

(a)   (b)

FIGURE 9-190
Comparison of VCC (a)
before and (b) after
constraint changes.

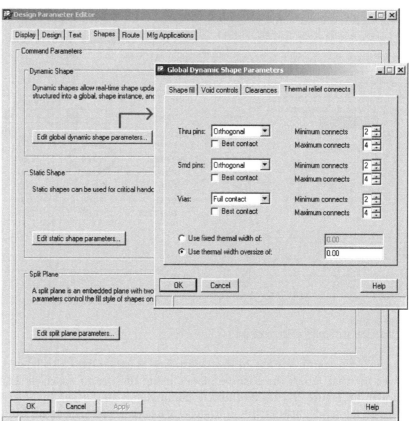

FIGURE 9-191
Other thermal relief
settings.

At the Global Dynamic Shape Parameters dialog box, you can control the rotation (Orthogonal, Diagonal, or Full contact, etc.), the number of connects (spokes), and set fixed or scaled spoke widths (relative to the trace width constraint).

One thing that is different about this plane and a typical negative plane is that the pads not connected to the planes are present, whereas with the negative planes, they are absent. Typically, it is desired that the unused pads on Plane layers are removed. Although the pads cannot be removed in the design mode, they can be removed during the artwork production.

Positive Plane Artwork Production

Artwork production is covered in detail in Chapter 10, but we touch on it here. The inner pads can be removed on positive planes during the artwork production, provided that three conditions are met: (1) the Inner layers option needs to be selected as Optional rather than Fixed in the Padstack Designer's Parameters tab (see Figure 9-187), (2) the Suppress unconnected pads option needs to be checked in the Artwork Control Form, and (3) the film Plot mode: needs to be processed as Positive artwork (also in the Artwork Control Form, as shown in Figure 9-192).

The IPC-2221A standard specifies that unconnected pads on routing planes should be maintained, but unconnected pads on Plane layers can be removed, so when using positive planes, suppress unconnected pads on only the Plane layers and leave the option unchecked for Routing layers.

Figure 9-193 shows the VCC layer photoplot artwork for this design (Silk-Screen and other layers not shown). The figure shows that the unused pads were removed as intended. Artwork verification is described in detail in Chapter 10.

Positive vs. Negative Plane File Sizes

It was mentioned at the beginning of this section that it is commonly held that the positive plane file sizes are larger than the negative ones. Let us take a look at the file sizes of the resulting artwork for this example. The GND plane artwork for this design in the positive image is 54 lines of Gerber code at 1.15 kB, while the negative artwork file is 128 lines of code at 2.34 kB. So, at least for this simple design, the positive artwork files are actually *smaller* than the negative files (as generated by PCB Editor anyway).

Pros and Cons of Using Positive vs. Negative Planes

In general there are two drawbacks to using positive planes. The first is that, in the padstack definition, the Inner layer must be set as Optional rather than Fixed for every padstack in the design for the artwork to be produced correctly. However, it is much simpler to do that than to make and assign thermal flashes for the padstacks. The second potential drawback is that constraints set for trace width and spacing affect the thermal relief geometry (possibly adversely), whereas thermal flashes on negative planes are fixed, and once they are engineered,

FIGURE 9-192
Artwork Control Form
for positive planes.

FIGURE 9-193
Gerber artwork for the
positive VCC plane.

they are unaffected by changes made in the Constraint Manager. But the latter potential drawback can be mitigated to some extent by using the Global Shape Parameters settings as described previously.

The bottom line though is that, when using OrCAD PCB Editor, positive planes are easier to implement than negative planes as long as the layer stack-up and thermal relief design do not have to be highly engineered.

DESIGN TEMPLATES

Making a Custom Capture Template

If you design a lot of projects that are similar to each other, setting up a project template in Capture can be a real time saver and can help eliminate errors by reusing known good project setups. Project templates can be used only when setting up a project through the Analog or Mixed A/D option from the New Project dialog box. However, since you can make a PCB design from either the Analog or Mixed A/D option or the PC Board Wizard option, you can still take advantage of creating your own Capture templates for PCB design projects.

To make a custom Capture project template, start a new Analog or Mixed A/D project; set it up with power supply and ground symbols, connectors, or whatever you want; then save it in the OrCAD/tools/Capture/templates/PSpice folder with the other templates. It will automatically be added to the templates list, so that you can select it the next time you start a new project.

Making a Custom PCB Editor Board Template

As with the Capture templates, if you design a lot of PCBs that are similar to each other, setting up a custom PCB Editor board file (.brd) can save a lot of time and eliminate many manufacturing errors by reusing known good PCB setups.

To make a custom start PCB Editor, begin with a new board design by selecting File → New from the menu. At the New Drawing dialog box, enter a name for the template and select Board from the Drawing Type: list, as shown in Figure 9-194. The board wizard will be demonstrated later, after a board template and technology file has been developed.

FIGURE 9-194
Open a new, blank board to start a board template.

When the board is opened, use the Design Parameters dialog box to set up the basic design parameters, including

Display tab:
    Display through holes
    Display thermals
    Size of DRC markers
Design tab:
    Design size
    Design extents

Shapes tab:
  Edit Global Dynamic shape parameters (RS274X for artwork format)
Route tab:
  Add connect defaults (trace width and angles)
  Create Fanout parameters
Manufacturing Applications tab:
  Edit Autosilk parameters

When the parameters have been set, click Apply then OK to dismiss the dialog box.

In addition to setting up previous parameters, the following items can also be set up:

- Default grid settings (Setup → Grids).
- A default layer stack-up using the Layout Cross Section dialog box.
- A board outline (select Setup → Outlines → Board Outline from the menu).
- Default artwork folders using the Artwork Control Form.
- A default, custom color scheme and layer visibility using the Color Dialog box.
- Design constraints using the Constraint Manager. Default settings might include the default trace width (Physical tab) and the default trace spacing (Spacing tab).

Once all of the design parameters and constraints have been established, save the board.

You can also save the Constraint Manager settings in a separate technology file that can be imported into any board. This is described in the next section.

To reuse the board template, assign it as the Input Board File: during the netlist creation in Capture, as shown in Figure 9-195. The new board will inherit all the template board properties.

Making a Custom PCB Editor Technology Template

If you often design boards with similar design constraints, you can save them in a board template. You can also save the Constraint Manager settings to a stand-alone technology file, so that, if you start a new design that is not based on the board template, you can still reuse the design constraints by importing the technology file.

To create a technology file, start from a board template or an active board design that has the constraints you want to be able to reuse. Open the Constraint Manager and select File → Export → Technology File... from the menu. Save the filename.tcf file in a folder that you can easily access for future designs.

When you start a new board and want to load the technology file, select File → Import → Technology File... from the Constraint Manager menu. The technology file will replace the existing constraints and layer stack-up with the setup saved in the technology file.

FIGURE 9-195
Use the board template during the initial netlist creation.

USING THE BOARD WIZARD

Over time you will likely develop several board characteristics that you have used but not every board uses all of them. For example you might have two or three board outlines you often use, but the layer stack-ups are different from one board to the next. Perhaps you have different technology files you use and have several mechanical symbols or drawing symbols that you need to select from, but they also vary from board to board. Rather than make a board template for every possible combination of these characteristics, you can save each of the characteristics and symbols in a library and use the Board Wizard to select specific characteristics and combine them into one new board design. The Board Wizard allows you to reuse your favorite board characteristics while keeping your template library more easily organized.

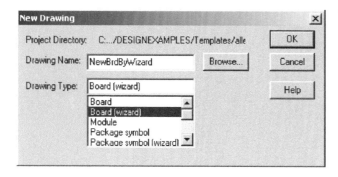

FIGURE 9-196
Select the Board (wizard) from the New Drawing dialog box.

When you start a new board, the wizard will look for templates and technology files in the current working directory where you start the process but will look in the symbols library for mechanical and drawing symbols. Before you begin make sure that any mechanical and drawing symbols you plan on using are in the symbols library and any board templates and technology files you will use are copied to the file path where you are setting up the new board.

To begin a new board using the wizard, start PCB Editor and select File → New from the menu. At the New Drawing dialog box, enter a name and path for the new board design and select Board (wizard) from the Drawing Type: list, as shown in Figure 9-196; click OK.

The Board Wizard dialog box will be displayed, which provides an introduction to the board wizard. Click the Next> button to continue.

The next step is to load a board template (*name*.brd) if you have one. To load the board template, select the Yes radio button then click the Browse... button to locate the file (see Figure 9-197). Remember that a copy of the board template needs to be located in the directory path where you are setting up the board, or the wizard will not be able to find it. At the Board Wizard Template Browser dialog box, select the board template on which you want to base the new board, then click OK to dismiss the Browser dialog box. Click the Next> button to continue to the next step.

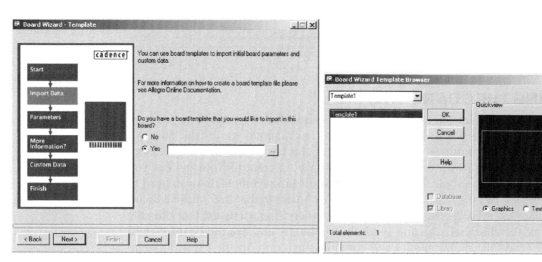

FIGURE 9-197
Loading a board template.

Next you can load a technology file (*name*.tcf). A technology file loaded by the wizard will override the technology file originally loaded with the board template. If you want to keep the original technology file, select the No radio button and click the Next> button to continue to the next step. Otherwise, to load a different technology file, select the Yes radio button and click the Browse... button to locate the file, as shown in Figure 9-198. Remember that a copy of the technology file needs to be located in the directory path where you are setting up the board, or the wizard will not be able to find it. At the Board Wizard Techfile Browser dialog box, select the technology file you want to use and click OK to dismiss the Browser dialog box. Click the Next> button to continue to the next step.

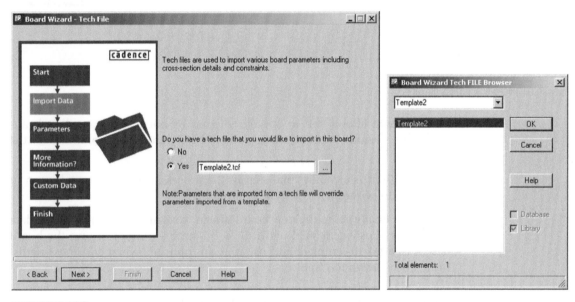

FIGURE 9-198
Loading a technology file.

In the next step, you can load a drawing file or mechanical symbol (Figure 9-199). If you have no drawing or mechanical symbols you want imported into the board, select the No radio button and click the Next> button to continue to the next step. Otherwise select the Yes radio button and click the Browse... button to locate the file. At the Board Wizard Mechanical Symbol Browser dialog box, select the symbol you want to add to the board then click OK to dismiss the Browser dialog box. Click the Next> button to continue to the next step.

With the next step, you can elect to import the selected files right away or wait until the last step in the wizard process, see Figure 9-200(a). Typically you can import the data right away. Click the Next> button to continue to the next step.

At the following step, Figure 9-200(b), you can specify the origin of the new board drawing. Click the Next> button to continue to the next step.

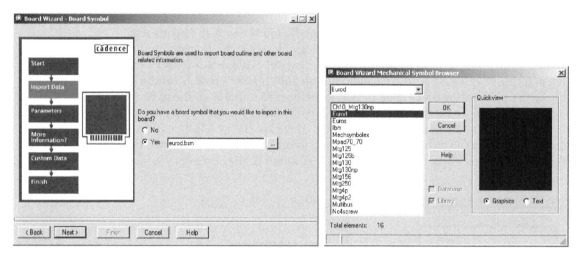

FIGURE 9-199
Loading a mechanical symbol drawing.

The next step allows you to specify whether or not to generate default artwork film, Figure 9-201(a). Click the Next> button to continue to the next step.

Figure 9-201(b) shows the final step in the process. Click the Finish button to complete the wizard and open the board design.

When you finally get to the new board design, you can either import logic data from a previously generated netlist (File → Import → Logic) or save the board design as a board template and use it as an input file when creating a netlist from a new design in Capture.

MOVING ON TO MANUFACTURING

The design examples included here are to provide an overview of the basic steps in the board design processes and introduce the various tools you can use to design different types of boards. These examples were *not* designed with manufacturability in mind, as that is another subject, which encompasses additional things to consider and would have made the examples overly cumbersome. Chapter 10 combines design for manufacturing topics introduced in Chapter 5 with design processes introduced here.

REFERENCES

Capture User's Guide.
IPC-2221A. Generic Standard on Printed Board Design. IPC/Association Connecting Electronic Industries, Northbrook, IL, May, 2003.
Ott, Henry W. *Noise Reduction Techniques in Electronic Systems.* 2nd ed. New York: Wiley, 1988, p. 129.
PSpice User's Guide.

FIGURE 9-200
Importing default data (a) and setting the drawing origin (b).

FIGURE 9-201
Setting artwork design parameters (a) and finishing the wizard (b).

CHAPTER 10

Artwork Development and Board Fabrication

Developing artwork for the board design was briefly introduced in Chapter 2. Here we look at this step in greater detail.

SCHEMATIC DESIGN IN CAPTURE

As with the previous examples, we use a simple circuit to minimize those details and allow us to focus on the board layout and manufacturing steps. The circuit (shown in Figure 10-1) consists of a 10-pin PCB connector (J1), a

FIGURE 10-1
Circuit design used for the fabrication process.

Table 10-1		Bill of Materials for the Manufacturing Design Example				
Item	Qty.	Ref.	Part	Symbol	Padstack	Flash
1	1	C1	0.1 µF	smdcap 1206	smd45rec71	N/A
2	1	J1	CON10	CH10_Conn10f	CH10_pad60cir36f	TR_90_60
3	2	M1, M2	MTG130	CH10_mtg130p	CH10_pad400cir130f	AB00
4	2	R1, R2	1k	CH10_res400f	CH10_pad72cir42f	TR_102_72
5	1	U1	7400	soic14	smd50_25	N/A
6	2		(NP mtg hole)	CH9_mtg130np	CH9_Hole130np	Circle 160
7			(VIA)		CH9_VIA	AB00

bypass capacitor (C1), a logic IC (U1), load resistors (R1 and R2), and two plated mounting holes that electrically connect the Ground plane to the PCB's mounting bracket. Once we move the design to PCB Editor, two more mounting holes will be added that are nonplated and electrically isolated from the board. The two types are used so that, when we generate the drill files, you will be able to see the differences between the two types of drill holes.

Table 10-1 shows the bill of materials for this design. Only the footprint symbol for U1 (soic14) is from the native PCB Editor symbols library. The footprint for C1 was made in the footprint design example from Chapter 8; the electrically connected mounting hole, mtg130p, is a custom symbol; and the remaining symbols are modifications of existing footprints. The footprint symbols are on the Web site for this book, and you need to copy them (and the associated padstacks and flash symbols) into PCB Editor's symbols folder. You also need the mtg130np mechanical symbol, which is included on the Web site in the same folder.

The footprint names for each of the components are shown on the schematic. *To display component information on the schematic*, double click the part to display the Property Editor spreadsheet (see Figure 10-2). Select the cell (row or column) that contains the information you want displayed. Click the Display... button at the top of the spreadsheet to show the Display Properties dialog box. Select one of the display formats and click OK. Click the Apply button at the top of the spreadsheet and close the spreadsheet.

Once the schematic is completed, generate a PCB Editor netlist as described in in Chapter 9, PCB Design Examples.

THE BOARD DESIGN WITH PCB EDITOR
Routing the Board

Figure 10-3 shows the routed board. The stack-up consists of the top and bottom Routing layers and the VCC and GND Plane layers. The stack-up was defined and the traces routed using the procedures described in the PCB Design Examples. The padstacks used in the footprint symbols for J1 and the two

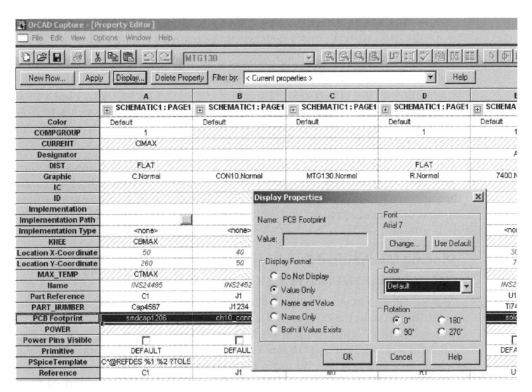

FIGURE 10-2
Use the Property Editor to display part information on the schematic.

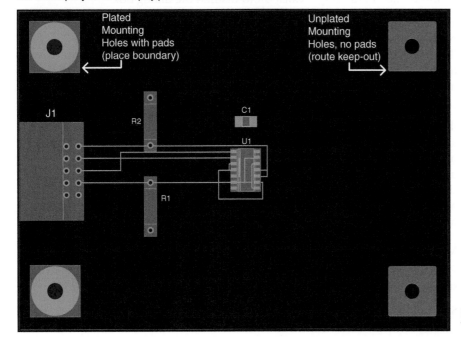

FIGURE 10-3
The routed board.

resistors (R1 and R2) have flash symbols already defined, so they should automatically show their thermal reliefs when the plane layers are visible.

Along with the components and routed traces, the board contains four mounting holes. The two on the left side of the board are the ones included on the schematic and are component symbols (CH9_mtg130p.psm) with plated padstack holes (CH9_pad400cir130f.pad). The f indicates the flash symbols have been predefined. They are placed on the board using the same procedure used for placing the other components.

The two on the right side of the figure are mechanical symbols (CH9_mtg130np.bsm) with nonplated padstack holes (CH9_Hole130np.pad) and are added to the board from within PCB Editor using the following steps. Make sure you have copied CH9_Hole130np.bsm and CH9_Hole130np.pad into your PCB Editor symbols library.

Placing Mechanical Symbols

MECHANICAL SYMBOLS FOR MOUNTING HOLES

To place the mechanical symbol mounting holes, select Place → Manually from the menu bar or select the Place Manual button on the toolbar. At the Placement dialog box (see Figure 10-4), select Mechanical symbols from the Placement List. Click the Advanced Settings tab and make sure that the Library option is checked. Select the Placement List tab again, and check the box beside the MTG130NP icon. Move your pointer to the design area; the symbol will be attached to it. Left click to place the mounting hole. Recheck the box for each hole you want to add to the board design. Click OK to dismiss the dialog box.

Since the plated holes are footprints, they contain a place boundary like any other component, but on this footprint place boundaries were added to both the Top and Bottom layers. The symbol was designed that way so no components are allowed to be located near the holes and cannot contact the mounting hardware (standoffs, washers, or nuts, say).

The unplated holes are mechanical symbols. They contain place boundaries on the top and bottom (not shown in the figure) for the same reason, but they also contain route keep-out areas (Route keepout/All). The keep-out areas are needed so that traces cannot be routed under mounting hardware or across the unplated hole (and subsequently drilled away). The DRC may not detect such a situation, because the Constraint Manager does not know about the mounting hardware that will be put into the hole, and there are no pads against which to maintain spacing rules.

MECHANICAL SYMBOLS FOR FIDUCIALS

Fiducials are marks or objects used for aligning the board with solderpaste stencils and pick and place machines.

Stencils are thin stainless steel plates or sheets through which solderpaste is squeegeed onto the board. The thickness of the stencil establishes the thickness

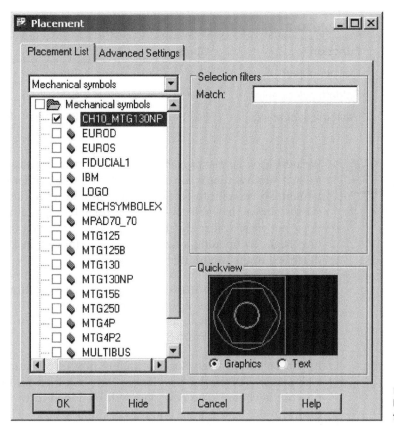

FIGURE 10-4
Using the Placement tool to add mounting holes.

(and therefore the amount) of the paste applied to the pads. The stencil manufacturer uses the pastemask file to fabricate the stencil.

Fiducial objects are placed at particular coordinates on both the board and the stencil, and optical equipment is used to align the stencil to the board during application of solderpaste.

During pick and place operations, the fiducials on the board are used to precisely identify the location and orientation of the board so that the components are accurately placed into the solderpaste.

On the circuit board the fiducial is typically a copper mark produced during the etch process. Because the fiducials are relied on optically, the soldermask is usually removed at the fiducial location. And since the fiducial needs to be visible through the stencil, an exact copy of the fiducial needs to be included with the pastemask pattern. A fiducial object then requires etch, soldermask, and pastemask elements.

Note: Many of the native PCB Editor footprint symbols do not contain pastemask elements, so you need to add them to the footprints if you plan to fabricate a stencil and use pick and place processes to populate your board.

Table 10-2	The Elements of a Fiducial	
Function	**Class**	**Subclass**
Copper object	Etch	Top
Soldermask opening	Board Geometry	Soldermask_ Top
Pastemask opening	Board Geometry	Pastemask_ Top

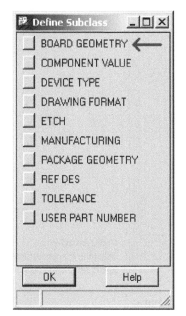

FIGURE 10-5
Defining a new subclass for a board pastemask.

We next look at *how to make a fiducial*. Ideally we want the fiducial to be a symbol that is an element of the board rather than a component. So we want to construct the symbol on the classes and subclasses shown in Table 10-2. Unfortunately a pastemask subclass does not exist under the Board Geometry class. So, we will make one. Begin by opening a new mechanical drawing in PCB Editor as described in the mounting hole example.

To define a new subclass, select Setup → Subclasses… from the menu. At the Define Subclass dialog box shown in Figure 10-5, select the class where you want to assign the new subclass. For this example click the box next to Board Geometry.

At the Define Non-Etch Subclass dialog box (see Figure 10-6), type PASTEMASK_TOP in the New Sublass: text box then hit the Enter key on your keyboard. The new class will be added to the list and displayed in the Options pane and the Color dialog box.

The geometry of a fiducial is typically specified by the user. In this example the fiducial will be a 40 × 40 mil square. The first step is to place the copper square. Change the All Etch grid setting (Setup → Grids) to 10 mils by 10 mils. Zoom in to the center of the drawing area (0, 0). Select the Etch/Top class and subclass from the Options pane. Select the Shape Place Rect tool and draw a 40 × 40 mil box centered at coordinate (0, 0).

Next we draw the soldermask opening. Change the Non-Etch grid setting (Setup → Grids) to 5 mils by 5 mils. Select the Board Geometry/Soldermask_Top class and subclass from the Options pane. Select the Shape Place Rect tool and draw a 50 × 50 mil box (5 mils larger than the pad on all sides) centered at coordinate (0, 0).

Finally the pastemask is defined. Repeat these procedures using the Shape Place Rect tool on the Board Geometry/Pastemask_Top class and subclass and draw a 40 × 40-mil box centered at coordinate (0, 0) directly over the copper pad. Save the drawing to create the .bsm file.

To place the fiducial, use the procedure described for placing a mechanical symbol. Place two or more fiducials on the outer perimeter of the board.

Generating Manufacturing Data

Manufacturing data include two general types of files: artwork files and NC files (for drilling and milling). These files are made using two or more distinct processes. We begin with the artwork files.

GENERATING THE ARTWORK

The artwork files describe elements of the design and include the Etch, Soldermask, Solder-Paste, and Silk-Screen layers.

Most of the manufacturing data are basically present in the board design but not in a format that manufacturers use. The artwork creation process performs that function.

Before we begin the process, one more element needs to be added to the board, a photoplot outline.

ADDING A PHOTOPLOT OUTLINE

The board outline generated by the Board Outline dialog box serves as a guide for the designer during the design process, but it does not contain manufacturing data for the manufacturer. That information is contained in a photoplot outline.

To draw a photoplot outline, select the Manufacturing class and Photoplot_Outline subclass from the Options pane. From the toolbar select the Add Rect tool (not the Shape Add Rect button). Draw a rectangle directly over the top of the board outline located on the Board Geometry/Outline layer. A photoplot outline is shown in Figure 10-7.

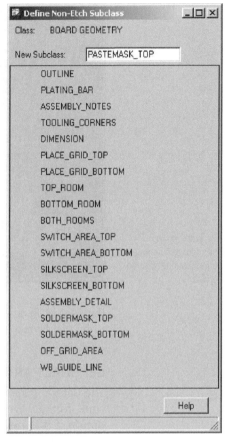

FIGURE 10-6
Define Non-Etch
Subclass dialog box.

OPTIONAL ARTWORK ITEMS

All the silk-screen elements in the design are associated with the individual components. You can add additional etch and silk-screen markings that are specific to the board. For example you can use objects (mechanical symbols) on the Top or Bottom layers to add a company logo or add silk-screen to display the board part number or manufacturing date, say, as shown in Figure 10-7.

Once the board is ready to go, the next step is to generate the artwork files.

Generating the Artwork Files

ARTWORK CONTROL FORM

The artwork files are generated using the Artwork Control Form dialog box (shown in Figure 10-8). *To display the* Artwork Control Form *dialog box*, select Manufacture → Artwork from the menu bar or click the Artwork button on the toolbar (if it is displayed). The Artwork Control Form contains two tabs, the Film Control tab and

FIGURE 10-7
Photoplot outline and optional manufacturing objects.

the General Parameters tab. The default settings may cause two warning boxes to be displayed. The procedure for clearing these warnings is given in the next section.

GENERAL PARAMETERS

The General Parameters tab is shown in Figure 10-8. Use the tab to set the artwork format. We will use the Excellon format, so select the Gerber RS274X option if it is not already checked. In the Decimal places: box, enter the same value that is in the Integer places: box. This will handle the warning shown in Figure 10-9(a). Select the artwork format.

To handle the second warning in Figure 10-9, *change the* Artwork Format *in the* Design Parameter Editor *to match the* Device *type in the* Artwork Control Form (Figure 10-8). To do this, select Setup → Design Parameters… from the menu. At the Design Parameter Editor dialog box, select the Shapes tab then click the Edit global dynamic shape parameters… button to display the dialog box (see Figure 10-10). Select the Void controls tab and choose Gerber RS274X from the Artwork format: list. Click OK twice to dismiss both dialog boxes.

FILM CONTROL

The Film Control tab shown in Figure 10-11 contains folders that in turn contain subclass items. To see the subclass items, click on the + symbol next to the folder you want to view. Initially the form contains folders related to the layer stack-up only. Folders for the other items (e.g., silk screen and soldermask) must be added.

A folder contains related items. For example recall that there are several types of silk-screen items (component reference designators, component outlines, board geometry text, etc.), which belong to different classes and subclasses.

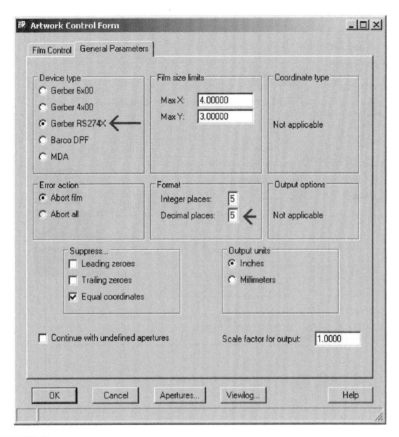

FIGURE 10-8
Selecting the artwork format.

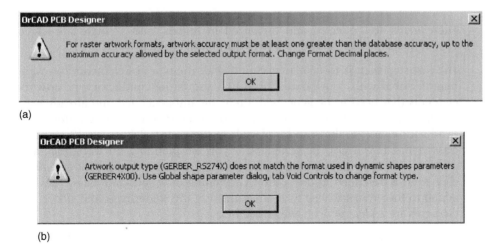

(a)

(b)

FIGURE 10-9
Artwork Control Form warnings: (a) decimal places, (b) artwork output type.

FIGURE 10-10
Setting the Artwork Format in the Design Parameter Editor.

These items are collected and organized in folders, and a single silk-screen Gerber file is made from the collective data in that folder.

The first step then is to take an account of all the types of elements you want manufactured with your board. Table 10-3 shows the conductor type artwork elements that need to be generated; these are autogenerated and already exist in the Artwork Control Form. Table 10-4 lists some of the most common non-conductor elements that can be placed on the board. Solderpaste (Pastemask) class elements can also be generated if you are planning on using pick and place machines to populate your board. Three classes contain pick and place elements, as described in the section Generating Pick and Place Files. They are added to the Artwork Control Form the same way the soldermask and silk-screen elements are added, which is described next.

For this design, the top side of the board needs a soldermask and silk screen, while the bottom side of the board needs only the soldermask. Since there is

FIGURE 10-11
The Artwork Control
Form with default layers.

Table 10-3	Conductor Artwork Elements
Class	**Subclass**
Etch	Any layer in the stack-up
Pin	
Via	

no silk screen on the bottom, we have no bottom silk-screen folder. So from the table, we have three folders: a Top-layer silk-screen folder, which includes the Board Geometry, Ref Des, and Package Geometry classes; a Top-layer solder-mask folder, which includes the Pin and Via classes; and a Bottom-layer solder-mask folder, which contains the same classes as the Top-layer soldermask (only on the Bottom layer).

Table 10-4	Nonconductor Artwork Elements	
	Subclass (*_Top and *_Bottom)	
Class	**Silk-screen**	**Soldermask**
Board Geometry	X	X
Component Value	X	
Component Device Type	X	
Component Reference Designator	X	
Component Tolerance	X	
Component User Part Number	X	
Package Geometry	X	X
Pin		X
Via		X

Additionally, we need a Gerber file for the board outline. Other types of Gerber files can be made, but for the time being, these are the only ones we need. We begin by adding the Top-layer soldermask folder.

ADDING ELEMENTS TO THE ARTWORK CONTROL FORM

When you add a folder to the control form, all classes that are visible in the design are added. So we begin by turning off all layers except for the classes and subclasses related to the Top-layer soldermask. Use the Color dialog box to control the visibility of the classes. *To turn off all the layers at once,* click the Off button (Global Visibility:) at the top of the dialog box, then click the Apply button at the bottom.

Next, turn on (put an X in the box) for the Soldermask_Top subclasses in each of the Package Geometry, Board Geometry, Pin, and Via classes. The Package and Board Geometry classes have nothing on them in this design, but it does not hurt to add them to get in the practice of including them for future designs.

Once these elements are the only ones visible, open the Artwork Control Form. We will add a new folder for the Top-layer soldermask items. *To add a new folder to the* Artwork Control Form, right click on any one of the existing folders and select Add from the pop-up menu (do not use the Add button at the bottom of the form). Enter a name for the folder in the new Film Control dialog box (see Figure 10-12). The name does not have to match any PCB Editor class or subclass name, but it can be helpful if it does or is similar. The new folder now will be listed in the form, and it will contain the subclasses visible in the design.

Using this procedure, make a new SoldermaskBottom folder and include the same classes as the Top-layer soldermask folder (VIA, PIN, PACKAGE, etc.).

FIGURE 10-12
Adding a new folder
to the artwork list.

Next we add a silk-screen object. As indicated in Table 10-4, many silk-screen objects can be added to the design. You can use the procedure just described to create the silk-screen films too. But there is another way to generate the silk-screen films using the Autosilk tool.

During the part placement steps, it is very possible that silk-screen objects can accidentally be removed (or be unable to be printed) on the soldermask because of openings in the mask, or they can become hidden under components or even printed on top of other silk-screen objects. Keeping the silk-screen objects under control can be a task. Fortunately PCB Editor has a tool that helps us keep track of them, Autosilk.

The Autosilk tool takes all of the individual silk-screen elements, combines them into one layer, and makes sure they are not under components, over

FIGURE 10-13
The Auto Silkscreen
dialog box.

padstacks, or too far away from their respective components. *To start the Autosilk process*, select Manufacture → Silkscreen from the menu bar. At the Auto Silkscreen dialog box (shown in Figure 10-13), you can select which elements you want added to the common Silk-screen files (one for Autosilk_Top and one for Autosilk_ Bottom). You can also set default values for some of the physical properties related to silk-screen objects. When the settings are complete, click the Silkscreen button. When the process is complete, the dialog box will dismiss itself and the results will be displayed in the command window. To view the new silkscreen, make sure the Manufacturing/Autosilk_Top (or _Bottom) subclass is visible. It is also helpful to make the new silk-screen a color different from the other ones.

If you do not notice a big difference, it could be that the new silk screen is directly on top of the original. This will happen if the Autosilk tool does not find anything wrong with the existing ones. To make sure the new silk screen is correct, turn off the original ones (all of them) and look at just the new one, which will be the sum of all of the others. Otherwise you should notice that some of the reference designators were moved away from fan-out vias and outside the place boundary shapes. Depending on how the autorouter routed your board, you may also notice that some of the component outlines are removed where vias are close to the component body. We now add the new silk screen to the artwork form.

If you previously made a silkscreen folder you can add the new autosilk data to the existing folder and remove the other, individual silkscreen elements.

To add an item to an existing folder, right click on any one of the existing items in that folder and select Add from the pop-up menu. At the Subclass Selection

FIGURE 10-14
Adding an artwork item to an existing folder.

dialog box (see Figure 10-14), check the boxes for the items you want to add and click the OK button. The items will be added to the folder. *To delete an item from a film folder*, right on the item and select Cut from the pop-up menu.

The last item to add is the board outline item. The board geometry outline drawn at the beginning of the board design process does not contain Gerber data and is for the designer's benefit. Gerber data for a board or design outline is contained in the photoplot outline.

Using the previous procedures, add a photoplot outline folder that contains the Manufacturing/Photoplot_Outline class and subclass. An example of the completed artwork list is shown in Figure 10-15.

We are almost ready to generate the Gerber files. Before we do, check to make sure that the Ground and Power layers are marked as Negative in the Plot Mode: area of the Artwork Control Form and you have entered a value in the Undefined Line Width: box (a number between 5 and 10 is usually good). If you forget to add a default line width value, any lines or text that have a width of 0 will be ignored (resulting in large blank areas on the silk screen). And recall from Chapter 8 that many of the silk-screen

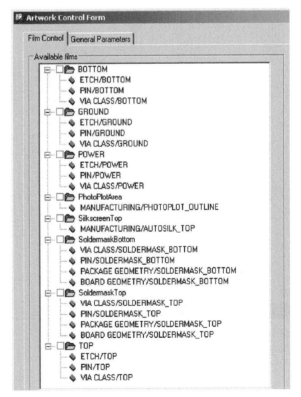

FIGURE 10-15
The complete artwork list.

outlines that come with the library use rectangles instead of lines, and rectangles always have a default width of zero.

Also, make sure that the Vector based pad behavior option is checked. This applies only to negative planes, but it affects how the unconnected pads on the plane layers will be represented in the artwork. Leaving this option checked removes unused pads from negative planes, which results in a larger (safer) clearance gap.

Next, we need to populate the aperture list. At the bottom of the Artwork Control Form, click the Apertures... button (see Figure 10-16). An Edit Aperture Wheels dialog box will be displayed with at least one entry. Click the Edit button to display the Edit Aperture Stations dialog box. If you do not see a list of entries, click the Auto button and select Without Rotation from the pop-up menu. After the aperture wheel list has been populated click OK twice to get back to the Artwork Control Form.

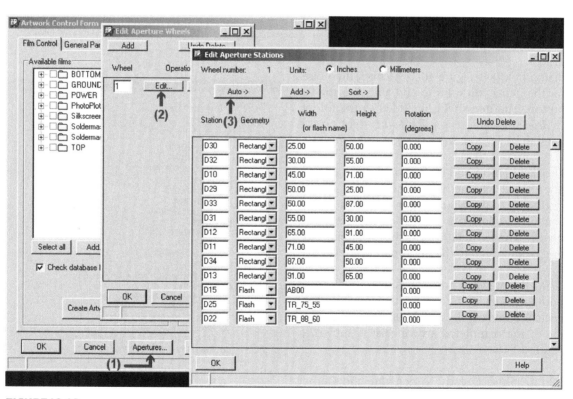

FIGURE 10-16
Populating the aperture list.

To generate the Gerber files, click the Select All button in the Artwork Control Form to check all the folders (or just check the ones you want—the others will not be written or overwritten if they already exist). Then click the Create Artwork button. The command window will report the results and you can read through the photoplot log to look for any errors. Check the working directory and make sure that the files are present. In this example eight files with the .art extension should have been generated. These are the Gerber files that will be sent to the board manufacturer.

That completes the artwork process. Now we move on to making the drill files.

FIGURE 10-17
The drill legend.

Generating Drill Files

Before generating the drill data, it is a good idea to take an inventory of the drill holes. You can do so using the drill legend and the Padstack Definition Report. The report provides all information about each padstack, but the drill legend is much easier to read.

To make a drill legend, zoom out a bit so you have room to place the drill legend. Select Manufacture → NC → Drill Legend… from the menu. The legend will be attached to your pointer. Left click to place it (typically above the board outline or off to one side, but any place is OK). The drill legend (shown in Figure 10-17) lists the holes by drill figure (specified in the .pad file as designed in Padstack Designer), and it shows the hole sizes, whether the hole is plated or not, and the total quantity of each type of hole. Notice too that, once the drill legend has been generated, the drill symbols are also displayed on each of the padstacks in the design.

DRILL CHART TOP to BOTTOM			
ALL UNITS ARE IN MILS			
FIGURE	SIZE	PLATED	QTY
'	14.0	PLATED	8
□	36.0	PLATED	10
◆	42.0	PLATED	4
○	130.0	PLATED	2
+	130.0	NON-PLATED	2

DRILL FILE FORMAT

As there are different formats for the artwork files, there are several types of drill formats. You need to know the format your board manufacturer requires and apply it to your design. *To specify the drill format,* select the Ncdrill Param button on the toolbar or select Manufacture → NC → NC Parameters… from the menu to display the NC Parameters dialog box (see Figure 10-18). At the NC Parameters dialog box, you can enter a header if your manufacturer asks for one, but PCB Editor does not care if you enter one. Enter your

FIGURE 10-18
The NC Parameters dialog box.

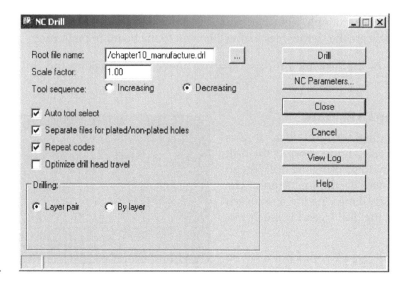

board manufacturer's specifications for the format (e.g., 2.4), zero suppression, Excellon format options, and the like. Once the values are set, click the Close button to save your settings.

The next step is to create the drill file. Select Manufacture → NC → NC Drill... from the menu to display the NC Drill dialog box shown in Figure 10-19. If you have not scaled your design drawing in any way, enter a scale factor of 1.

If your board manufacture prefers a particular tool sequence, you can select that here; otherwise use the default setting.

Select the Auto tool select option. This option causes PCB Editor to look for a tool list. If one does not exist, it will create one automatically. It also adds drill tool header information to Excellon formatted files. If the Auto tool select option is not enabled, no tool list is created and the drill fill contains only drill locations and manual stops where the CNC machine operator is required to indicate the drill tool from a manually generated tool list.

If you know for sure that your board manufacturer can process drill files that contain both plated and nonplated hole information, you can leave the Separate files for plated/non-plated holes option unchecked. Otherwise check the box to make two distinct drill files. Nonplated hole drill files have np in the filename, as described later, whereas the plated holes do not.

The last two options (repeat codes and optimized head travel) also depend on your board manufacturer's preferences. If the manufacturer does not specify one way or the other, check them both.

In the Drilling: option box, select the Layer pair option. The By layer option is used for back-drilling and blind/buried-via drilling operations and is explained briefly later.

Click the Drill button to create the drill file(s). For this design with the settings chosen in the NC Drill dialog box, two drill files are generated, chapter10_manufacture-1-4.drl and chapter10_manufacture-1-4-np.drl. The naming convention is *filename*-startlayer-endlayer.drl. The np in the second file indicates that the file is for the nonplated drill holes. These *.drl files are the NC files your board manufacturer will use during the board fabrication.

If you chose the By layer option in Figure 10-19, the same files would be generated, but additional drill files would also be generated, where a separate drill file is generated for each layer in the stack-up. The drill file names would then be, for a four-layer board for example, *filename*-bl-1-2.drl, *filename*-bl-2-3.drl…, *filename*-bl-3-4.drl, in addition to the one file named *filename*-1-4.drl. The bl in the file name indicates that the by layer option was checked.

MANUALLY GENERATING A TOOL LIST

Instead of using the default nc_tools_auto.txt file, you can specify your own tool list. To do so, you must check the Auto tool select option and create a tool list file, which must have the file name nc_tools.txt. *To generate a tool list*, open a text editor such as Notepad and save it as nc_tools.txt. A tool list is shown in Figure 10-20. Each line begins with the diameter of the drill hole. The P or N indicates whether the hole is plated or nonplated. The T*nn* is the tool number, and the numbers following that indicate the + and − tolerance. Save the tool list file into the working directory, and PCB Editor will automatically look for it there.

FIGURE 10-20
The nc_tools drill tool file.

Generating Route Path Files

For most designs and most manufacturers, the artwork and drill files provide enough information for them to fabricate your board. But, under certain circumstances, you may also need to generate an NC route path file for the milling operation used to cut your board from a larger panel or for cutting slots or irregularly shaped drill holes in the board.

As described in Chapter 4, the PCB manufacturers use standard panel sizes to fabricate boards. To help reduce costs, several designs may be placed on a single panel during fabrication and cut apart by a milling machine. The cutting path that the machine takes is called the *NC route path*. Normally you need not worry about it, since you do not know how the board manufacturer will utilize the panel space. However, if, for example, you want to relay to your manufacturer specific cutting requirements or you buy your boards in panels for mass production pick and place processes and separate the pieces yourself, you need to generate an NC route file.

Generating an NC route file typically requires two steps, but additional steps can be employed, which will be discussed here for completeness. The first required step is to create an ncroutebits.txt tool file and the second is to place an NC route outline. The optional steps are placing cut marks at the board corners and placing cutting path direction indicators on the route path outline.

ESTABLISHING AN NC ROUTE TOOL LIST FILE

Unlike the NC drill process, you have to specify a routing tool list. Setting up an NC route tool list is similar to making the drill tool list described previously. A sample is shown in Figure 10-21. Use a simple text editor (such as Notepad) and open a blank file. The contents have a line that includes the bit diameter (in inches if your design is using English units or millimeters for metric units) and a tool number for each cutting tool that will be used in the design, as defined by the line widths on the Ncroute_Path subclass. For this example the bit diameter is 0.100 in. (100 mils). Save the file into your working directory with the filename ncroutebits.txt and then close the file.

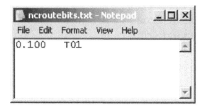

FIGURE 10-21
The ncroutebits tool file.

CUT MARKS

The actual cutting information is contained in the route path outline described later and not in the cut marks. But we begin with this optional step because cut marks can be used as a guide in placing the route path outline, and it will help make sense of the later steps.

FIGURE 10-22
Configuring the cut marks.

We specified that the bit used to cut around the board is 100 mils in diameter, so the cut marks and the placement of the route outline will be based on that. *To place cut marks on your board*, select Manufacture → Cut Marks from the menu bar. At the Cut Mark Options dialog box (see Figure 10-22), enter 100 as the Line width:, because that is the diameter of the cutting bit. Next, assuming that the edge of the bit will cut right along the board edge, the center of the cut path has to be offset by one half of the bit width (50 mils) from the board so that the bit does not cut into the board itself. Therefore, enter 50 in the Offset: option. The line length is not critical, as it simply determines the visual length of the mark. A line length of twice the width is a good rule of thumb. If your board is not a simple rectangle, then you should select both inside and outside corners in the Corner type box, otherwise the outside only is adequate. When you are ready to place the marks, click the Apply button. PCB Editor automatically places cut marks around the board outline on the Board Geometry/Cut_Marks class and subclass. An example is shown in Figure 10-23.

DRAWING THE CUT PATH

The next step is to place an outline along the cutting path. Begin by selecting the Ncroute_Path subclass under the Board Geometry class in the Options pane. Select the Add Line tool from the toolbar. Set the width of the cut in the Line width: box in the Options pane (100 mils). The center of the cut marks will serve as a drawing guide. Left click and release at the center of the cut mark to begin drawing the cutting path. Continue around the board, drawing the cut path

FIGURE 10-23
Drawing a cut path
around the board outline.

through the center of the cut marks. When you return to the starting point left click to end the last segment at the starting point and then right click and select Done from the pop-up menu.

Note: You must use the Add Line tool (not a rectangle) to draw the cut path so that you can specify the cut width. PCB Editor uses the width of the line to determine which cutting tool to use from the ncroutebits file when it generates the route file (*.rou). If you have only one cut path (and width) and only one tool in the ncroutebits file, PCB Editor will assume you mean to use that bit for that cut. But if you have multiple cuts (e.g., for edge slots etc.) and multiple bits, then you have to specify the line width and have the correct bits listed in the bit file.

Figure 10-23 shows a small 1 near the starting point. This is an indicator used to tell the NC machine where to start. A 2 at one of the adjacent corners defines the direction that the cut should be made (not the order the outline was drawn). The numbers are optional, but we will place them as a demonstration.

To place cutting order text objects, select the Add Text tool from the toolbar and the Ncroute_Path subclass under the Board Geometry class in the Options pane (same as for the cutting path outline). Left click near the corner you would like to have the cutting tool start (it does not have to be the corner you started the outline). Enter a 1 in the text block, right click, and select Done from the pop-up menu. That defines the starting point. Use the same procedure to place a 2 at the next corner. The second number defines the direction of the cut path. You can place starting and direction indicators on as many distinct cutting paths as there are on your board. You do not have to place a number at every corner of a closed path unless your board requires a special cutting process.

Generating the Route File

The last step in this example is to generate the NC route file. To do so, select Manufacture → NC → NC Route... from the menu. At the NC Route dialog box

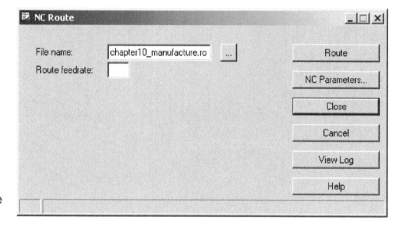

FIGURE 10-24
The NC Route dialog
box for starting the route
process.

(see Figure 10-24), enter a feed rate if you know what it is or leave it blank. You can change the NC Parameters if necessary; we will use the same ones specified during the drill setup. Click the Route button to continue. When the route process is complete, the route log will be displayed. The NC route file will be named *boardname*.rou.

Verifying the Artwork

Before you send your artwork files to the board manufacturer, it is a good idea to review them.

To use PCB Editor to review your artwork files, open a new, blank board design in PCB Editor. To do so select File → New from the PCB Editor menu bar. At the New Drawing dialog box, select Board from the Drawing Type: selection list. Enter a name for the board and choose a destination using the Browse… button then click the OK button. A new, blank design will be opened.

Use the Design tab in the Design Parameter Editor dialog box (Setup → Design Parameters… from the menu) to setup the drawing size needed for the artwork files. Use the Layout Cross Section dialog box (Xsection button on the toolbar) to set up etch layers for each conductor artwork file you intend to import (e.g., Top, Ground).

Next, select File → Import → Artwork… from the menu (see Figure 10-25).

At the Load Cadence Artwork dialog box (see Figure 10-26), select the class and subclass for the artwork you want to import then use the Filename: Browse… button to locate that artwork file. Click the Load button.

The artwork will be attached to your pointer as an outline with the artwork origin as the insertion point. Click the P button at the bottom of the design window to display the Pick dialog box. Enter 0 0 and click the Pick button. The origin of the artwork file will be placed at the origin of the current drawing.

Figure 10-27 shows an example of what the Top layer looks like when importing the Top.art file. The traces (clines) are now just lines and no net information is associated with them. The pads are shapes and have a dummy net property attached to them, so DRC markers are placed wherever a line crosses a pad (even if it is not supposed to). Since we are just reviewing the artwork, you can disable the DRC markers by turning off that layer in the Color Dialog panel (go to the Stack-Up folder and disable the entire DRC column). Alternatively you can set up layer subclasses in the manufacturing class and import the files there, where the DRC checker will ignore them.

When you import a negative plane layer, it is displayed as an actual negative image, not the way PCB Editor displays it when you are in the design mode. In the negative view, what you see is what will be removed during the etch process.

When you import the silk-screen file, you can import it into the Board Geometry class or into the Manufacturing class, but if you created the silk screen using the autosilk feature it is best to import it into the Manufacturing class.

The types of artwork files that you can import are limited to the basic etch and nonetch objects. You cannot import machining information (e.g., drilling and milling) or photoplot information. If you want to review that information, you need to use a CAM tool.

USING CAD TOOLS TO 3-D MODEL THE PCB DESIGN

Before you send the Gerber files off to the board house to be manufactured, you might want to know if the board will be physically compatible with the house mounting hardware and enclosure (i.e., will it fit?). You can export the design information from PCB Editor to a .DXF file, which you can open with most popular CAD drawing packages.

Graphically speaking the design has a tremendous amount of information, a lot of

FIGURE 10-25
Importing artwork into PCB Editor.

FIGURE 10-26
Loading an artwork file.

FIGURE 10-27
The imported Top-layer
artwork.

which does not need to be exported to a drawing file. To select layers to plot, use the Colors Dialog panel to turn on only the layers you want to include in the drawing.

To export a PCB design as a .DXF file, select File → Export → DXF... from the menu. At the DXF Out dialog box (Figure 10-28), enter a name in the file name box. Click the Layer conversion file entry Browse... button. At the dialog box, click the Open button to accept the default file name or enter a new name and then click Open.

The conversion file needs to be set up. To do so, click the Edit button to display the conversion file editing dialog box shown in Figure 10-29. You can specify layers and names or map the PCB Editor classes directly to the new DXF layers. The options shown in the figure map them directly. To execute the conversion, click the Map button. The class and subclass names will be combined into one name in the DXF layer boxes. Click OK to dismiss the conversion dialog box and return to the DXF Out dialog box.

In the Data Configuration section, you can leave everything unchecked, but the objects in the DXF file will be two dimensional. If you want a 3-D representation, then at a minimum you need to check the Export symbols and padstacks as BLOCKS option. If you also check the Fill options, flat objects (traces and silk screens) will be filled in instead of just an outline. This will make the DXF drawing look richer, but it significantly increases the memory usage and drawing regeneration time.

Once the remaining options are set in the DXF Out dialog box, click the Export button.

When the conversion is complete, the DXF file can be opened with a CAD program.

FIGURE 10-28
Setting up the DXF Out dialog box to generate data for a CAD drawing.

Figure 10-30 shows a 3-D model of the design using a popular drawing program (the colors were modified with the CAD program). The CAD program can then be used to add the enclosure and mounting hardware to verify form, fit, and function. Dimensions and other manufacturing information can easily be added to the drawing file to fully describe manufacturing and assembly instructions.

The height information that the CAD program uses is derived from the PACKAGE_HEIGHT_MAX property of the component's place-bound outline in PCB Editor. You can view the property by selecting Edit → Properties… from the menu and left clicking on the component. The information shown in Figure 10-31 is from the place-boundary outline of J1. As mentioned earlier, place boundaries can be drawn using filled rectangles or shapes. In some CAD programs, filled rectangles behave as solid objects whereas shapes behave as wire frame objects.

FABRICATING THE BOARD

Not all board houses are the same. While most follow certain fabrication standards, not all have the same fabrication capabilities (e.g., number of layers, copper thickness, and drill sizes). There are also differences in design submission policies and minimum order and billing practices.

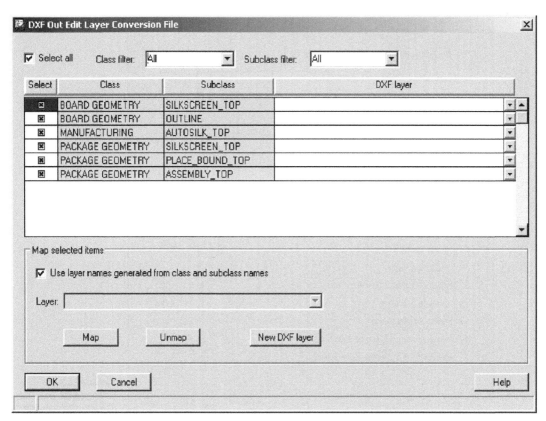

FIGURE 10-29
Mapping PCB Editor Classes into DXF layers.

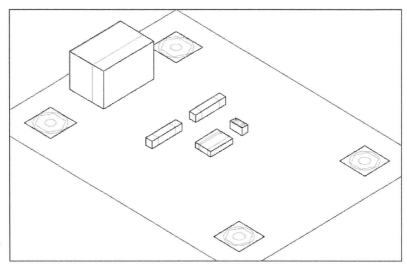

FIGURE 10-30
The 3-D CAD model of the PCB.

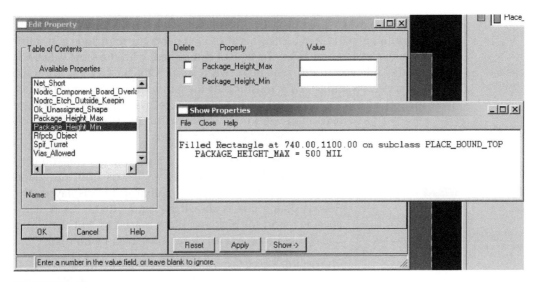

FIGURE 10-31
Set the height information with the Edit Property dialog box in PCB Editor.

Many board houses use an online file submission and quote process, but not all do. Some have you send the Gerber files by e-mail, and after someone looks at them with CAM software (such as GerbTool), the company sends you a quote via e-mail.

An example of the generated Gerber files that would be sent to the board manufacturer is shown in Figure 10-32.

Name ▲	Size	Type
BOTTOM.art	2 KB	ART File
chapter10_manufacture-1-4.drl	1 KB	DRL File
chapter10_manufacture-1-4-np.drl	1 KB	DRL File
chapter10_manufacture.rou	1 KB	ROU File
GROUND.art	7 KB	ART File
Outline.art	1 KB	ART File
POWER.art	4 KB	ART File
SilkscreenTop.art	7 KB	ART File
SoldermaskBottom.art	2 KB	ART File
SoldermaskTop.art	2 KB	ART File
TOP.art	3 KB	ART File

FIGURE 10-32
A list of artwork files.

Receipt Inspection and Testing

Regardless of which house fabricates your boards, you should perform an inspection of the boards, especially if it is a first run. You may also want to do some initial electrical tests (continuity and isolation), especially on the power and ground, before you place any components on the board. IPC-2515 and IPC-6011 outline many types of inspection and acceptance tests.

The next step is to populate the boards. The next section describes how to generate a file describing your board and parts that pick and place machines can use to populate the board.

GENERATING PICK AND PLACE FILES

There are many types of pick and place machines, each with its own programming and setup requirements. As an example here, we just look at the basics most machines require. At a minimum the machine needs to know which parts are to be placed, where they are to be placed, and in what rotation. When placing large quantities of parts on a board, it is also often necessary to include part numbers, as there may be more than one type of component with the same value. For example the only way to differentiate between a size 1206 smd, 0.1%, 1-k resistor and a size 1206 smd, 1%, 1-k resistor is by the part numbers. So we will customize the pick and place list to include the components' part numbers so that the assembler will know which part to place in which tray on the machine and how to program it.

FIGURE 10-33
The Reports selection dialog box.

The pick and place part list is a report. In general you generate reports by selecting Tools → Reports from the menu or the Reports button on the toolbar if it is displayed. The Reports dialog box is shown in Figure 10-33. There is no specific report for pick and place machines, but the Placed Component Report comes close, so we begin with that.

The default Placed Component Report is shown in Figure 10-34. This report provides the reference designator (REFDES), the symbol X and Y coordinates, and the rotation, but no part numbers, and it contains other information that we do not care about.

We can create a report that will provide the part number, but this information must already be assigned to each part. An easy way to do that is to add the part number to the part in Capture. *To add a part number to a component*, select the component (or components) on the schematic, right click and select Edit Properties… from the pop-up menu. Enter the part numbers in the PART_NUMBER cells in the spreadsheet, as shown in Figure 10-35. Perform an ECO to update the part number information to the board design in PCB Editor (if it has not already been done).

FIGURE 10-34
The default Placed Component Report.

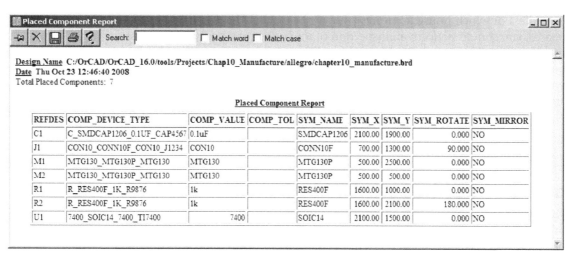

Design Name C:/OrCAD/OrCAD_16.0/tools/Projects/Chap10_Manufacture/allegro/chapter10_manufacture.brd
Date Thu Oct 23 12:46:40 2008
Total Placed Components: 7

Placed Component Report

REFDES	COMP_DEVICE_TYPE	COMP_VALUE	COMP_TOL	SYM_NAME	SYM_X	SYM_Y	SYM_ROTATE	SYM_MIRROR
C1	C_SMDCAP1206_0.1UF_CAP4567	0.1uF		SMDCAP1206	2100.00	1900.00	0.000	NO
J1	CON10_CONN10F_CON10_J1234	CON10		CONN10F	700.00	1300.00	90.000	NO
M1	MTG130_MTG130P_MTG130	MTG130		MTG130P	500.00	2500.00	0.000	NO
M2	MTG130_MTG130P_MTG130	MTG130		MTG130P	500.00	500.00	0.000	NO
R1	R_RES400F_1K_R9876	1k		RES400F	1600.00	1000.00	0.000	NO
R2	R_RES400F_1K_R9876	1k		RES400F	1600.00	2100.00	180.000	NO
U1	7400_SOIC14_7400_TI7400		7400	SOIC14	2100.00	1500.00	0.000	NO

	A	B	C	D
	⊞ SCHEMATIC1 : PAGE1	⊞ SCHEMATIC1 : PAGE1	⊞ SCHEMATIC1 : PAGE1	⊞ SCHEMATIC1 : PAGE1
Part Reference	C1	J1	R1	U1A
PART_NUMBER	Cap4567	J1234	R9876	Ti7400
PCB Footprint	smdcap1206	ch10_conn10f	ch10_res400f	soic14

FIGURE 10-35
Using Capture to add part numbers to components.

Make sure that the parts have the part number property attached to them. To check, select the Show Element tool from the toolbar and select one of the parts. The PART_NUMBER property should be visible in the Show Element text box that pops up.

The next step is to create a custom report. *To make a new report*, display the Reports dialog box (see Figure 10-33). Click the New/Edit button at the bottom. At the Extract UI dialog box (see Figure 10-36), select the Data Fields tab, select COMPONENT from the database list, and select REFDES in the Available Fields list to add it to the Current Configuration window. Next, select the Properties tab, scroll down the Available Properties window until you find PART_NUMBER then click it to add it to the list. Return to the Data Fields tab and look for SYM_CENTER_X and SYM_CENTER_Y and add them to the list. Also add SYM_ROTATE to the list.

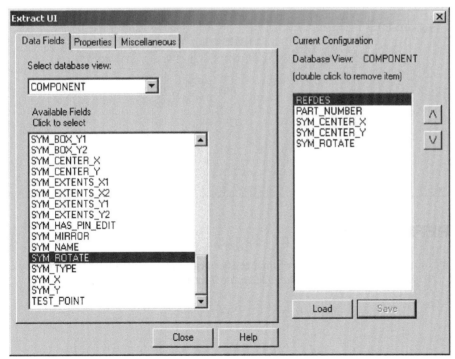

FIGURE 10-36
Making a custom report in PCB Editor.

FIGURE 10-37
The Custom report shown in the Available Reports list.

The final list is shown in Figure 10-36. The order that the properties are listed in the dialog box is the order that they will appear in the report. If you need change the order, use the up and down arrows to the right of the list to move a selected item.

To save this report, click the Save button and enter a name in the dialog box that pops up. Click the Close button to dismiss the Extract UI dialog box.

To generate the new report, scroll all the way down to the bottom of the Available Reports list in the Reports dialog box (Figure 10-37), and your report will be listed.

Double click the new report to select it then click the Report button. The report is shown in Figure 10-38. You can save the report as an html file or copy the data from the report and paste it into a text or Excel file, which the pick and place machine programmer can use to set up the machine to populate your board.

FIGURE 10-38
A custom report for pick and place machines.

REFDES	PART_NUMBER	SYM_CENTER_X	SYM_CENTER_Y	SYM_ROTATE
J1	J1234	475.00	1500.00	90.000
M1		500.00	2500.00	0.000
M2		500.00	500.00	0.000
R1	R9876	1600.00	1200.00	0.000
R2	R9876	1600.00	1900.00	180.000
C1	Cap4567	2100.00	1900.00	0.000
U1	Ti7400	2100.00	1500.00	0.000

REFERENCES

IPC-2515.
IPC-6011. July 1996. Generic Performance Specification for Printed Boards. Northbrook, IL: IPC/Association Connecting Electronic Industries.

Appendices

ANSI STANDARDS

ANSI B94.11-197 (see also ASME B94.1 IM-1993) Twist Drills

ANSI Y32.2-1975 (see also IEEE Std 315-1975) Graphic Symbols for Electrical and Electronics Diagrams

ASME STANDARDS

B18.2.8-1999, Clearance Holes for Bolts Screws and Studs

B18.6.3-2003, Machine Screws and Machine Screw Nuts

ASA 618.11-1961, Miniature Screws

Y 14.5 M, Dimensioning and Tolerancing

ASME B94.1 IM-1993, Twist Drills

IEEE STANDARDS

IEEE Std 315-1975 (ANSI Y32.2-1975), Graphic Symbols for Electrical and Electronics Diagrams

IPC STANDARDS

IPC-1902/IEC 60097, Grid Systems for Printed Circuits

IPC-2141, Controlled Impedance Circuit Boards and High Speed Logic Design

IPC-2221A, Generic Standard on Printed Board Design

IPC-2222, Sectional Design Standard for Rigid Organic Printed Boards

IPC-2223A, Sectional Design Standard for Flexible Printed Boards

IPC-2224, Sectional Standard for Design of PWBs for PC Cards

IPC-2225, Sectional Design Standard for Organic Multichip Modules (MCM-L) and MCM-L Assemblies

IPC-2226, Sectional Design Standard for High Density Interconnect (HDI) Printed Boards

IPC-2251, Design Guide for the Packaging of High Speed Electronic Circuits

IPC-2515A, Sectional Requirements for Implementation of Bare Board Product Electrical Testing Data Description [BDTST]

IPC-2615, Printed Board Dimensions and Tolerances

IPC-4101A, Specification for Base Materials for Rigid and Multilayer Printed Boards

IPC-7351, Generic Requirements for Surface Mount Design and Land Pattern Standard

IPC-9252, Guidelines and Requirements for Electrical Testing of Unpopulated Printed Boards

IPC-A-600G, Acceptability of Printed Boards

IPC-A-610, Acceptability of Electronic Assemblies

IPC-HDBK-610, Handbook and Guide to IPC-A-610

IPC-CM-770E, Guidelines for Printed Board Component Mounting

IPC-D-322, Guidelines for Selecting Printed Wiring Board Sizes Using Standard Panel Sizes

IPC-D-350, Printed Board Description in Digital Form

IPC-D-356B, Bare Substrate Electrical Test Data Format

IPC/JPCA-2315, Design Guide for High Density Interconnects (HDI) and Microvias

IPC/JPCA-6801, Terms and Definitions, Test Methods, and Design Examples for Build-up/High Density Interconnect (HDI) Printed Wiring Boards

IPC-SM-780, Component Packaging and Interconnecting with Emphasis on Surface Mounting

IPC/WHMA-A-620, Requirements and Acceptance for Cable and Wire Harness Assemblies

JEDEC STANDARDS

Master Index for JEDEC Publication No. 95 (see also Appendix B for specific standards)

MILITARY STANDARDS

MIL-HDBK-198A, Capacitors, Selection and Use Of

MIL-HDBK-199A, Resistors, Selection and Use Of

MIL-HDBK-5961A, List of Standard Semiconductor Devices

MIL-STD-1276G, Leads for Electronic Component Parts

MIL-STD-275E, Printed Wiring for Electronic Equipment (superseded by IPC-2221A)

Appendix B
Partial List of Packages and Footprints and Some of the Footprints Included in OrCAD Layout

Table B-1   A List of Package Abbreviations

Abbreviation	Full name
BDIP (SDIP)	Butt-mounted dual inline package (surface DIP, std pitch)
BGA	Ball grid array
BQFP	Bumper quad flat package
CBGA	Ceramic column ball grid array
CFP	Ceramic flat packages
CGA	Column grid array
CQFP	Ceramic quad flat packages
DIMM	Dual inline memory module
DIP	Dual inline package
DO	Diode outline
DPAK	Discrete packaging (type 1, TO-252)
D2PAK	Discrete packaging (type 2, TO-263)
D3PAK	Discrete packaging (type 3, TO-268)
LCC/LCCS	Leadless chip carrier/leadless ceramic chip carrier
LGA	Land grid array
MELF	Metal electrode face
MSOP	Micro (mini) small outline package
MLP	Micro leadframe package (no lead)
PGA	Pin grid array
PLCC	Plastic leaded chip carriers
PLCCR	Plastic leaded chip carriers rectangular
PLCCS	Plastic leaded chip carriers square
PQFP	Plastic quad flat package
QBCC	Quad bottom chip carrier

Continued

Table B-1	A List of Package Abbreviations (Continued)
Abbreviation	**Full name**
QFP	Quad flat packages
QFN (QFPNL)	Quad flat no lead package
QLCCC	Quad leadless ceramic chip carrier (see LCC/LCCS)
SIMM	Single inline memory module
SDIP	Shrink dual inline package
SOD	Small outline discrete (or diode)
SOIC (SOJ)	Small outline integrated circuit, J-lead
SOIC (SOG)	Small outline integrated circuit, gull wing
SON	Small outline nonleaded
SOP	Small outline package
SOT	Small outline transistor
SSOT	Shrink small outline transistor
SQFP	Shrink quad flat package
SSOP	Shrink small outline package
TO	Transistor outline
TQFP	Thin quad flat package
TSOP	Thin small outline package
TSSOP	Thin shrink small outline package

Table B-2	List of Common Discrete Component Packages	
Name or type	**Name, case, or size**	**Standards**
Resistor, chip	0402, 0805, 1206, etc.	IEC 60115-B, JIS C 5201-B,
		EIAJ RC-2134B
Capacitor, tantalum (molded)	A	EIA 3216-18
	B	EIA 3528-21
	C	EIA 6032-28
	D	EIA 7343-31
	E	EIA 7260-38
	R	EIA 2012-12
	T	EIA 3528-12
	V	EIA 7343-20
	X	EIA 7343
	Y	EIA 7340
MELF (DL-41, LL-34)	Metal electrode face	EIC 10H01, LL-34
SOD, SC-76 (molded)	Small outline diode	JEDEC DO215-D, EIAJ SC-76, EIAJ SC-76
SMA (molded)	SMT diode outline	JEDEC DO214-D (variation AC)
SMB (molded)	SMT diode outline	JEDEC DO214-D (variation AA)
SMC (molded)	SMT diode outline	JEDEC DO214-D (variation AB)
DO-213	Diode outline (~MELF)	JEDEC DO213-D
DO-214 (see SMA, etc.) (See also SOT, DPAK, D2PAK packages)	Diode outline (molded)	JEDEC DO214-D
Inductor, chip	0805, 1206, etc.	(see Resistor, chip)
Inductor, molded		IMC-2220 (see Vishay Web site)
Inductor, wire wound	Power SMD inductor	MSS5131 (see Coilcraft Web site)

Table B-3	Discrete Package (DPAK)		
Name	**Package outline**	**JEDEC DWG No.**	**Variation—pitch**
DPAK (TO-252)		TO252-E	AA–0.090 in. (3-lead)
			AB–0.090 in. (3-lead)
			AC–0.090 in. (3-lead)
			AD–0.045 in. (5-lead)
D2PAK (TO-263)		TO263-D	AA–0.100 in. (4-lead)
			AB–0.100 in. (3-lead)
			BA–0.067 in. (6-lead)
			BB–0.067 in. (5-lead)
			CA–0.050 in. (8-lead)
			CB–0.050 in. (7-lead)
D3PAK (TO-268)		TO268-A	AA–5.45 mm (4-lead)

Table B-4 Some of the Small Outline Transistor (SOT/SSOT/SC) Packages

Name	Package outline	JEDEC DWG No.	Variation—pitch
SOT23-3 (3-lead) SC-59 (3-lead) SSOT-3 (1.9 mm) (3-lead)		TO-236	AA–0.95 mm (a) AB–0.95 mm (a) –1.90 mm (b)
SOT23-5 (5-lead) EIAJ SC-74A (5-lead) SOT-26 (6-lead) SOT23-8 (8-lead)		MO178-C MO193-C	AA–0.95 mm (5-lead) AB–0.95 mm (6-lead) BA–0.65 mm (8-lead)
SOT223-3 (4-lead) SOT223-4 (5-lead)		TO261-C	AA–2.30 mm (4-lead) AB–1.50 mm (5-lead)
SOT-89 (2- & 3-lead)		TO243-C	AB–3.0 mm (2-lead) AA–1.5 mm (3-lead)
SOT-143 (4-lead) SOT-343 (4-lead)		TO253-D (EIAJ SC-61B)	AA–1.92 mm –1.30 mm
SOT-353 (5-lead) SC-88 (5-lead) SC70 (6-lead) SC-74 (8-lead) SSOT-n (5, 6, 8-leads)		MO059-B MO203-B	AA–0.65 mm (5-lead) AB–0.65 mm (6-lead) BA–0.50 mm (8-lead)

	Table B-5	Some of the Small Outline Integrated Circuit (SOIC/SOP/SO) Packages		
Name	**Package outline (no. leads)**		**JEDEC DWG No.**	**Variation—pitch**
MSOP Body width 2.3 mm, 2.8 mm, 3.0 mm	(8, 10)		MO187-E	AA–0.65 mm (8-lead)
				AA–T–0.65 mm (8-lead)
				DA–0.65 mm (8-lead)
				CA–0.50 mm (8-lead)
				BA–0.50 mm (10-lead)
				BA–T–0.50 mm (10-lead)
SOIC	(8, 14, 16)		MS012-E	Width at lead– 0.236 in. (6.0 mm)
Narrow body (0.150 in.) (3.8 mm)				Pitch–0.050 in. (1.27 mm) Lead width– 0.016 in. (0.40 mm) Lead spacing– 0.034 in. (0.87 mm)
SOIC Wide body (0.300 in.) (7.5 mm)	(14, 16, 18, 20, 24, 28)		MS013-E	Width at lead–0.403 in. (10.3 mm) Pitch–0.050 in. (1.27 mm) Lead width– 0.016 in. (0.4 mm) Lead spacing– 0.034 in. (0.87 mm)

Table B-5	Some of the Small Outline Integrated Circuit (SOIC/SOP/SO) Packages (Continued)

Name	Package outline (no. leads)	JEDEC DWG No.	Variation—pitch
SSOP Narrow body (0.150 in.) (3.8 mm)	(14, 16, 18, 20, 24, 28)	MO137-C	Width at lead– 0.236 in. (6.0 mm) Pitch–0.025 in. (0.635 mm) Lead width– 0.010 in. (0.25 mm) Lead spacing– 0.015 in. (0.385 mm)
SSOP Wide body (0.300 in.) (7.5 mm)	(28, 48, 56, 64)	MO118-B	Width at lead–0.410 in. (10.4 mm) Pitch–0.025 in. (0.635 mm) Lead width– 0.010 in. (0.25 mm) Lead spacing– 0.015 in. (0.4 mm)
SOP	(44 - 90...)	MO174-A MO175-A MO180-B	Various package configurations

Table B-6	Some of the Common Through-Hole Packages	
Package	**Name**	**JEDEC standard(s)**
DIP	Dual inline package	MS001-D, TO250-A,
	(0.100 in. pitch)	MO043-A
DO-15 (DO-41 glass/plastic)	Diode outline	DO204B-D
DO-35	Diode outline	DO-204-AH
IPAK (TO-251)	Transistor outline, flange mount	TO251-D
I2PAK (TO-262)	Transistor outline, flange mount	TO262-A
TO-205AF/TO-39	Transistor outline	TO205-E
TO-3P/TO-41/TO-247AD	Transistor outline	TO204-C
TO-92	Transistor outline	TO226-G
TO-92; TO-18, lead form STD	Transistor outline	TO226-G, TO206-B
TO-126	Transistor outline	(Similar to TO-220)
TO-218AC	Transistor outline	TO218-E
TO-220 (and variations)	Transistor outline	TO220-K, TO262-A
TO-226AE	Transistor outline	TO226-G
TO-247, 2L, 3L	Transistor outline	TO247-E_01
TO-264	Transistor outline	TO264-B

Table B-7	List of Ball Grid Array Standards
JEDEC doc.	**JEDEC title**
MO-151	Elevated to MS-034A. MS-034A was revised and became MS-034B, 2/20/03.
MO-156	Square ceramic ball grid array (BGA) family, 1.00, 1.27, and 1.50 mm pitch, CBGA.
MO-157	Rectangular ceramic BGA family, CBGA.
MO158-D	The addition of 47.5, 50.0, 52.5, and 55.0 mm body variations with 1.27 and 1.00 mm ball pitch to column grid array registration.
MO163-B	Replaced by MS-028-A.
MO-192	Low profile square ball grid array family.
MO-195	Thin, fine pitch ball grid array family, 0.5 mm pitch.
MO-205	Low profile, fine pitch BGA family, 0.80 mm pitch (rectangular).
MO-207	Square and rectangular die-size, ball grid array family.
MO-210	Thin, fine pitch, rectangular ball grid array family, 0.80 mm pitch.
MO-211	Die size ball grid array, fine pitch, thin/very thin/extremely thin profile.
MO-216	Thin profile, square, and rectangular BGA family, for 1.00 and 0.80 mm pitch.
MO-219	Low profile, fine pitch ball grid array (FBGA) registration, 0.80 mm pitch (square and rectangle).
MO-221	Extremely thin, two row cavity down, 0.50 mm pitch BGA family.
MO-222	Ceramic BGA rectangular.
MO-225	Addition of variations AB and BC to VFBGA.
MO-228	Square, dual pitch and FBGA family.
MO-233	Mixed pitch (0.80 and 1.00 mm), rectangular die size, fine dual pitch ball grid array (DSBGA) family.
MO-234B	Low profile rectangular ball grid array family.
MO-237E	DDR2 SDRAM DIMM package with 1.00 mm contact centers.
MO-242B	Rectangular die-size, stacked ball grid array family, 0.80 mm pitch.
MO-246C	Rectangular, fine pitch, thin ball grid array, 0.65 mm pitch.

Continued

Table B-7	List of Ball Grid Array Standards (Continued)
JEDEC doc.	**JEDEC title**
MO-261A	Thick and very thick, fine pitch, rectangular ball grid array family, 0.80 mm pitch.
MO-264A	Rectangular die-size, stacked ball grid array family, dual pitch.
MO-266A	Very thin, fine pitch, stackable BGA family, 0.50 mm pitch.
MO-273A	Upper pop package, square, fine pitch BGA, 0.65 and 0.50 mm pitch.
MO-275-A	Low profile, fine pitch ball grid array family, square.
MO-280A	Ultra thin and very, very thin profile, fine pitch BGA family.
MO-028-C	Addition of rectangular BGA variations to BGA family.
MS-034-D	Elevation of registration MO-151, plastic BGA, 1.0, 1.27, 1.5 mm pitch, with increased ax dimensions to allow for thicker packages.

Note: Because there are so many types and variations, only some of the JEDEC standards and titles are listed here.

Table B-8	List of Quad Flat Packs Standards
JEDEC doc.	**JEDEC title**
MO-134-A	Ceramic quad flatpack family (CQFP), 0.50 mm lead pitch with ceramic nonconductive tie bar.
MO-143-C	Replaced—see MS-029-A.
MO-148-A	Multichip module (MCM) ceramic quad flatpack family, S-CQFP.
MO-188-B	Power PQFP with heat slug.
MO-189-A	Plastic QFP/heat slug (H-LQP/G) 2.00 mm thick/2.00 mm footprint.
MO-198-A	3-tier family, PQFP-B.
MO-204-B	Plastic quad flatpack (PQFP) outline with exposed heat sink, thermally enhanced PQFPs.
MO-220K	Thermally enhanced plastic very thin and very, very thin fine pitch quad flat no lead package.
MO-239-B	Thermally enhanced plastic very thin dual row fine pitch quad flat no lead package.
MO-241-B	Dual in-line compatible, thermally enhanced, plastic very thin fine pitch, quad flat no lead package family.

Table B-8	List of Quad Flat Packs Standards (Continued)
JEDEC doc.	**JEDEC title**
MO-243-A	Thermally enhanced plastic very thin and very very thin fine pitch bumped quad flat no lead package.
MO-247C	Plastic quad no lead staggered multirow packages.
MO-248E	Thermally enhanced plastic ultra thin and extremely thin fine pitch quad flat no lead package.
MO-250-A	New family of thermally enhanced plastic very thin and very very thin pitch bumped quad flat no lead packages.
MO-251-A	Thermally enhanced plastic very thick, quad flat no lead package.
MO-254-A	Thermally enhanced plastic low and thin profile fine pitch quad flat no lead package.
MS026-D	Standard–low/thin profile plastic quad flat package, 2.00 mm footprint, optional heat.
TO271-A	4 lead quad flat pack.

Table B-9	List of Some Quad Flat Packs–No Lead (QFN) Standards
JEDEC doc.	**JEDEC title**
MO-220K	Thermally enhanced plastic very thin and very, very thin fine pitch quad flat no lead package.
MO-241-B	Dual in-line compatible, thermally enhanced, plastic very thin fine pitch, QFN package family, includes addition of very thin profile variations.
MO-247C	Plastic quad no lead staggered multirow packages.
MO-255-B	Plastic very, very thin, ultra thin, and extremely thin, fine pitch quad flat small outline, nonleaded package family.
MO-257-B	Plastic fine pitch quad no lead staggered two row thermally enhanced package family.
MO-262A	0.50 mm pitch very thin and very, very thin flange-molded thermally enhanced (top side) QFNs.
MO-263A	0.50 and 0.40 mm pitch very thin and very, very thin flange-molded QFNs.
MO-265A	Thermally enhanced plastic very thin fine pitch quad flat no lead package, including corner terminals.
MO-267B	Punch-singulated, fine pitch, square, very thin, lead-frame-based quad no lead staggered dual-row QFN package family.

Table B-10	List of Some Leadless Chip Carrier Standards
JEDEC doc.	**JEDEC title**
MO041-C	0.050 in. center leadless rectangular chip carrier type E, variations AA–AF
MO042-A	0.050 in. center leadless rectangular chip carrier type F.
MO044-A	Leaded ceramic chip carrier 0.050 in. center, 68 and 84 terminals, item 11.11-138.
MO047-B	Plastic chip carrier (PCC) family, 0.050 in. lead spacing, square, item 11.11-242
MO052-A	Replaced—see MS-016-A. PCC family 0.050 in. lead spacing, rectangular.
MO056-A	Ceramic 0.025 in. center chip carrier.
MO057-A	Ceramic 0.020 in. center chip carrier.
MO062-A	148 pin leadless ceramic chip carrier, 0.025 in. pitch.
MO075-A	0.050 in. center nonhermetic leadless chip carrier quad series, square.
MO076-A	0.050 in. center nonhermetic leadless chip carrier SO series, rectangular
MO107-A	Ceramic multilayer leaded chip carrier, 0.050 in. pitch, J-bend leadform, 20 mil min.
MO110-A	Round lead, J-form square body, 0.050 in. pitch center ceramic chip carrier.
MO111-A	Family of round leads, 0.050 in. pitch, gull-wing lead-form, center ceramic chip carrier.
MO126-B	Leadless small outline ceramic chip carrier, 0.400 in. body, 0.050 in. pitch, 28, 32, 26 leads, variations AA–AC.
MO129-A	Top brazed ceramic leaded chip carrier (0.020 in. lead pitch) with plastic nonconductive tie bar.
MO130-A	Top brazed ceramic leaded chip carrier (0.015 in. lead pitch) with plastic nonconductive tie bar.
MO131-A	Top brazed ceramic leaded chip carrier (0.025 in. lead pitch) with plastic nonconductive tie bar.
MO144-A	Leadless small outline ceramic chip carrier, 0.350 in. body, 0.050 in. pitch, R-CDCC-N.
MO147-A	Small outline J-lead ceramic chip carrier, 0.415 in. body, 0.050 in. lead spacing.
MO217-B	Very very thin quad bottom terminal chip carrier family with addition of variations AE, AF, AG, BE, BF, and BG.

Table B-10	List of Some Leadless Chip Carrier Standards
JEDEC doc.	**JEDEC title**
MS002-A	0.050 in. leadless chip carrier, type A, variations AA–AH.
MS003-A	0.050 in. leadless chip carrier, type B, variations BA–BH.
MS004-B	0.050 in. center leadless chip carrier, type C, variations CA–CH.
MS005-A	0.050 in. center leadless chip carrier, type D, variations DA–DH.
MS006-A	0.050 in. center leaded chip carrier, 24 terminal leaded, type A.
MS007-A	0.050 in. center lead chip carrier, type A, variations AA–AH.
MS008-A	0.050 in. center lead chip carrier, type B, variations BA–BH.
MS009-A	0.040 in. center leadless chip carrier packages, variations AA–AJ.
MS014-A	Single-layer chip carrier family, 0.040 in. terminal spacing, ceramic, variations AA–AJ.
MS016-A	PCC family, 1.27 mm/0.050 in. lead spacing, rectangular.
MS018-A	Square plastic chip carrier family, 1.27 mm/0.050 in. pitch.

Logic family		Transition time (ns)		Rank (fastest)
		RT	FT	
BiCMOS	ABT	1.6	1.4	14
	BCT	0.7	0.7	12
	LVT	2.7	2.8	29
CMOS	AC	1.7	1.5	15
	ACT	1.7	1.5	25
	ACQ	2.4	2.4	16
	ACTQ	2.5	2.4	27
	AHCT	2.4	2.4	26
	C	35	25	39
	FCT	1.5	1.2	13
	HC	3.6	4.1	33
	HCT	4.6	3.9	34
	LCX	2.9	2.4	28
	LV	3.0	3.0	30
	LVQ	3.5	3.2	31
	LVX	4.8	3.7	35
	VCX	2.0	2.0	20
	VHC	4.1	3.2	32
ECL	10K	2.2	2.2	22
	10KH	1.7	1.7	17
	100K	0.6	0.6	11
	300K	0.5	0.5	10
	E	0.38	0.38	8
	EP	0.11	0.11	6

Continued

Logic family		Transition time (ns)		Rank (fastest)
		RT	FT	
	LVEL	0.22	0.22	3
	EL	0.23	0.23	5
	SiGe-2.5V	0.02	0.02	1
	SiGe-3.3V	0.3	0.03	2
GaAs		0.3	0.1	4
LVDS		0.3	0.3	7
SSTL		0.3	0.5	9
TTL	74nn	8.0	5.0	36
	ALS	2.3	2.3	24
	AS	2.1	1.5	18
	F	2.3	1.7	21
	FR	2.1	1.5	19
	H	7.0	7.0	37
	L	35	30	40
	LS	15	10	38
	S	2.5	2.0	23

Partial data from IPC-2251 and Coombs 2001.

Appendix D
Drill and Screw Dimensions

| Table D-1 | English Sizes |

		Clearance hole (drill diameter)								
		Close fit		Normal fit		Loose fit				
Screw size no. or diam.	Threads per inch	Drill bit gauge	Min (in.)	Drill bit gauge	Min (in.)	Drill bit gauge	Min (in.)	Head diam. typ (in.)	Washer diam. typ (in.)	Nut size (in.)
0	80	No. 51	0.067	No. 48	0.079	$3/32$	0.104	0.116	0.188	$5/32$
1	72 or 64	No. 46	0.081	No. 43	0.092	No. 37	0.114	0.141	0.219	$5/32$
2	64 or 56	$3/32$	0.094	No. 38	0.105	No. 32	0.126	0.167	0.250	$3/16$
3	56 or 48	No. 36	0.106	No. 32	0.119	No. 30	0.140	0.193	0.312	$3/16$
4	48 or 40	No. 31	0.120	No. 30	0.130	No. 27	0.156	0.219	0.375	$1/4$
5	44 or 40	$9/64$	0.141	$5/32$	0.160	$11/64$	0.184	0.245	0.406	$1/4$
6	40 or 32	No. 23	0.154	No. 18	0.174	No. 13	0.197	0.270	0.438	$5/16$
8	36 or 32	No. 15	0.180	No. 9	0.200	No. 3	0.225	0.322	0.445	$11/32$
10	32 or 24	No. 5	0.206	No. 2	0.225	B	0.250	0.373	0.500	$3/8$
$1/4$	28 or 20	$17/64$	0.266	$9/32$	0.286	$19/64$	0.311	0.492	0.625	$7/16$
$5/16$	24 or 18	$21/64$	0.328	$11/32$	0.349	$23/64$	0.373	0.615	0.688	$9/16$
$3/8$	24 or 16	$25/64$	0.391	$13/32$	0.411	$27/64$	0.438	0.740	0.813	$5/8$

Table D-2	Metric Sizes (mm)					
Screw size	Pitch (mm)	Clearance hole fit			Head, nut size	Washer diameter
		Close	Normal	Loose		
M 1.6	0.4	1.7	1.8	2.0	2.9	3.2
M 2.0	0.4	2.2	2.4	2.6	3.6	4.0
M 2.5	0.5	2.7	2.9	3.1	4.5	5.0
M 3.0	0.5	3.2	3.4	3.6	5.4	6.0
M 4.0	0.7	4.3	4.5	4.8	7.2	8.0
M 5.0	0.8	5.3	5.5	5.8	9.0	10.0
M 6.0	1.0	6.4	6.6	7.0	10.8	12.0
M 8.0	1.3	8.4	9.0	10.0	14.4	16.0
M 10	1.5	10.5	11.0	12.0	18.0	20.0

COMPONENT PACKAGE TYPES AND MOUNTING (SMD)

IPC-7351, Section 8.0, pp. 44–70.

IPC-D-330, Section 5.

(see also Land Patterns)

COMPONENT PLACEMENT, SPACING, AND ORIENTATION

General Considerations

IPC-1902 (grid resolution considerations).

IPC-2221A, Section 7.2.3, p. 51; Section 8.0, p. 55.

IPC-7351, Section 3.4, p. 20 (SMD).

IPC-AJ-820.

IPC-CM-770E, Section 8.1, p. 44.

IPC-D-330, Fig. 11-64, p. 65.

Coombs' Printed Circuits Handbook, 5th ed., Clyde F. Coombs, Jr., New York: McGraw-Hill, 2001, Section 49.4.

Capacitors/Bypassing/Placement

Avoiding Passive-Component Pitfalls, Doug Grant and Scott Wurcer, Analog Devices Application Note AN-348.

Biasing and Decoupling Op Amps in Single Supply Applications, Charles Kitchin, Analog Devices Application Note AN-581.

EMC and the Printed Circuit Board: Design, Theory, and Layout Made Simple, Mark I. Montrose, IEEE Press Series on Electronics Technology, New York: IEEE Press, 1999, p. 126.

High-Speed Digital Design: a Handbook of Black Magic, Howard Johnson, Martin Graham, Englewood Cliffs, NJ: Prentice-Hall, 1993, pp. 281–288.

Noise Reduction Techniques in Electronic Systems, 2nd ed., Henry W. Ott, New York: Wiley, 1988, p. 129.

Printed Circuit Board Design Techniques for EMC Compliance, 2nd ed., Mark I. Montrose, IEEE Press Series on Electronics Technology, New York: IEEE Press, 2000, Chap. 3.

Fiducials

IPC-2221A, Section 3.4.4, p. 22.

IPC-CM-770E, Section 8.4, p. 46.

Wave Soldering Considerations

IPC-2221A, Section 8.1.3, p. 57.

IPC-7351, Section 3.1, pp. 9–16 (SMD footprint design).

Coombs' Printed Circuits Handbook, 5th ed., Clyde F. Coombs, Jr., New York: McGraw-Hill, 2001, Section 43.7.8

Design Rule Checking

IPC-CM-770E, Section 7.2, p. 35.

Coombs' Printed Circuits Handbook, 5th ed., Clyde F. Coombs, Jr., New York: McGraw-Hill, 2001, Section 19.5.1.

LAND PATTERNS (FOOTPRINT DESIGN)

SMD

IPC-7351, Section 3.1, pp. 9–16 (SMD footprint design); Table3-15, p. 16 (IPC SMD naming convention); Land Pattern Viewer (free demo version from IPC Web site).

IPC-SM-780, Section 6.7.2, p. 66.

Through Hole

(see Lead Relationships)

LAYER STACK-UPS

Advanced Circuits Web site (www.4pcb.com, search for "controlled impedance").

Coombs' Printed Circuits Handbook, 5th ed., Clyde F. Coombs, Jr., New York: McGraw-Hill, 2001, Chap. 13, Fig. 13.9; Chap. 27.

DP83865 Gig PHYTER® V10/100/1000 Ethernet Physical Layer Design Guide, National Semiconductor Application Notes, August 2003.

High-Speed Digital Design: a Handbook of Black Magic, Howard Johnson, Martin Graham, Englewood Cliffs, NJ: Prentice-Hall, 1993, Chap. 5, pp. 212, 217–221.

H. Ott's Web site (http://hottconsultants.com/techtips/pcb-stack-up-3.html).

Signal Integrity Issues and Printed Circuit Board Design, Douglas Brooks, Prentice Hall Modern Semiconductor Design Series, 2003, p. 308.

LEAD RELATIONSHIPS

Lead-to-Hole Ratio

IPC-AJ-820, Section 2.5.2, p. 2-11.

IPC-SM-780, Section 8.6.1.5, p. 103.

Coombs' Printed Circuits Handbook, 5th ed., Clyde F. Coombs, Jr., New York: McGraw-Hill, 2001, Section 42.2.1.

Lead Bend Radius and Spacing

IPC-2221A, Fig. 8-9, p. 61.

IPC-AJ-820, Section 2.5.1.4, p. 2-11; Section 2.5.3.1, p. 2-13.

IPC-CM-770, Section 11.1.8, p. 67.

IPC-D-330, Section 11, Fig. 11-8, p. 12.

IPC-SM-780, Section 8.4.2, p. 94.

MATERIALS

Copper

IPC-2221A, Section 4.4, p. 20; Table 4-3, p. 21.

IPC-4101A.

Prepreg/Laminate

IPC-4101A.

Coombs' Printed Circuits Handbook, 5th ed., Clyde F. Coombs, Jr., New York: McGraw-Hill, 2001, Chap. 8, Chap. 10, Section 10.3.

Plating

IPC-2221A, Tables 10-1, 10-2, p. 77.

Coombs' Printed Circuits Handbook, 5th ed., Clyde F. Coombs, Jr., New York: McGraw-Hill, 2001, Sections 28.5, 29.4; Chaps. 30, 31.

Solder

(see Soldering and Assembly)

Dielectric (Relative Permittivity) Properties

IPC-2221A, Table 4-1, p. 18; Table 6-2, p. 45.

IPC-4101A.

EMC and the Printed Circuit Board: Design, Theory, and Layout Made Simple, Mark I. Montrose, IEEE Press Series on Electronics Technology, New York: IEEE Press, 1999, Table 6-2.

Noise Reduction Techniques in Electronic Systems, 2nd ed., Henry W. Ott, New York: Wiley, 1988, Table 4-1, p. 125.

Printed Circuit Board Design Techniques for EMC Compliance, 2nd ed., Mark I. Montrose, IEEE Press Series on Electronics Technology, New York: IEEE Press, 2000, p. 111.

Signal Integrity—Simplified, Eric Bogatin, Upper Saddle River, NJ: Prentice Hall, 2004, p. 349.

MOUNTING HARDWARE AND SPECIFICATIONS

ASME B1-10M Unified Miniature Screw Threads.

ASME B18.2.4.1M-2002 Metric Hex Nuts—Style 1.

ASME B18.2.4.2M-2005 Metric Hex Nuts—Style 2.

ASME B18.2.8-1999, Clearance Holes for Bolts, Screws, and Studs.

ASME B18.6.3-2003 Machine Screws and Machine Screw Nuts (inch series).

ASME B18.6.7M-2000 Machine Screws (metric series).

ASME B18.12-2001 Glossary of Terms for Mechanical Fasteners.

ASME B18.13-1996 Screw and Washer Assemblies (inch series).

ASME B18.13.1M-1999 Screw and Washer Assemblies (metric series).

ASME B18.21.1-1999 Lock Washers (inch series).

ASME B18-21-2M-1999 Lock Washers (metric series).

ASME B18.22.1-1981 Plain Washers (inch series).

ASME B18.22M-1981 Plain Washers (metric series).

ASME B94.11M-1993 Twist Drills (drill bit sizes—inch and metric).

STANDARDS

IPC-D-330, Section 1.

(see Appendix A)

PCB

Fabrication

Coombs' Printed Circuits Handbook, 5th ed., Clyde F. Coombs, Jr., New York: McGraw-Hill, 2001, Chaps. 13, 14.

Fabrication Allowances

IPC-7351, Section 3.1, p. 9–16 (SMD footprint design); Tables 3-17, 3-18; Fig. 3-17, p. 28.

IPC-CM-770E, Tables 8-1, 8-2, p. 50.

Coombs' Printed Circuits Handbook, 5th ed., Clyde F. Coombs, Jr., New York: McGraw-Hill, 2001, Section 17.1.

Borders

IPC-2222, Section 5.2.1.1.

General Design Information

ANSI/IPC-D-322 (standard panel sizes).

IPC-2615 (PCB dimensions and tolerances).

IPC-AJ-820, Section 2.0, pp. 2-2–2-8.

IPC-D-330, Section 2.

DFM in PWB Fab: a Review of Predictive Engineering Benefits, Happy T. Holden, Hewlett Packard Co., Loveland, CO.

PCB layout application notes, Maxim, Dallas, Web site: www.maxim-ic.com/appnotes10.cfm/ac_pk/35/ln/en.

Planning PCB Design: for Fun and Profit, Happy Holden, TechLead Corp., Evergreen, CO.

Performance Classes

IPC-2221A, Section 1.6.2, p. 1.

IPC-7351, Section 1.2.1, p. 2.

IPC-CM-770E, Section 1.2.1, p. 1.

Assembly Types

IPC-2222, Section 1.5.1, Fig. 1-1, p. 1.

IPC-AJ-820, Section 1.3.2, Fig. 1-1, p. 1-5.

IPC-CM-770E, Sections 1.2.2, 1.2.4, p. 1.

Density Levels

IPC-7351, Section 1.4, p. 2.

Producibility Levels

IPC-2221A, Section 1.6.3, p. 2.

IPC-7351, Section 1.3.1, p. 2.

IPC-CM-770E, Section 1.2.2, p. 1.

Power Distribution Concepts

IPC-2221A, Figs. 6-1, 6-2, 6-3, p. 38.

Sizes/Panels

IPC-2221A, Fig. 3-5, p. 14; Fig. 5-1, p. 28.

Testing

IPC-2221A, Sections 3.5–3.5.6, p. 7-12.

PLATED THROUGH HOLES

Types: Through, Blind, Buried, Mounting Holes

IPC-2221A, Section 9.2, p. 75.

IPC-7351, Section 1.5, p. 3.

Fabrication of Through Holes

Coombs' Printed Circuits Handbook, 5th ed., Clyde F. Coombs, Jr., New York: McGraw-Hill, 2001, Chap. 11, Sections 21.2, 22.4.3.

Aspect Ratio

IPC-2221A, Table 4-3, p. 21.

IPC-2222, Section 9.3, p. 20.

IPC-D-330, Section 6.18, Tables 6-29 through 6-32, p. 53.

Lead-to-Hole Ratio

Coombs' Printed Circuits Handbook, 5th ed., Clyde F. Coombs, Jr., New York: McGraw-Hill, 2001, Section 42.2.1.

(see also Lead Relationships)

Hole-to-Land Ratio

IPC-2221A, Table 4-3, p. 21.

Land Requirements

IPC-2221A, Section 9.1.1, p. 73.

Annular Ring Requirements

IPC-2221A, Section 9.1.2, p. 73.

IPC-7351, Section 3.4.7.4, p. 27.

Coombs' Printed Circuits Handbook, 5th ed., Clyde F. Coombs, Jr., New York: McGraw-Hill, 2001, Sections 42.2.2, 48.8.1.

Clearance from Planes

IPC-2221A, Section 6.3.1, Table 6-1 (column B1), p. 42.

IPC-2222, Section 9.1.3, p. 17.

Nonfunctional Lands

IPC-2222, Section 9.1.4, p. 18.

Unsupported Holes

IPC-2222, Section 9.2.1, p. 18.

Fan-Outs

IPC-SM-780, Section 6.7.1.1, p. 64.

Via (Through-Hole) Guidelines and Fabrication

IPC-7351, Section 3.4.6, p. 24.

IPC-CM-770E, Section 8.6, p. 47.

Coombs' Printed Circuits Handbook, 5th ed., Clyde F. Coombs, Jr., New York: McGraw-Hill, 2001, Section 23.8

Filled/Plugged Holes

IPC-2222, Section 9.2.2.3, p. 19.

RISE AND FALL TIMES OF LOGIC FAMILIES

IPC-2251, Table 5-4, p. 29.

Coombs' Printed Circuits Handbook, 5th ed., Clyde F. Coombs, Jr., New York: McGraw-Hill, 2001, Chap. 13, Table 13.2.

EMC and the Printed Circuit Board: Design, Theory, and Layout Made Simple, Mark I. Montrose, IEEE Press Series on Electronics Technology, New York: IEEE Press, 1999, Table 3-1.

Noise Reduction Techniques in Electronic Systems, 2nd ed., Henry W. Ott, New York: Wiley, 1988, p. 275.

(see also Appendix C)

SCHEMATICS
Symbols
ANSI 315-1975.

Diagrams
IPC-2221A, Section 3.3, p. 6.

SIGNAL INTEGRITY

A Practical Guide to High-Speed Printed-Circuit-Board Layout, John Ardizzoni, Analog Devices: Analog Dialogue, Applications Note.

Analysis on the Effectiveness of High Speed Printed Circuit Board Edge Radiated Emissions Based on Stimulus Source Location, Mark I. Montrose, Jin Hong-Fang, Er-Ping Li.

Analysis on the Effectiveness of High Speed Printed Circuit Board Edge Termination Using Discrete Components Instead of Implementing the 20-H Rule, Mark I. Montrose, Jin Hong-Fang, Er-Ping Li.

Combating Noise Effects in Adaptive Cable Equalizer Designs, James Mears National Semiconductor Applications Note.

EMC and the Printed Circuit Board: Design, Theory, and Layout Made Simple, Mark I. Montrose, IEEE Press Series on Electronics Technology, New York: IEEE Press, 1999.

High-Speed Digital Design: A Handbook of Black Magic, Howard Johnson, Martin Graham, Englewood Cliffs, NJ: Prentice-Hall, 1993.

Printed Circuit Board Design Techniques for EMC Compliance, 2nd ed., Mark I. Montrose, IEEE Press Series on Electronics Technology, New York: IEEE Press, 2000.

QPI-AN1, PICOR general application note on QPI Family.

Signal Integrity Issues and Printed Circuit Board Design, Douglas Brooks, Prentice Hall Modern Semiconductor Design Series, Upper Saddle River, NJ: Prentice Hall, 2003.

Cross Talk

EMC and the Printed Circuit Board: Design, Theory, and Layout Made Simple, Mark I. Montrose, IEEE Press Series on Electronics Technology, New York: IEEE Press, 1999, Chap. 7.

High-Speed Digital Design: A Handbook of Black Magic, Howard Johnson, Martin Graham, Englewood Cliffs, NJ: Prentice-Hall, 1993, p. 191.

Printed Circuit Board Design Techniques for EMC Compliance, 2nd ed., Mark I. Montrose, IEEE Press Series on Electronics Technology, New York: IEEE Press, 2000, p. 131.

Ground Bounce/Rail Collapse

Signal Integrity—Simplified, Eric Bogatin, Upper Saddle River, NJ: Prentice Hall, 2004, pp. 163, 249.

Propagation Speed/Time Delay

High-Speed Digital Design: A Handbook of Black Magic, Howard Johnson, Martin Graham: Prentice-Hall, 1993, p. 178.

Signal Integrity—Simplified, Eric Bogatin, Upper Saddle River, NJ: Prentice Hall, 2004, pp. 210–211, 214–216, 255.

Reflections/Ringing

EMC and the Printed Circuit Board: Design, Theory, and Layout Made Simple, Mark I. Montrose, IEEE Press Series on Electronics Technology, New York: IEEE Press, 1999, pp. 188–194.

High-Speed Digital Design: A Handbook of Black Magic, Howard Johnson, Martin Graham, Englewood Cliffs, NJ: Prentice-Hall, 1993, p. 160.

Signal Integrity Issues and Printed Circuit Board Design, Douglas Brooks, Prentice Hall Modern Semiconductor Design Series, Upper Saddle River, NJ: Prentice Hall, 2003, Chap. 10.

Guard Rings and Traces

Printed Circuit Board Design Techniques for EMC Compliance: A Handbook for Designers, 2nd ed., M.I. Montrose, New York: IEEE Press, 2000, pp. 137–141.

Signal Integrity—Simplified, E. Bogatin, Upper Saddle River, NJ: Prentice Hall, 2004, pp. 452–459.

Signal Integrity Issues and Printed Circuit Board Design, D. Brooks, Upper Saddle River, NJ: Pearson Education, 2003, p. 233.

Coombs' Printed Circuits Handbook, 5th ed., Clyde F. Coombs, J., New York: McGraw-Hill, 2001, P. 15.6.

High-Speed Digital Design: A Handbook of Black Magic, H. Johnson, M. Graham, Englewood Cliffs, NJ: Prentice-Hall, 1993, pp. 201–204, 262.

Noise Reduction Techniques in Electronic Systems, 2nd ed., H.W. Ott, New York: Wiley, 1988, p. 343.

Rich, A. Shielding and Guarding, How to Exclude Interference-Type Noise, What to Do and Why to Do It—A Rational Approach. Analog Devices Application Note AN-347: www.analog.com/UploadedFiles/Application_Notes/41727248AN_347.pdf.

Barnes, J. R. Designing Electronic Systems For ESD Immunity. Conformity®: February 2003 (0302designing.pdf).

Terminations

EMC and the Printed Circuit Board: Design, Theory, and Layout Made Simple, Mark I. Montrose, IEEE Press Series on Electronics Technology, New York: IEEE Press, 1999, p. 217.

High-Speed Digital Design: A Handbook of Black Magic, Howard Johnson, Martin Graham, Englewood Cliffs, NJ: Prentice-Hall, 1993, Chap. 6.

Printed Circuit Board Design Techniques for EMC Compliance, 2nd ed., Mark I. Montrose, IEEE Press Series on Electronics Technology, New York: IEEE Press, 2000, p. 143.

Ground Areas (Split Planes, Moated Grounds)

EMC and the Printed Circuit Board: Design, Theory, and Layout Made Simple, Mark I. Montrose, IEEE Press Series on Electronics Technology, New York: IEEE Press, 1999, pp. 107–111, 119.

Printed Circuit Board Design Techniques for EMC Compliance, 2nd ed., Mark I. Montrose, IEEE Press Series on Electronics Technology, New York: IEEE Press, 2000, pp. 152–170.

Signal Integrity Issues and Printed Circuit Board Design, Douglas Brooks, Prentice Hall Modern Semiconductor Design Series, Upper Saddle River, NJ: Prentice Hall, 2003, Chap. 16.

Guard Shields

Coombs' Printed Circuits Handbook, 5th ed., Clyde F. Coombs, Jr., New York: McGraw-Hill, 2001, Chap. 17, Section 17.2.3.2.

High-Speed Digital Design: A Handbook of Black Magic, Howard Johnson, Martin Graham, Englewood Cliffs, NJ: Prentice-Hall, 1993, p. 201.

Noise Reduction Techniques in Electronic Systems, 2nd ed., Henry W. Ott, New York: Wiley, 1988, p. 106.

Printed Circuit Board Design Techniques for EMC Compliance, 2nd ed., Mark I. Montrose, IEEE Press Series on Electronics Technology, New York: IEEE Press, 2000, pp. 137, 141.

Shielding and Guarding, Alan Rich, Analog Devices Application Note AN-347.

Noise Reduction (General)

An Applications Guide for Op Amps, National Semiconductor Application Note AN-20, February 1969.

An IC Amplifier User's Guide to Decoupling, Grounding, and Making Things Go Right for a Change, Paul Brokaw, Analog Devices Application Note AN-202.

Common Mode Ground Currents, *High-Speed Digital Design, Online Newsletter* Vol. 7, Issue 02, Sigcon.

EMC and the Printed Circuit Board: Design, Theory, and Layout Made Simple, Mark I. Montrose, IEEE Press Series on Electronics Technology, New York: IEEE Press, 1999, p. 41.

Grounding for Low- and High-Frequency Circuits, Paul Brokaw and Jeff Barrow, Analog Devices Application Note AN-345.

Grounding Rules for High-Speed Circuits, Don Brockman and Arnold Williams, Analog Devices Application Note AN-214.

Noise Reduction Techniques in Electronic Systems, 2nd ed., Henry W. Ott, New York: Wiley, 1988, Chap. 6 (shielding).

Op Amp Selection Guide for Optimum Noise Performance, Glen Brisebois, Linear Technologies Design Ideas, *Linear Technology Magazine*, December 2005.

PCB Design Guidelines That Maximize the Performance of TVS Diodes, Jim Lepkowske, ON Semiconductor Application Note AND8232/D.

Power Supply Effects on Noise Performance, Nicholas Gray, National Semiconductor Application Note AN-1261, September 2002.

Power Supply Ripple Rejection and Linear Regulators: What's All the Noise About? Gabriel A. Rincón-Mora and Vishal Gupta, Texas Instruments Planet Analog application note (www.planetanalog.com/printableArticle. jhtml;jsessionid = CRXB).

Shielding and Guarding: How to Exclude Interference-Type Noise, Alan Rich, Analog Devices Application Note AN-347.

EMI/EMC

Coombs' Printed Circuits Handbook, 5th ed., Clyde F. Coombs, Jr., New York: McGraw-Hill, 2001, Section 17.2.3.2.

EMC and the Printed Circuit Board: Design, Theory, and Layout Made Simple, Mark I. Montrose, IEEE Press Series on Electronics Technology, New York: IEEE Press, 1999.

Printed Circuit Board Design Techniques for EMC Compliance, 2nd ed., Mark I. Montrose, IEEE Press Series on Electronics Technology, New York: IEEE Press, 2000.

Signal Integrity Issues and Printed Circuit Board Design, Douglas Brooks, Prentice Hall Modern Semiconductor Design Series, Upper Saddle River, NJ: Prentice Hall, 2003, Chaps. 8, 9.

SOLDERING AND ASSEMBLY

IPC J-STD-001D (soldering requirements).

IPC-SM-780, Section 8.1, p. 84.

Assembly Issues and Processes

IPC-2221A, Section 8.0, p. 55.

IPC-7351, Section 7.4, Fig. 7-1, p. 40.

IPC-CM-770E, Sections 6.3, 7.1, p. 37.

IPC-D-330, Section 11.

IPC/WHMA-A-620.

Coombs' Printed Circuits Handbook, 5th ed., Clyde F. Coombs, Jr., New York: McGraw-Hill, 2001, Chap. 13, Section 13.3; Chaps. 41, 42, 43.

Reflow Soldering

IPC-CM-770E, Sections 6.2.2.3, 6.2.2.4, p. 35.

Coombs' Printed Circuits Handbook, 5th ed., Clyde F. Coombs, Jr., New York: McGraw-Hill, 2001, Section 43.6.

Wave Soldering

IPC-CM-770E, Section 6.2.2.1, p. 35, and Section 8.1.2.

Coombs' Printed Circuits Handbook, 5th ed., Clyde F. Coombs, Jr., New York: McGraw-Hill, 2001, Section 43.7.

Components, Placing/Spacing

IPC-2221A, Section 8.1.2, p. 55.

Components, Orientation

IPC-2221A, Fig. 8-1, p. 57.

Components, Mounting

IPC-2221A, Section 8.1.9.1, p. 58.

Reliability

Coombs' Printed Circuits Handbook, 5th ed., Clyde F. Coombs, Jr., New York: McGraw-Hill, 2001, Section 53.3.

THERMAL MANAGEMENT

IPC-2221A, Section 7.0, p. 48; Fig. 7-1 (heat sink and spacing), p. 51.

IPC-D-330, Section 4.

ADN2530—11.3 Gbps, Active Back-Termination, Differential VCSEL Driver, Analog Devices data sheet (333411103ADN2530.pdf).

Analytical Modeling of Thermal Resistance in Bolted Joints, S. Lee, M. M. Yovanovich, S. Song, and K. P. Moran, HTD-Vol. 263, Enhanced Cooling Techniques for Electronics Applications ASME 1993.

Evaluation of Power MOSFET Thermal Solutions for Desktop and Mobile Processor Power, Tim McDonald and John Ambrus, International Rectifier Corp.

How to Select a Heat Sink, Seri Lee, Aavid Thermal Technologies, Inc., Application Note.

JEDEC Standard 51.

LFPAK Thermal Design Guide—LFPAK Thermal Performance Junction to Ambient, Philips Application Note, Rev. 01, June 16, 2003.

LT1806/LT1807 data sheet (18067f[1].pdf), Linear Technology Corp.

Monitor Heat Dissipation in Electronic Systems by Measuring Active Component Die Temperature, Maxim, Dallas, Semiconductor Application Note 3500, March 18, 2005.

Package Thermal Characteristics, Actel Corp. Application Note, February 2005.

Thermal Considerations of QFN and Other Exposed-Paddle Packages, Maxim Integrated Products Application Note HFAN-08.1, November 2001 (4hfan081. doc 11/20/2001).

Thermal Design Considerations, Philips Semiconductors IC Packages, Chap. 6 (IC26_CHAPTER_6_2000.pdf).

Thermal Performance of National's LLP Package, Highlighting the LM2750 Regulated Voltage Converter, Brian J. Conner, National Semiconductor Application Brief 111.

Thermal Resistance v PCB Area Comparison, Zetex Application Note.

TRACE ROUTING

Trace Spacing (Voltage Withstanding)

IPC-2221A, Table 6-1 (trace to trace), p. 42–43; Fig. 8-4 (trace to mounting hardware), p. 59.

IPC-AJ-820, Fig. 2-12 (trace to mounting hardware), p. 2-13.

Coombs' Printed Circuits Handbook, 5th ed., Clyde F. Coombs, Jr., New York: McGraw-Hill, 2001, Section 49.3.2.

Trace Spacing (Cross Talk/3-W Rule)

EMC and the Printed Circuit Board: Design, Theory, and Layout Made Simple, Mark I. Montrose, IEEE Press Series on Electronics Technology, New York: IEEE Press, 1999, p. 207.

High-Speed Digital Design: A Handbook of Black Magic, Howard Johnson, Martin Graham, Englewood Cliffs, NJ: Prentice-Hall, 1993, pp. 215, 219.

Printed Circuit Board Design Techniques for EMC Compliance, 2nd ed., Mark I. Montrose, IEEE Press Series on Electronics Technology, New York: IEEE Press, 2000, pp. 131, 136.

Signal Integrity—Simplified, Eric Bogatin, Upper Saddle River, NJ: Prentice Hall, 2004, pp. 421–425.

Trace Width—Current Handling

IPC-2221A, Fig. 6-4, pp. 40–41.

Trace Width—Transmission Lines

IPC-2221A, Fig. 6-5, pp. 44–47.

IPC-2251, Section 5.5–5.5.5.4, pp. 31–36.

High-Speed Digital Design: a Handbook of Black Magic, Howard Johnson, Martin Graham, Englewood Cliffs, NJ: Prentice-Hall, 1993, p. 214.

Microstrip

IPC-2221A, Section 6.4.1, p. 44.

Embedded Microstrip

IPC-2221A, Section 6.4.2, p. 44.

Stripline

IPC-2221A, Section 6.4.3, p. 44.

Asymmetric Stripline

IPC-2221A, Section 6.4.4, p. 46.

(see also Transmission Lines)

TRANSMISSION LINES

Characteristic Impedance (Microstrip, Stripline Topologies, Etc.)

IPC-2141, Sections 4.0–4.7, pp. 11–16.

IPC-2251, Sections 5.5–5.5.5.4, pp. 31–36.

IPC-2221A, Fig. 6-5, pp. 44–47.

IPC-D-330, Sections 4, 6.5.4.2.

Coombs' Printed Circuits Handbook, 5th ed., Clyde F. Coombs, Jr., New York: McGraw-Hill, 2001, Chap. 16.

EMC and the Printed Circuit Board: Design, Theory, and Layout Made Simple, Mark I. Montrose, IEEE Press Series on Electronics Technology, New York: IEEE Press, 1999, pp. 171–179.

High-Speed Digital Design: A Handbook of Black Magic, Howard Johnson, Martin Graham, Englewood Cliffs, NJ: Prentice-Hall, 1993, pp. 186–188.

Printed Circuit Board Design Techniques for EMC Compliance, 2nd ed., Mark I. Montrose, IEEE Press Series on Electronics Technology, New York: IEEE Press, 2000, pp. 101–190.

Signal Integrity Issues and Printed Circuit Board Design, Douglas Brooks, Prentice Hall Modern Semiconductor Design Series, Upper Saddle River, NJ: Prentice Hall, 2003, pp. 185, 203, 265.

Signal Integrity—Simplified, Eric Bogatin, Upper Saddle River, NJ: Prentice Hall, 2004, pp. 220–233, 258–260, 274.

(see also Trace Routing)

Critical Length

EMC and the Printed Circuit Board: Design, Theory, and Layout Made Simple, Mark I. Montrose, IEEE Press Series on Electronics Technology, New York: IEEE Press, 1999, p. 195.

High-Speed Digital Design: A Handbook of Black Magic, Howard Johnson, Martin Graham, Englewood Cliffs, NJ: Prentice-Hall, 1993, p. 166.

Noise Reduction Techniques in Electronic Systems, 2nd ed., Henry W. Ott, New York: Wiley, 1988, p. 308.

Signal Integrity—Simplified, Eric Bogatin, Upper Saddle River, NJ: Prentice Hall, 2004, p. 297.

Propagation Delay Time

IPC-2141, Section 3.4.7.1, p. 6.

IPC-2251, Section 5.4, p. 30.

Printed Circuit Board Design Techniques for EMC Compliance, 2nd ed., Mark I. Montrose, IEEE Press Series on Electronics Technology, New York: IEEE Press, 2000, p. 111.

Signal Integrity—Simplified, Eric Bogatin, Upper Saddle River, NJ: Prentice Hall, 2004, pp. 214–216.

Reflections/Impedance Mismatches

IPC-2251, Section 5.6.2, p. 36.

Coombs' Printed Circuits Handbook, 5th ed., Clyde F. Coombs, Jr., New York: McGraw-Hill, 2001, Chap. 16.

Noise Reduction Techniques in Electronic Systems, 2nd ed., Henry W. Ott, New York: Wiley, 1988, p. 169.

Signal Integrity—Simplified, Eric Bogatin, Upper Saddle River, NJ: Prentice Hall, 2004, pp. 278–283, 302.

Sharp Corners

High-Speed Digital Design: a Handbook of Black Magic, Howard Johnson, Martin Graham, Englewood Cliffs, NJ: Prentice-Hall, 1993, p. 174.

Signal Integrity—Simplified, Eric Bogatin, Upper Saddle River, NJ: Prentice Hall, 2004, p. 315.

Index

Printed and bound by CPI Group (UK) Ltd, Croydon, CR0 4YY

03/10/2024

01040325-0020